NUMERICAL INITIAL VALUE PROBLEMS IN ORDINARY DIFFERENTIAL EQUATIONS

Prentice-Hall Series in Automatic Computation
George Forsythe, editor

ANSELONE, *Collectively Compact Operator Approximation Theory and Applications to Integral Equations*
ARBIB, *Theories of Abstract Automata*
BATES AND DOUGLAS, *Programming Language/One, 2nd Ed.*
BAUMANN, FELICIANO, BAUER, AND SAMELSON, *Introduction to ALGOL*
BLUMENTHAL, *Management Information Systems*
BOBROW AND SCHWARTZ, editors, *Computers and the Policy-Making Community: Applications to International Relations*
BOWLES, editor, *Computers in Humanistic Research*
CRESS, DIRKSEN, AND GRAHAM, *FORTRAN IV with WATFOR and WATFIV*
DANIEL, *The Approximate Minimization of Functionals*
DESMONDE, *Computers and Their Uses, 2nd Ed.*
EVANS, WALLACE, AND SUTHERLAND, *Simulation Using Digital Computers*
FIKE, *Computer Evaluation of Mathematical Functions*
FORSYTHE AND MOLER, *Computer Solution of Linear Algebraic Systems*
GAUTHIER AND PONTO, *Designing Systems Programs*
GEAR, *Numerical Initial Value Problems in Ordinary Differential Equations*
GORDON, *System Simulation*
GREENSPAN, *Lectures on the Numerical Solution of Linear, Singular and Nonlinear Differential Equations*
HARTMANIS AND STEARNS, *Algebraic Structure Theory of Sequential Machines*
HEAPS, *An Introduction to Computer Languages*
HULL, *Introduction to Computing*
JOHNSON, *System Structure in Data, Programs, and Computers*
LORIN, *Parallelism in Hardware and Software: An Introduction to Considerations in Real and Apparent Concurrency*
MARTIN, *Design of Real-Time Computer Systems*
MARTIN, *Future Developments in Telecommunications*
MARTIN, *Programming Real-Time Computer Systems*
MARTIN, *Systems Analysis for Data Transmission*
MARTIN, *Telecommunications and the Computer*
MARTIN, *Teleprocessing Network Organization*
MARTIN AND NORMAN, *The Computerized Society*
MATHISON AND WALKER, *Computers and Telecommunications: Issue in Public Policy*
MCKEEMAN, HORNING, AND WORTMAN, *A Compiler Generator*
MINSKY, *Computation: Finite and Infinite Machines*
MOORE, *Interval Analysis*
PYLYSHYN, editor, *Perspectives on the Computer Revolution*
SALTON, *The SMART Retrieval System: Experiments in Automatic Data Processing*
SAMMET, *Programming Languages: History and Fundamentals*
STERLING AND POLLACK, *Introduction to Statistical Data Processing*
STROUD, *Approximate Calculation of Multiple Integrals*
TAVISS, editor, *The Computer Impact*
TRAUB, *Iterative Methods for the Solutions of Equations*
VARGA, *Matrix Iterative Analysis*
VAZSONYI, *Problem Solving by Digital Computers with PL/1 Programming*
WILKINSON, *Rounding Errors in Algebraic Processes*

TABLE OF THEOREMS

TABLE OF LEMMAS

NUMERICAL INITIAL VALUE PROBLEMS IN ORDINARY DIFFERENTIAL EQUATIONS

C. William Gear

Department of Computer Science
University of Illinois

Prentice-Hall, Inc. Englewood Cliffs, New Jersey

Current Printing (last digit):
10 9 8 7 6 5 4 3 2

13-626606-1
Library of Congress Catalog Card Number: 75-152448
Printed in the United States of America

PRENTICE-HALL INTERNATIONAL, INC., *London*
PRENTICE-HALL OF AUSTRALIA, PTY., LTD., *Sydney*
PRENTICE-HALL OF CANADA, LTD., *Toronto*
PRENTICE-HALL OF INDIA PRIVATE LIMITED., *New Delhi*
PRENTICE-HALL OF JAPAN, INC., *Tokyo*

To my children, Jodi and Chris

FOREWORD

The successful solution of a realistic problem in applied mathematics requires the fusion of four distinct ingredients: (1) knowledge of the subject area of the problem; (2) knowledge of the relevant mathematics; (3) knowledge of the relevant computer science; (4) a talent for selecting just what part of all this knowledge will actually solve the problem, and ignoring the rest.

Professor Gear is one of those rare individuals with these four capabilities. He has made himself familiar with subject areas that spawn differential equations. He has a good grounding in mathematical analysis. He has demonstrated a capacity as a systems programmer and teacher of systems programming, and is as capable in symbol manipulation as he is in computing with numbers. He has created powerful programs (e.g., DIFSUB on pp. 158–166) for solving difficult practical problems. In particular, he knows a lot about solving stiff differential equations, a difficult area that has received much attention in recent years.

Because of Professor Gear's fine credentials, I am delighted that this book is joining my Prentice-Hall Series in Automatic Computation, and I believe the reader will find it both readable and very useful.

GEORGE E. FORSYTHE

PREFACE

This book is the outcome both of teaching graduate level courses on the subject matter and of the application of many of the methods in large automatic processing systems concerned with aspects of computer aided design and simulation. As such it is addressed to two audiences: the men with problems to solve and the student numerical analysts. However, it is my belief that the goals of these two groups should not be greatly different. Both groups should be concerned with the understanding of all of the properties that determine the best method for a given problem.

The progression of many numerical methods has been that they originate with the engineer (scientist, astronomer, etc.) who solves his problem that was previously unmanageable, progress through the mathematician-numerical analyst who proves that they do indeed solve the problem under some conditions, and finish with the numerical analyst-computer scientist who concerns himself with their effective implementation and utilization on the computer. As automatic programming systems provide problem oriented languages that make the computer more accessible to the non-professional user (that is, not part of the numerical analyst, computer scientist, or research engineering professions), it becomes more important that the program be foolproof, or rather, problem proof.

In this book I have tried to gather together methods, mathematics and implementations, and to provide guidelines for their use on problems. The material is separated into chapters on this basis so that the motivations for methods can be presented clearly before the mathematics of errors clouds the scene. The results shown in the chapters on theory are quoted in the development of methods so that the reader only interested in results can skip these chapters entirely, and continue to the chapters on the selection of methods and implementation.

The reader is assumed to have a background equivalent to a beginning course in numerical analysis at the advanced undergraduate level in which the topics of interpolation, solution of nonlinear equations, and some basic matrix theory have been covered. No previous knowledge of numerical techniques for differential equations is assumed, but a basic knowledge of differential equations is required. The chapters on theory require a greater knowledge of matrix theory, such as obtained from an undergraduate course on linear algebra.

Chapters 4 and 10 deal with the theory of one-step and multistep methods, respectively. The material was chosen to provide the justification for many of the approaches used in current programs (some of which are given in the text), where that material was of reasonable proportions. In addition, I have tried in Chapter 10 to set down the theory of multivalue methods for both first and higher order equations.

Chapter 11 deals mainly with the subject of stiff equations, which has become increasingly important as computers are applied in large simulation and design situations. It treats the subject pragmatically and does not discuss the theory for two reasons. First, the major problems are unsolved (for example, reasonably unrestrictive sufficient conditions for error bounds independent of the stiffness to be obtained). Secondly, those results that do exist are still sufficiently new that it is too early to judge their merit.

A book such as this owes debts to many people, a large number of whom I have only met via their published work. I would like to give special acknowledgement to a number of people. Professors G. Dahlquist and P. Henrici did more than most people to put the mathematics in order, and provided me with much reading. Professors W. Gautschi, W. Kahan, and B. Parlett reviewed the manuscript and made many useful suggestions. Particularly stimulating environments were provided for my own work by Dr. W. Givens at Argonne National Laboratory, Professor W. Miller at the Stanford Linear Accelerator Center, and Professor G. Forsythe at Stanford University. I would also like to quote loosely from my advisor of graduate student days, Professor A. H. Taub, who said a lot of things I didn't believe until later. In particular, he pointed out that the academic life was a good one, and that the least one could do to repay one's debt to society was to set one's thoughts on paper so that the useful among them might benefit others.

Thanks are also due to the University of Illinois for the sabbatical leave during which this book was written, and Stanford University and its Linear Accelerator Center for providing a home for me in that time. Much of the testing of early versions of the subroutine in Chapter 9 was done by Mrs. K. Ratliff of Urbana. Messrs. S. Eisenstaat and M. Saunders of Stanford and many other students both at Stanford and Urbana made many helpful comments (and served as guinea pigs for many of the problems). To all these people I am grateful, and especially to B. Hurdle who typed the manuscript accurately and at record speed.

C. W. GEAR

Stanford, California

CONTENTS

1 INTRODUCTION

1.1 THE PROBLEM TO BE SOLVED

The theoretical physicist or chemist spends much of his time working with models of the world around him. A model is often given as a mathematical description of variables, some of which certainly exist and are measurable (pressure, for example), others of which may only be hypothesized (the characteristics of a new particle, for example). Frequently these models will be sets of ordinary differential equations in which the independent variable is time and the dependent variables are the physical variables. Occasionally the independent variable may be one of the physical variables due either to an interchange of variables, or the solution of a system in equilibrium state which is therefore independent of time, but we will call the independent variable t time throughout this book. The methods to be discussed are not dependent on any special property of time, so that the reader should be prepared to substitute x for space or whatever meets his fancy.

When a model has been constructed, two things may happen to it. It may be necessary first to "test" it. This will involve the experimentalist in his laboratory. The behavior of the model must be compared with the observed behavior in the experiment. Secondly, the theoretician may wish to predict the behavior of the real world by studying the behavior of the model. Except in those few cases in which the equations can be integrated directly, they will finish up on the desk of the numerical analyst with a request to "get some numbers." Let us suppose that the pair of equations

$$\begin{aligned} y' &= (p - tq)y - z \\ z' &= y \end{aligned} \tag{1.1}$$

has just been put in front of us, where y' means dy/dt. What questions must we ask the physicist and what must we ask ourselves before we "put it on the computer"?

Before starting an integration, we must get enough information to specify a problem. Equations (1.1) contain two unknown constants p and q. We will call these *parameters*, and we must know their value. If the problem is basically to determine these parameters by comparing experimental results with numerical integrations for various values of p and q, we may be faced with many integrations, and may need to investigate better methods. A discussion of such problems will be delayed until a later chapter; we will first assume that all parameters are specified.

Consider the equations

$$\begin{aligned} y' &= z \\ z' &= -y \end{aligned} \tag{1.2}$$

These have the solution

$$\begin{aligned} y &= A \sin (t + \alpha) \\ z &= A \cos (t + \alpha) \end{aligned} \tag{1.3}$$

There are two constants of integration A and α. In general, a system of N first order equations has N constants of integration. If we are to provide numerical answers, that is, calculate the values of $y(t)$ and $z(t)$ for a number of values of t, enough information must be given to specify these constants of integration. If the values of the dependent variables y and z are specified for a value of t which we will call the *initial value* t_0, then the solution will in general be determined. (We will assume that a transformation has been made so that $t_0 = 0$. This does not affect the problem or the methods to be discussed.) The problem of determining values of y and z at future times t is then called an *initial value problem*. Alternatively, values of some of the dependent variables may be specified at a number of different values of t. This is called a *boundary value problem*. The most common form of this problem is the *two-*

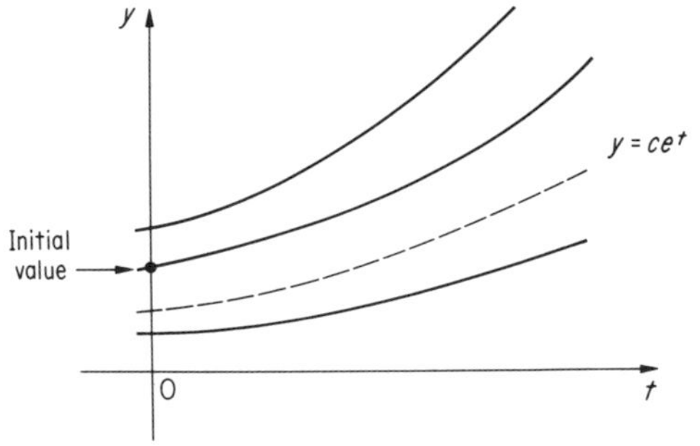

Fig. 1.1 Family of solutions of a first order differential equation.

point boundary value problem, where function values are given for two values of t, say 0 and b. If there were N first order differential equations and M of the dependent variables were specified at $t = 0$, then $N - M$ would have to be specified at $t = b$. Under some conditions, this will lead to a solution.

The integral of a system of differential equations is a family of curves. The integral of $y' = y$, for example, is $y = ce^t$, which is the family shown in Figure 1.1. The choice of an initial value serves to select one of the curves of the family. If there is more than one constant of integration, it is difficult to draw the family of solutions. However, we can illustrate the two-point boundary value problem by considering the solutions (1.3) of Eqs. (1.2). If a value of y is specified for $t = 0$, then different initial values of z at $t = 0$ will give rise to a smaller family of different solutions as shown in Figure 1.2. If y is specified at another value of $t = b$, it serves to select the member of the family required. If, in this example, we had chosen $y(0) = 0$ and $b = \pi$, we would have obtained the family shown in Figure 1.3 for different initial values of z. The specification of $y(\pi)$ does not lead to a well-defined problem in this case. If $y(\pi) = 0$, there are infinitely many solutions; if $y(\pi) \neq 0$, there are no solutions. Thus, we see that a boundary value problem does not always have a unique answer. The same is true for initial value problems, but fortunately it is possible to state a reasonable criterion for initial value problems to have a unique solution. We will only discuss methods for initial value problems, since the class of methods for the two-point boundary value problems is different.

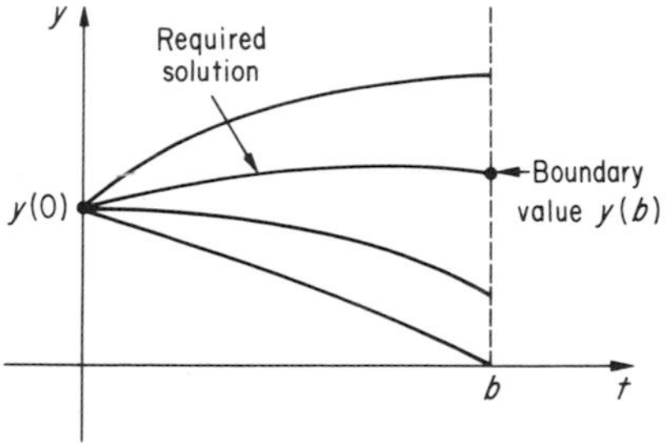

Fig. 1.2 Two-point boundary value problem.

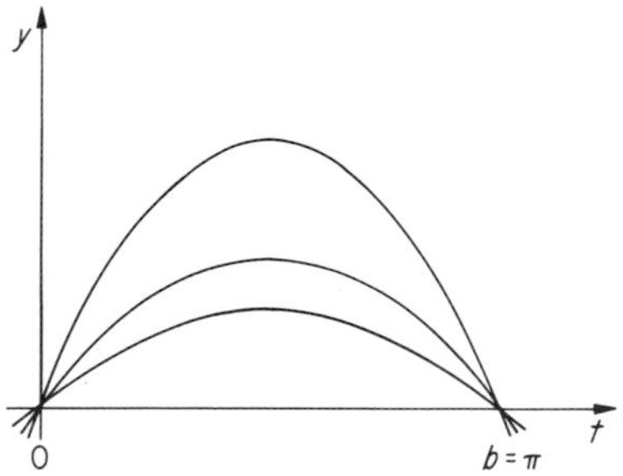

Fig. 1.3 Two-point boundary value problem without a unique solution.

Having gotten enough information to specify the solution desired, we must now ask our physicist what accuracy he requires in his solution. This must be used to determine the method to be applied but it should also be used to check the accuracy of the initial values and parameter values supplied. For now, we will assume that the parameters were given exactly, but let us consider the effect of errors in the initial values. The solution family for $y' = y$ was shown in Figure 1.1. An initial error will cause the wrong member to be chosen. Suppose that

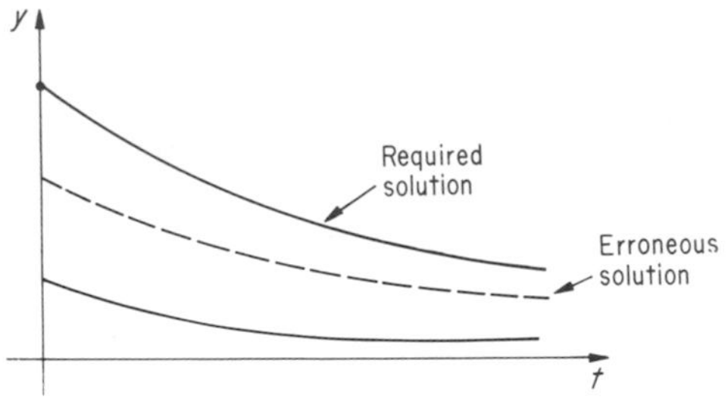

Fig. 1.4 Stable family of solutions.

the dashed rather than the heavy solid curve is selected because of initial value error. As time increases, the difference between the curves increases, in this case by the factor e^t. We call this phenomenon *instability of the equation.* If, on the other hand, we have the equation $y' = -y$, we get the family shown in Figure 1.4. In this case, the error decreases as t increases. This is called *stability of the equation.* If the initial values are given at $t = 0$, and the integration continues to $t = b$, the final error is e^{-b} times the initial error. For the differential equation $y' = \lambda y$, where λ is given, the error growth is $e^{\lambda b}$. If $\lambda \leq 0$, the initial error is not increased, so the equation is stable. If $\lambda > 0$, the equation is unstable.

It is apparent that the accuracy we can obtain is partially limited by the accuracy of the initial values. In a similar manner, the accuracy will be limited by errors in the parameters. Part of the integration task may be to determine the error in the solution due to these errors. If we were able to integrate the equations explicitly, we could do this by examining the family of solutions for all possible initial values. However, if an explicit solution cannot be found, and a numerical solution is used, the numerical method will introduce additional errors. The highest accuracy that can be obtained is bounded by the initial value errors.

Although we may think that we have enough information to solve the equations, we must ask ourselves about the existence of solutions before beginning a numerical integration. Many numerical methods will return some numbers, but it is obvious that they do not have any meaning if the problem has no solution. Specifically, we must be sure that the equations have a unique solution. A standard theorem states:

THEOREM 1.1

If $y' = f(y, t)$ is a differential equation such that $f(y, t)$ is continuous in the region $0 \leq t \leq b$, and if there exists a constant L such that

$$|f(y, t) - f(y^*, t)| \leq L|y - y^*|$$

for all $0 \leq t \leq b$ and all y, y^* (this is called the Lipschitz condition and L is called the Lipschitz constant), then there exists a unique continuously *differentiable function* $y(t)$ *such that*

$$y'(t) = f(y(t), t) \tag{1.4}$$

and $y(0) = y_0$, *the initial condition.* A proof of this can be found in most books on ordinary differential equations. [For example, see Ince (1956), Chapter 3.]

There is no requirement that $f(y, t)$ be differentiable, but if it is, then the Lipschitz condition guarantees that $|\partial f/\partial y| \leq L$. Conversely, if f is differentiable with respect to y, and $|\partial f/\partial y| \leq L$, f satisfies the Lipschitz condition. This is frequently the easiest way of verifying that it is satisfied.

In some cases the Lipschitz condition is not satisfied, but additional restraints on the solution may make it unique. Consider, for example, the equation

$$y' = \sqrt{1 - y^2} = f(y, t)$$

At $y = 1$, $\partial f/\partial y$ is unbounded. In fact, the family of solutions is $y = \sin(t + \alpha)$ shown in Figure 1.5. $y = \pm 1$ are also solutions of a rather special kind. They form the *envelope* of the family of solutions; that is, they are curves that are everywhere tangential to at least one member of the solution family.

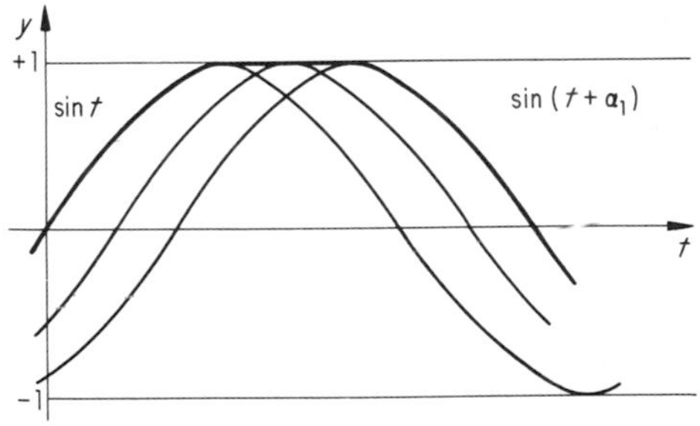

Fig. 1.5 Family of solutions with an envelope singularity.

Consequently, the thick curve shown consisting of a portion of $\sin t$, a portion of $y = 1$, and a portion of $\sin(t + \alpha_1)$ is also a solution for any α_1. It has a continuous first derivative. Thus, for any starting point $-1 \leq y_0 \leq 1$, there exist infinitely many solutions. However, if we also demand a continuous second derivative, there exists a unique solution. In many physical problems we can expect a number of continuous derivatives, and this fact can be used both to guarantee a solution and to assist in the selection of methods. Another example that illustrates both this problem and the fact that an initial value problem may have no solution is the equation

$$y' = \frac{y}{t}$$

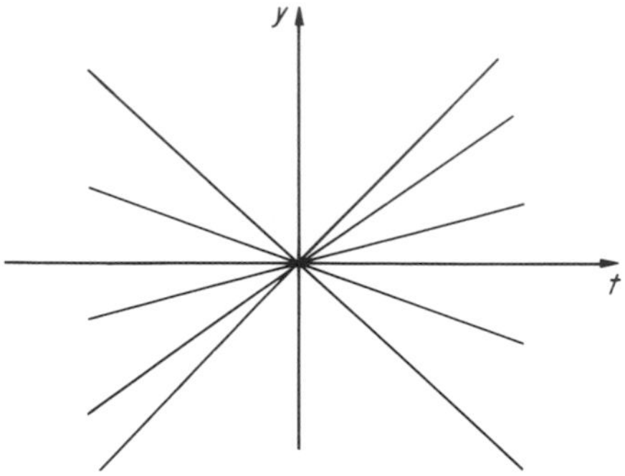

Fig. 1.6 Family of solutions with a point singularity.

The family of solutions is $y = ct$, as shown in Figure 1.6. Since they all pass through $y = t = 0$, initial values cannot be given at the singular point $t = 0$.

In addition to assuring ourselves that the problem has a solution, we must also assure ourselves that it is *well-posed.* By this we mean that small perturbations in the stated problem will only lead to small changes in the answers. This is obviously a useful condition because a numerical approximation to the solution may well introduce perturbations such that a different problem is being solved; it is desirable that the solution can be made as accurate as needed by keeping these perturbations small. Fortunately, the Lipschitz condition is a sufficient condition for an ordinary initial value problem to be well-posed. We can show this by considering the perturbed problem

$$\begin{aligned} z' &= f(z, t) + \delta(t) \\ z(0) &= y_0 + \epsilon_0 \end{aligned} \tag{1.5}$$

where $\delta(t)$ and ϵ_0 are the small perturbations. Let ϵ be the *norm*† $\|\epsilon_0, \delta\|$ defined as $\max\,[|\epsilon_0|, \max_{0 \le t \le b} |\delta(t)|]$. If $\epsilon(t)$ is the difference between the changed solution z and the true solution y, we get, by subtracting (1.4) from (1.5),

$$\epsilon'(t) = f(z, t) - f(y, t) + \delta(t) \qquad |\epsilon(0)| = |\epsilon_0| \le \epsilon$$

Therefore,

$$|\epsilon'(t)| \le |f(z, t) - f(y, t)| + |\delta(t)| \le L|\epsilon(t)| + \epsilon$$

By integration we get

$$|\epsilon(t)| \le \frac{\epsilon}{L}[(L + 1)e^{Lt} - 1]$$

Consequently, the largest change in the solution of the perturbed problem is

†A *norm* is a real positive number which is the measure of the size of something. We will introduce various norms for vectors and matrices later. The important property here is that $\|\epsilon_0, \delta\| = 0$ implies that $\epsilon_0 = 0$ and $\delta(t) = 0$.

bounded by

$$\max_{0\le t\le b} |\epsilon(t)| \le ||\epsilon_0, \delta|| \cdot \frac{1}{L}[(L+1)e^{Lb} - 1] = k\epsilon \tag{1.6}$$

where k is independent of ϵ.

Thus we formally define well-posed by:

DEFINITION 1.1

The ordinary differential equation (1.4) is well-posed with respect to the initial condition y_0 if there exist strictly positive constants k and $\tilde{\epsilon}$ such that for any $\epsilon \le \tilde{\epsilon}$ the perturbed problem (1.5) satisfies

$$|z(t) - y(t)| \le k\epsilon$$

whenever $|\epsilon_0| < \epsilon$, $|\delta(t)| < \epsilon$ for all $0 \le t \le b$.

We can then state:

THEOREM 1.2

If $f(y, t)$ satisfies a Lipschitz condition, (1.4) is well-posed with respect to any initial condition.

In many problems we cannot get a Lipschitz condition for all y, but only in a region of the y-space. [For example, if $f(y, t) = y^2$, $\partial f/\partial y$ exists and is bounded in any finite region.] Theorem 1.1 can be used to guarantee a unique solution while y remains in that region. The proof of Theorem 1.2 is valid as long as both z and y remain in the region, so it may be necessary to limit the maximum perturbation by $\tilde{\epsilon}$. For example, perturbations to $y' = 1/y^2$, $y(0) > 0$ are bounded as long as the perturbation does not reduce y below 0. Consequently, it is well-posed with respect to any positive initial value y_0, but when y_0 is close to 0, $\tilde{\epsilon}$ is small and k is large.

1.2 NUMERICAL APPROXIMATION OF THE SOLUTION

There are two basic approaches to the numerical approximation of solutions of differential equations. One is to represent an approximate solution by the sum of a finite number of independent functions, for example, a truncated power series or the first few terms of an expansion in *orthogonal functions*.† These methods are usually better suited to hand computations, although there has been a lot of work on the application of Chebyshev polynomials to ordinary differential equations.

†A set of functions $\{\phi_n\}$ is *orthogonal* with respect to an interval $[a, b]$ and a weight function $\omega(t)$ if $\int_a^b \omega(t)\phi_n(t)\phi_m(t)\,dt = 0$ for $m \ne n$. Many orthogonal functions have useful properties for numerical approximation purposes. See discussions in A. Ralston (1965), p. 93.

The second approach, the one we are going to examine in this book, is the *difference method.* The solution is approximated by its value at a sequence of discrete points called the *mesh points.* In much of our discussion we will assume that these points are equally spaced and call them $t_i = ih$, where h is the spacing between adjacent points. The end point will usually be called $t_N = b$, so that $N = b/h$. However, the mesh spacing, or step size h, will be seen to affect the error introduced, and what may be a good size of h in one region of the interval will not be good elsewhere. Consequently, we may use a variable step size, in which case we will have

$$t_{i+1} = t_i + h_i, \qquad t_0 = 0$$

A difference method is also called a step by step method and provides a rule for computing the approximation at step i to $y(t_i)$ in terms of the values of y at t_{i-1} and possibly preceding points. We will call this approximation y_i. Ideally, the solution could be represented by its actual value at each mesh point, so that it could be approximated to high accuracy by interpolation between the mesh points. However, two problems interfere with this ideal: First, the exact solution of the differential equation is not, in general, known and cannot be calculated, so the solution to a different problem which can be calculated is sought (the difference between these two solutions will be called *truncation error*); and secondly, numbers cannot be represented exactly in the numerical processes involved (the change introduced by this mechanism will be called *round-off error*). Consequently, the difference method solution will be represented by a finite number of finite precision numbers containing two sources of error, round-off and truncation. The difference methods, also called discrete variable methods, are generally more suited for the automatic computation of general nonlinear problems than are series expansion methods, and are the methods most frequently used in common computer subroutine libraries.

When we think of approximating a solution numerically, we naturally are concerned with how accurate we can make the numerical solution to the actual solution. When we pick a method it may depend on one or more parameters, for example, the step size h (or $\max_i (h_i)$ if the step is variable) or the number of terms in a series expansion. We would like to know how to pick these parameters to achieve any desired accuracy. It is possible that there is an error below which it is not possible to go. *At this point we loosely define the concept of convergence to mean that any desired degree of accuracy can be achieved for any problem satisfying a Lipschitz condition by picking a small enough h.* This definition will be made more precise when specific classes of methods are discussed. Since as h decreases the number of points and hence the amount of calculation increases, we would expect the effect of round-off errors to increase because there are more of them. Therefore, in defining convergence, we must require that the computations indicated in the method be performed exactly. In practice, this means that additional digits are carried in the computations as h decreases.

We previously assured ourselves that the problem was *well-posed* so that the effects of errors are bounded. We also need to know that small changes in the initial values only produce bounded changes in the numerical approximations provided by the method. We term this concept *stability*, which we loosely define in the following sense for discrete variable methods: *If there exists an* $h_0 > 0$ *for each differential equation such that a change in the starting values by a fixed amount produces a bounded change in the numerical solution for all* $0 \leq h \leq h_0$, *then the method is stable.* This definition will be made more precise when specific classes of methods are discussed. We see that stability is related to a method as well-posed is related to a problem. Note that stability does not require convergence, although we will show that the converse is true. Thus, the "method" $y_n = y_{n-1}$, $n = 1, 2, \ldots, N$ is stable, but not convergent for any but the trivial problem $y' = 0$.

The concepts of stability and convergence are concerned with the limiting process as $h \longrightarrow 0$. In practice, we must compute with a finite number of steps, and we are really concerned with the size of errors for such nonzero h. In particular, we want to know if the errors we introduce at each step (truncation and round-off) have a small or large effect on the answer. We, therefore, attempt to define *absolute stability* as follows: "A method is absolutely stable for a given step size h and a given differential equation if the change due to a perturbation of size δ in one of the mesh values y_n is no larger than δ in all subsequent values y_m, $m > n$." Unfortunately, this definition is hopelessly dependent on the problem, so we utilize the idea of a "test equation." We will define absolute stability for the differential equation $y' = \lambda y$, where λ is a complex constant,† and say that *the* ***region of absolute stability*** *is that set of values of h (real nonnegative) and* λ *for which a perturbation in a single value* y_n *will produce a change in subsequent values which does not increase from step to step.*

†A linear system of equations $\mathbf{y}' = A\mathbf{y}$ can be reduced to a set of test equations if A is diagonalizable. If $SAS^{-1} = \Lambda$ is a similarity transformation of A to the diagonal matrix Λ with diagonal entries λ_i, we can write $\mathbf{z} = S\mathbf{y}$, and get

$$S^{-1}\mathbf{z}' = AS^{-1}\mathbf{z}$$

or

$$\mathbf{z}' = SAS^{-1}\mathbf{z} = \Lambda\mathbf{z}$$

This is a set of independent equations each of the form $z_i' = \lambda_i z_i$. In a general nonlinear equation we identify the λ_i with the eigenvalues of the Jacobian $\partial\mathbf{y}'/\partial\mathbf{y}$. These values determine the local behavior of the system to a first approximation. These eigenvalues λ_i may, of course, be complex.

1.3 AN ILLUSTRATION—THE EULER METHOD

It is instructive to examine the simplest method, the Euler method, because it is easy to analyze and is often the basis of constructive existence proofs. In the Euler method, the value of the dependent variable at one point is calculated by straight line extrapolation from a previous point. We consider the single equation

$$y' = f(y, t)$$

and suppose that $y(0) = y_0$ is given. $y'_0 = f(y_0, 0)$ can be calculated and from this we can calculate an approximation to $y(h)$ by using the first two terms of a Taylor's series.

$$y(h) \cong y_0 + hy'(0)$$

We let $t_1 = h$, and define our approximation to $y(t_1)$ as y_1. Thus

$$y_1 = y_0 + hf(y_0, t_0)$$

In general, we define the next value in terms of the current value y_n and t_n by

$$y_{n+1} = y_n + hf(y_n, t_n) \tag{1.7}$$

where

$$t_n = nh$$

At each step we usually cross over onto another member of the family of solutions as shown in Figure 1.7, so we expect the accuracy of the solution to be strongly linked to the stability of the equations. If the equations are very stable, then errors in the early stages will have little effect. On the other hand, if they are unstable, the early errors will have a larger effect.

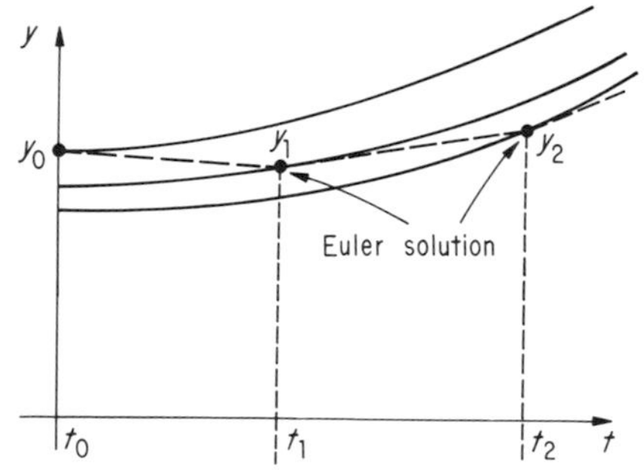

Fig. 1.7 Euler method graphically displayed.

What if the error of the Euler method for a given *step size h* is too large? Can we reduce it? In other words, does the method converge as $h \longrightarrow 0$? The answer is given in the following theorem.

THEOREM 1.3

If $f(y, t)$ satisfies a Lipschitz condition in y and is continuous in t for $0 \leq t \leq b$ and all y, if $h = t/n$, if the sequence y_i, $i = 1, \ldots, n$ is defined by (1.7), and if $y_0 \longrightarrow y(0)$, then *$y_n \longrightarrow y(t)$ as $n \longrightarrow \infty$ uniformly in t, where $y(t)$ is the solution of (1.4) with initial value $y(0)$.*

We will call y_0 the *starting value* to distinguish it from the initial value $y(0)$. In practice we can only expect the starting value used in numerical computations to approach the initial value as h decreases and as we use more precision in our computation. In this theorem we are assuming that (1.7) is solved without round-off errors.

Although we asked for a Lipschitz condition for all y, we only need it in a closed region R that contains the solution $y(t)$ in its interior. The step can then be made small enough to keep the numerical solution also in R so that the condition applies. The fact that the numerical solution and $y(t)$ both remain in a closed region is required in most theorems to infer boundedness of f and its derivatives from any continuity used. The proof of Theorem 1.3 will be discussed below. It consists of deriving a bound for the error

$$e_n = y_n - y(t_n)$$

and showing that this bound can be made arbitrarily small. By making additional assumptions about the function f and about the solution y, we can get better error bounds. If a bound depends only on knowledge of the equation but not on knowledge of the solution $y(t)$, it is called an *a priori bound.* If, on the other hand, a knowledge of the properties of the solution is required, its error bound is called an *a posteriori bound.* Error bounds are usually much larger than the actual errors incurred in a numerical integration, so we sometimes make error *estimates* of an *asymptotic* form. That is, we find a function $e(t)$ such that

$$\frac{e_n}{e(t_n)} = 1 + 0(h) \quad \text{as} \quad h \longrightarrow 0\dagger$$

Error estimates are usually of the a posteriori type.

Proof of Theorem 1.3: To get an a priori bound let us write

$$y(t_{n+1}) = y(t_n) + hf(y(t_n), t_n) + d_n \tag{1.8}$$

d_n is called the *local truncation error.* It is the amount by which the solution fails to satisfy the difference method. Subtracting this from (1.7), we get

$$e_{n+1} = e_n + h(f(y_n, t_n) - f(y(t_n), t_n)) + d_n \tag{1.9}$$

†By $0(h)$ we mean any function of h such that there exists constants h_0 and k independent of h for which $|0(h)| \leq kh$ for all $|h| \leq h_0$.

Let us write

$$f(y_n, t_n) - f(y(t_n), t_n) = e_n L_n \tag{1.10}$$

Therefore,

$$e_{n+1} = e_n(1 + hL_n) + d_n$$

This is a difference equation† for e_n. The error e_0 is known, so it can be solved if we know L_n and d_n. We have a bound of the Lipschitz constant L for $|L_n|$. Suppose we also have $D \geq |d_n|$. Then we have

$$|e_{n+1}| \leq |e_n|(1 + hL) + D \tag{1.11}$$

This difference equation occurs frequently in this work, and leads to:

Lemma 1.1

If $|e_n|$ satisfies (1.11) and $0 \leq nh \leq b$, then

$$\begin{aligned} |e_n| &\leq D\frac{(1 + hL)^n - 1}{hL} + (1 + hL)^n|e_0| \\ &\leq \frac{D}{hL}(e^{Lb} - 1) + e^{Lb}|e_0| \end{aligned} \tag{1.12}$$

The first inequality in (1.12) follows by induction. It is trivially true for $n = 0$. Assuming it is true for n, we have from (1.11)

$$\begin{aligned} |e_{n+1}| &\leq D\frac{(1 + hL)^n - 1}{hL}(1 + hL) + D + (1 + hL)^{n+1}|e_0| \\ &= D\frac{(1 + hL)^{n+1} - (1 + hL) + hL}{hL} + (1 + hL)^{n+1}|e_0| \\ &= D\frac{(1 + hL)^{n+1} - 1}{hL} + (1 + hL)^{n+1}|e_0| \end{aligned}$$

Hence (1.12) is true for $n + 1$. The second inequality in (1.12) follows from the fact that $nh \leq b$, and for $hL \geq 0$, $1 + hL \leq e^{Lh}$, so that $(1 + hL)^n \leq e^{Lnh} \leq e^{Lb}$, proving the lemma.

To continue the proof of Theorem 1.3 we need to investigate D, the bound on the local truncation error. From (1.8),

$$-d_n = y(t_{n+1}) - y(t_n) - hf(y(t_n), t_n)$$

By the mean value theorem, we get for $0 \leq \theta \leq 1$

$$\begin{aligned} |d_n| &= |hf(y(t_n + \theta h), t_n + \theta h) - hf(y(t_n), t_n)| \\ &\leq h|f(y(t_n), t_n + \theta h) - f(y(t_n), t_n)| \\ &\qquad + h|f(y(t_n + \theta h), t_n + \theta h) - f(y(t_n), t_n + \theta h)| \\ &\leq h|f(y(t_n), t_n + \theta h) - f(y(t_n), t_n)| + hL|(y(t_n + \theta h) - y(t_n))| \end{aligned} \tag{1.13}$$

†Difference equations are treated in many texts, such as P. Henrici (1963), Chapter 3. This is a particularly simple difference equation. More complex ones will be discussed in later chapters.

The last term can be treated by the mean value theorem to get a bound

$$L\theta h^2|y'(\xi)| \leq h^2 LZ$$

where $Z = \max|y'(t)|$, which exists because of the continuity of y and f in a closed region. The treatment of the first term in (1.13) depends on our hypotheses. The proof can be continued with no additional hypotheses. See, for example, Henrici (1962), Section 1.2.4. However, if we are prepared to assume that $f(y, t)$ also satisfies a Lipschitz condition in t (as will happen in at least a piecewise manner in practice), we can bound the first term of (1.13) by $K\theta h^2$, where K is the Lipschitz constant for f as a function of t. Consequently,

$$|d_n| \leq h^2(K + LZ) = D$$

so from (1.12) we get

$$|e_n| \leq h\frac{K + LZ}{L}(e^{Lb} - 1) + e^{Lb}|e_0| \tag{1.14}$$

Thus, the numerical solution converges as $h \longrightarrow 0$ if $|e_0| \longrightarrow 0$.

1.3.1 Error Estimates

In proving convergence above, we derived the error bound (1.14) for the solution. This demonstrated that the error behaves like $0(h)$ if $\partial f/\partial t$ exists, is continuous, and is bounded. Generally the function f will be differentiable, and the bounds K, L, and Z can be calculated, at least for a finite region of the y space within which the actual and numerical solutions can be shown to lie. However, the error bound so obtained may not be very good because the largest values of $|y'|$, $|\partial f/\partial y|$, and $|\partial f/\partial t|$ will have to be chosen. If we have some knowledge of the solution, and assume that its second derivative is continuous and bounded by a known quantity, say C, we can get a better bound. We first express d_n by using a Taylor's series expansion at t_n with a remainder term to get

$$-d_n = y(t_{n+1}) - y(t_n) - hf(y(t_n), t_n) = \frac{h^2}{2}y''(\xi)$$

for

$$\xi \in (t_n, t_{n+1})$$

Therefore,

$$|d_n| \leq \frac{Ch^2}{2}$$

and

$$|e_n| \leq \frac{h}{2}\frac{C}{L}(e^{Lb} - 1) + e^{Lb}|e_0| \tag{1.15}$$

This is an a posteriori bound because it depends on a knowledge of the second derivative of the solution, but by writing

$$y'' = (y')' = \frac{df}{dt} = \frac{\partial f}{\partial t} + \frac{\partial f}{\partial y}y'$$

we can convert it back to an a priori bound when y'' exists and is continuous.

It is difficult to improve the bound given in (1.15) for the error, but we can instead look for an estimate of the error. Since we know that if y'' exists the error behaves as $0(h)$, we try to express it in the form $e_n = h\delta_n$ and get an estimate of the size of δ_n. To do this we will assume that y has a continuous third derivative. Then we can write

$$d_n = -\frac{1}{2}h^2y''(t_n) - \frac{h^3}{6}y'''(\xi)$$

where $\xi \in (t_n, t_{n+1})$, and Eq. (1.9) can be written as

$$e_{n+1} = e_n + h(f(y_n, t_n) - f(y(t_n), t_n)) - \frac{1}{2}h^2y''(t_n) - \frac{h^3}{6}y'''(\xi) \tag{1.16}$$

The parenthesized term can be expressed as

$$e_n\frac{\partial f}{\partial y}(y(t_n), t_n) + \frac{1}{2}e_n^2\frac{\partial^2 f}{\partial y^2}(\bar{\xi}, t_n)$$

by a Taylor's series with a remainder term. If we replace e_n in (1.16) by $h\delta_n$, we get

$$\delta_{n+1} = \delta_n + h\left(\delta_n\frac{\partial f}{\partial y} - \frac{1}{2}y''\right) + \frac{1}{2}e_n^2\frac{\partial^2 f}{\partial y^2}(\bar{\xi}) - \frac{h^2}{6}y'''(\xi) \tag{1.17}$$

All derivatives are evaluated at t_n, $y(t_n)$ except where indicated otherwise. We have from (1.15) that $|e_n| < hK_1$, hence (1.17) gives

$$\delta_{n+1} = \delta_n + h\left(\delta_n\frac{\partial f}{\partial y} - \frac{y''}{2} + hc_n\right) \tag{1.18}$$

where

$$|c_n| \leq \frac{K_1^2}{2}\max\left|\frac{\partial^2 f}{\partial y^2}\right| + \frac{1}{6}\max|y'''| \leq K_2$$

Notice that (1.18) is the numerical solution, by Euler's method, of the differential equation

$$\frac{d\delta}{dt} = g(t)\delta - \frac{1}{2}y'' + hc(t), \qquad \delta(0) = \frac{e_0}{h} \tag{1.19}$$

where $c(t)$ is any function with values c_n at $t = t_n$, and $g(t) = \partial f/\partial y$ evaluated for $y = y(t)$. The actual solution of (1.19) is equal to the numerical solution (1.18) plus $0(h)$, as is shown by (1.14), while the actual solution of (1.19) is the solution of

$$\frac{d\delta}{dt} = g(t)\delta - \frac{1}{2}y'', \qquad \delta(0) = \frac{e_0}{h} \tag{1.20}$$

plus a bounded multiple of $h \max c(t) \leq kK_2$ since the problem is well-posed. (See Theorem 1.2.) Therefore, we have shown the following error estimate:

THEOREM 1.4

The error in Euler's method has the form

$$e_n = h\delta(t_n) + 0(h^2) \tag{1.21}$$

where $\delta(t)$ is the solution of (1.20) if y is continuously differentiable three times.

1.3.2 Comparison of Error Estimates with Actual Errors

We now have the bounds (1.14) and (1.15) and the estimate (1.21). Let us compare them with the actual error in some examples. Consider $y' = y$, $y(0) = 1$. The solution is $y(t) = e^t$, so $y(1) \cong 2.71828$. Euler's method gives $y_{n+1} = y_n + hy'_n = y_n(1 + h)$. Therefore, $y_N = (1 + h)^N$. If $t_N = 1$, $y_N = (1 + h)^{1/h}$. The error bound (1.14) with $e_0 = 0$ gives

$$|e_N| \leq h2.71828(2.71828 - 1) = 4.67077h$$

while (1.15) gives

$$|e_N| \leq h1.35914(2.71828 - 1) = 2.33539h$$

The error estimate of (1.21) gives

$$e_N \cong h\delta(t)$$

where δ is the solution of

$$\delta'(t) = \delta(t) - \tfrac{1}{2}e^t, \qquad \delta(0) = 0$$

Consequently, $\delta(t) = -\frac{1}{2}te^t$, and $\delta(1) = -1.35914$. These are shown for several values of h in Table 1.1.

We can see that as $h \longrightarrow 0$ the error approaches the estimate and that the bounds are unduly pessimistic. Now let us look at

$$y' = -y, \qquad y(0) = 1$$

The solution is $y = e^{-t}$, so the largest value of $|y'|$ and $|y''|$ in [0, 1] is 1.

Table 1.1 ACTUAL ERRORS VERSUS BOUNDS AND ESTIMATES FOR $y' = y$

h	y_N	e_N	$e_{N/h}$	*Bound* (1.14)/h	*Bound* (1.15)/h	*Estimate* (1.21)/h
1	2	−0.71828	−0.71828	4.67077	2.33539	−1.35914
$\frac{1}{2}$	2.25	−0.46828	−0.93656	4.67077	2.33539	−1.35914
$\frac{1}{4}$	2.44141	−0.27688	−1.10750	4.67077	2.33539	−1.35914
$\frac{1}{8}$	2.56578	−0.15250	−1.21998	4.67077	2.33539	−1.35914
$\frac{1}{16}$	2.63793	−0.08035	−1.28565	4.67077	2.33539	−1.35914
$\frac{1}{32}$	2.67699	−0.04129	−1.32134	4.67077	2.33539	−1.35914
$\frac{1}{64}$	2.69734	−0.02094	−1.33997	4.67077	2.33539	−1.35914

Consequently, the bounds (1.14) and (1.15) give 1.71828h and 0.85914h, respectively. Since $y'' = e^{-t}$, δ is given by

$$\delta'(t) = -\delta(t) - \tfrac{1}{2}e^{-t}, \qquad \delta(0) = 0$$

so $\delta(t) = -\frac{1}{2}te^{-t}$ and $\delta(1) = -0.18394$. A comparison of errors for various step sizes is shown in Table 1.2. We see from this that the estimate from Eq. (1.21) gives a smaller number than the actual error, but again, as the step decreases, the actual error approaches the estimated error.

Table 1.2 ACTUAL ERRORS VERSUS BOUNDS AND ESTIMATES FOR $y' = -y$

h	y_N	e_N	$e_{N/h}$	*Bound* (1.14)/h	*Bound* (1.15)/h	*Estimate* (1.21)/h
1	0	−0.36788	−0.36788	1.71828	0.85914	−0.18394
$\frac{1}{2}$	0.25	−0.11788	−0.23576	1.71828	0.85914	−0.18394
$\frac{1}{4}$	0.31641	−0.05147	−0.20589	1.71828	0.85914	−0.18394
$\frac{1}{8}$	0.34361	−0.02427	−0.19416	1.71828	0.85914	−0.18394
$\frac{1}{16}$	0.35607	−0.01181	−0.18889	1.71828	0.85914	−0.18394
$\frac{1}{32}$	0.36206	−0.00582	−0.18637	1.71828	0.85914	−0.18394
$\frac{1}{64}$	0.36499	−0.00289	−0.18515	1.71828	0.85914	−0.18394

The additional error introduced by a single step of a method is usually called the *local truncation error*. It is the difference between the value given by the method and the value of the solution of the differential equation which passes through the value at the beginning of the step. In the Euler method we can find it by comparing

$$y(t_{n+1}) = y(t_n) + hy'(t_n) + \tfrac{1}{2}h^2y''(\xi)$$

with the Euler formula. The local truncation error is the additional term $\frac{1}{2}h^2y''(\xi)$. Note that this, apart from the h^2 factor, is essentially the inhomogeneous term in Eq. (1.20) for the error estimate. The truncation error in one step of the Euler method affects the next step in the same way that an initial value error would if it had moved the numerical solution onto the same new member of the family of solutions. In later chapters we will consider methods in which an error introduced at one step may have effects in several later steps because information is used from several different values of t in order to get a good approximation at a new value of t. In this case, the effect of the local error will enter into the equivalent of Eq. (1.20) in a similar form.

1.3.3 Stability

A change in one of the calculated values from y_n to z_n will cause us to solve

$$z_{m+1} = z_m + hf(z_m, t_m)$$

instead of (1.7). Subtracting (1.7) from this and setting $e_n = z_n - y_n$, we get

$$|e_{m+1}| \leq |e_m| + hL|e_m|$$

or

$$|e_N| \leq (1 + hL)^{N-n}|e_n|$$
$$\leq e^{bL}|e_n|$$

which is a bounded multiple of the introduced error $|e_n|$ and is independent of h. Hence the method is stable.

To examine absolute stability, we consider the equation $y' = \lambda y$. For this,

$$y_{n+1} = y_n + \lambda h y_n = (1 + \lambda h)y_n$$

Consequently, Euler's method is absolutely stable in the region $|1 + \lambda h| \leq 1$, which is a unit circle in the complex λh-plane centered at $(-1, 0)$. We can see the effect of absolute stability very clearly in the following example:

$$y' = -1000(y - t^2) + 2t$$
$$y(0) = 0$$

Calculate $y(1)$

This was done by Euler's method with step sizes of 10^{-m}, $m = 0, 1, 2, \ldots, 5$, on a machine with 10 significant digits of accuracy. The answers were as listed in Table 1.3.

Table 1.3 EFFECT OF ABSOLUTE STABILITY

h	N	$y(1)$
1	1	0
0.1	10	$0.90438207503 \times 10^{16}$
0.01	100	overflow
0.001	1000	0.99999900001
0.0001	10000	0.99999990000
0.00001	100000	0.99999998997

Overflow means that the largest number retainable in the computer was exceeded, in this case, 10^{38}. This occurred because $\partial f/\partial y$ for this problem is -1000, so with $h = 0.01$, the error is amplified by $(1 - \frac{1000}{100}) = -9$ at each step. In 100 steps, the error in the first step is increased by nearly 10^{100} in size. Once $|1 + h\lambda|$ was less than one, the answers were reasonable, and converged to the correct answer. The error for a step size of 10^{-k} is very close to 10^{-k-3}, showing the linear dependence on h *within the region of absolute stability.*

Error bounds (1.14) and (1.15) are valid regardless of absolute stability, but for this problem $L = 1000$ and $C = 2$, hence (1.15) only yields the bound

$$|e_n| \leq \frac{h}{1000}(e^{1000} - 1)$$

which is too large to be of practical value. In this particular case we can show that if $e_n = y_n - t_n^2$,

$$e_{n+1} = e_n - 1000he_n - h^2$$

so that

$$e_N = \frac{h}{1000}((1 - 1000h)^N - 1)$$

For $h \leq .001$, $N = h^{-1}$,

$$|e_N| \cong \frac{h}{1000}$$

but for $h \geq .01$, $N = h^{-1}$,

$$|e_N| \cong \frac{(1000)^{N-1}}{N^{N+1}}$$

which shows the critical dependence on absolute stability.

1.3.4 Round-Off Errors

We have assumed that all the numerical calculations used in the Euler method were performed exactly. In practice, *round-off errors* occur due to the finite number of digits used in the calculation. Henrici (1962) examines round-off for fixed point machines in great detail and the reader can refer there for a statistical treatment. We will examine the effects of floating point round-off errors briefly. Each time that the Euler formula is evaluated, an additional term r_n is added due to numerical inaccuracy. Thus we have

$$y_{n+1} = y_n + hf(y_n, t_n) + r_n \tag{1.22}$$

and r_n acts in the same way as an additional local truncation error. At each step it may move the solution onto a different family as shown in Figure 1.7. This can be expressed along with the truncation error in Eq. (1.20) if there is a function $r(t)$ such that $r(t_n) = r_n$. Let us assume the existence of such a function. Then the analysis that leads to (1.20) would show that

$$e_n = h\delta(t_n) + 0(h^2)$$

where $\delta(t)$ is the solution of

$$\delta' = g(t)\delta - \tfrac{1}{2}y'' + h^{-2}r(t) \tag{1.23}$$

The h^{-2} appears in (1.23) in the same way that h^2 was dropped from the local truncation error $\frac{1}{2}h^2y''$. This shows us that $e_n = 0(h^{-1}|r| + h)$, where $|r|$ is a measure of the dependence of $r(t)$ on h.

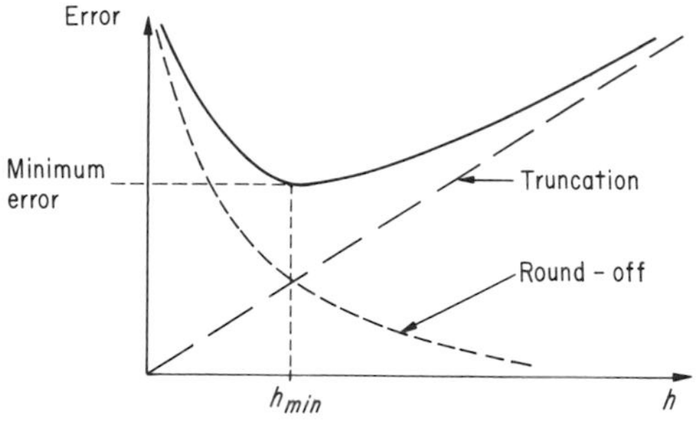

Fig. 1.8 Error as a function of h.

If $|r|$ is independent of h, as would happen if the precision is kept fixed, the total error will initially decrease as h decreases and the truncation error decreases, then increase as round-off becomes significant. This is shown in Figure 1.8.

In most integrations performed on a digital computer, we do not need to achieve the minimum error, so an h well in excess of $h_{\min}$ can be used. However, if more accuracy is required, so that h has to be reduced below $h_{\min}$, it will be necessary to increase the precision so as to reduce the size of $h^{-1}|r|$.

If a worst-case round-off of ϵ were to occur every time, then $r(t) = \epsilon$ and $e_n = 0(h^{-1})$. In other words, in this case, the numerical solution does not converge as $h \longrightarrow 0$. A moment's reflection will convince the reader that this is reasonable. If the amount of error introduced at each step is fixed, and the number of operations increases, the total error introduced can be expected to increase. In practice, computations are done in floating point. Thus, the number $hf(t_n, y_n)$ is calculated to a given number of significant digits. However, the addition of $hf(t_n, y_n)$ to y_n will cause an error of size ϵy_n, where ϵ is in the range of $[-\frac{1}{2}, \frac{1}{2}]10^{-k}$ and k is the number of significant digits retained, assuming that rounding is done.† If the method is to converge in practice, it means that the number of digits of accuracy in $hf(t_n, y_n)$ must be increased as h decreases so that the value of this error is $o(h)$,‡ and the precision in the calculation $y_n + hf(t_n, y_n)$ must be increased so that $10^{-k} = o(h)$. Thus, for each ten-fold decrease in h, more than one decimal digit of precision must be added.

Fortunately, the round-off error is not constant and does not take the worst case every time. We often assume that it is randomly distributed over its range with a *rectangular distribution*, that is, any value is just as likely as

†Although this should be done, there are a number of machines that do not do this as a matter of course. Many FORTRAN compliers on the IBM 7090 series machines give a truncated answer to multiplication, for example, rather than a rounded one. The effects can be significant on a lengthy integration since the error ϵ then lies in the range $[0, 1]10^{-k}$.

‡The notation $o(h)$ means a function of h which goes to zero faster than h, thus $o(h)/h \longrightarrow 0$ as $h \longrightarrow 0$.

any other value. Under this assumption, we can say something about the effect of round-off error over a number of steps. If the $g(t)$ appearing in (1.22) is zero, then the effect of round-off can be found by simply summing each of the round-off errors. The sum of N errors in the range $[-\frac{1}{2}, \frac{1}{2}]$ will lie in the interval $[-N/2, N/2]$. However, we also ask "how far can we expect the answer to be from the center of this interval?" This question is usually answered by evaluating the *standard deviation*, which is the square root of the average value of $(R - \mu)^2$, where R is the answer and μ is the mean of R. In this case, μ is 0. If the errors at each step are independent of each other, then the standard deviation of a sum of N random variables each rectangularly distributed in the range $[-\epsilon/2, \epsilon/2]$ is $\epsilon\sqrt{N/12}$. This tells us that, although the error may be as bad as $N\epsilon/2$, we do not expect it to be much worse than $\epsilon\sqrt{N/12}$. In fact, it can be shown that for large N, there is a 99.73% probability that the error will be less than three standard deviations.†

If the round-off error is *biased*, that is, if it has a mean that is not zero, then the mean of the sum of N such errors has a mean that is N times as large. Thus, machines that chop off the extra digits after operations such as multiplication (confusingly, this is called truncation) can cause much larger errors. Consequently, multiplication on the IBM 7094 produces an error in the range $[0, 2^{-27}]$ with a mean of 2^{-28} relative to the value of the answer. After N steps, the mean is $2^{-28}N$, which grows linearly with N.

The calculations in Table 1.1 were performed to eight significant digits, so that round-off was not a problem. Table 1.4 shows the effect of performing the same calculations to only three significant decimal digit accuracy. In one column we see the answers with truncation performed (y_T), while in the next column correct rounding was performed at each step to get y_R. The errors in the two answers are shown in the next two columns. We see that they first decrease as the truncation of the Euler method decreases, then

Table 1.4 ROUND-OFF ERRORS

	Numerical solutions		*Errors*		*Truncation error e from table 1.1*	*Round-off errors*	
h	y_T	y_R	$e_T = 2.72 - y_T$	$e_R = 2.72 - y_R$		$e_T - e$	$e_R - e$
$\frac{1}{2}$	2.25	2.25	0.47	0.47	0.47	0	0
$\frac{1}{4}$	2.43	2.44	0.29	0.28	0.28	0.01	0
$\frac{1}{8}$	2.43	2.48	0.29	0.24	0.15	0.14	0.09
$\frac{1}{16}$	2.41	2.55	0.31	0.17	0.08	0.23	0.09
$\frac{1}{32}$	2.33	2.56	0.39	0.16	0.04	0.33	0.12
$\frac{1}{64}$	1.64	1.78	1.08	0.94	0.02	1.06	0.92

†See P. Henrici (1962), Section 1.5.2.

increase as round-off becomes significant. The actual truncation error e from Table 1.1 is shown, so that the final two columns show the contribution of the round-off error.

1.3.5 The Perturbation Due to the Numerical Approximation

For many problems the differential equation only represents an approximation to the physical problem. There may be omitted terms, errors in the coefficients, etc. Hence the physical system is behaving according to

$$w'(t) = f(w(t), t) + s(t)$$
$$w(0) = y(0) + d_0$$

where d_0 and $s(t)$ represent perturbations to the equation due to inaccuracies of the problem statement. In this section we will show that the numerical solution is equivalent to the solution of the differential equation with a different perturbation. That is, $y_n = z(t_n)$, where z is given by

$$z'(t) = f(z(t), t) + r(t)$$
$$z(0) = y(0) + e_0 = y_0$$

Furthermore, $\|r(t)\|$ can be made arbitrarily small by choosing a small enough h and sufficient precision. Hence the perturbations due to numerical inaccuracies can be made arbitrarily small compared to the perturbations due to problem inaccuracies.

Since the numerical solution is only specified on a finite point set, there are many different functions $z(t)$ which agree with the numerical solution on that set. We are interested in one for which the *residual* $r(t) = z'(t) - f(z(t), t)$ is small.

Let us write $y(t; x, \tau)$ for the solution of $y' = f(y, t)$ which passes through the point (x, τ) in the (y, t)-plane. (We will deal only with points inside a region in which f satisfies continuity and Lipschitz conditions.) We consider the solution of the differential equation that passes through (y_n, t_n). If a Euler step of size τ is taken from (y_n, t_n), we define the *local error* to be

$$d(\tau; y_n, t_n) = y_n + \tau f(y_n, t_n) - y(t_n + \tau; y_n, t_n) \tag{1.24}$$

It is the amount by which the Euler solution differs from the true solution passing through (y_n, t_n). It is, in other words, the local truncation error for the solution $y(t; y_n, t_n)$. [See Eq. (1.8).] Since we have seen that the local truncation error behaves like h^2 for smooth solutions, let us write

$$d(\tau; y_n, t_n) = \tau^2 T(\tau; y_n, t_n) \tag{1.25}$$

and assume that T is a differentiable function of τ. (It will be if y''' is continuous.)

The numerical method generates the solution

$$y_{n+1} = y_n + hf(y_n, t_n) + R$$

where R is the rounding error. Let us now define† $z(t)$ in the interval $(t_n, t_n + h)$ by

$$\begin{aligned} z(t_n + \tau) = y_n + \tau f(y_n, t_n) + \frac{\tau R}{h} \\ + \left(\frac{h}{\tau} - 1\right)(y_n + \tau f(y_n, t_n) - y(t_n + \tau; y_n, t_n)) \end{aligned} \tag{1.26}$$

Note that $z(t_n) = y_n$ and $z(t_n + h) = y_{n+1}$, so that the $z(t)$ so defined is continuous, passes through the numerical solution, and is differentiable everywhere except on a finite point set. It is differentiable from the right everywhere and its derivative is continuous except on a finite point set.‡ Now let us examine the residual of this $z(t)$.

$$r(t_n + \tau) = z'(t_n + \tau) - f(z(t_n + \tau), t_n + \tau)$$

Substitute from (1.24), (1.25), and (1.26) to get

$$\begin{aligned} r(t_n + \tau) &= \frac{d}{d\tau}\left[y(t_n + \tau; y_n, t_n) + \frac{\tau R}{h} + h\tau T(\tau; y_n, t_n)\right] \\ &\quad - f(z(t_n + \tau), t_n + \tau) \\ &= \frac{d}{d\tau}y(t_n + \tau; y_n, t_n) + \frac{R}{h} + hT(\tau; y_n, t_n) \\ &\quad + h\tau\frac{d}{d\tau}T(\tau; y_n, t_n) - f(z(t_n + \tau), t_n + \tau) \end{aligned}$$

which gives

$$\begin{aligned} r(t_n + \tau) = \frac{R}{h} + hT(\tau; y_n, t_n) + h\tau\frac{d}{d\tau}T(\tau; y_n, t_n) \\ + f(y(t_n + \tau; y_n, t_n), t_n + \tau) - f(z(t_n + \tau), t_n + \tau) \end{aligned} \tag{1.27}$$

Thus the numerical solution is given by the solution of the differential equation

$$z'(t) = f(z(t), t) + r(t)$$

where $r(t)$ is defined by (1.27). $r(t)$ can be bounded by use of the Lipschitz condition and bounds on the derivatives. Note that for the Euler method

$$\begin{aligned} \tau^2 T(\tau; y_n, t_n) &= y_n + \tau f(y_n, t_n) - y(t_n + \tau; y_n, t_n) \\ &= -\frac{\tau^2}{2}\frac{d^2}{dt^2}y(t_n; y_n, t_n) - \frac{1}{6}\tau^3\frac{d^3}{dt^3}y(\xi_\tau; y_n, t_n) \end{aligned} \tag{1.28}$$

†This is, unfortunately, one of those functions that is "pulled out of a hat" and then can be seen to provide the required answer. It is possible to argue from the result back to this function or something similar, but to do so obscures the important issues.

‡It is possible to define a $z(t)$ that is continuously differentiable everywhere but it only complicates the definition and does not aid in the analysis.

while by differentiation

$$2\tau T(\tau; y_n, t_n) + \tau^2 \frac{dT}{d\tau}(\tau; y_n, t_n)$$

$$= f(y_n, t_n) - \frac{d}{dt} y(t_n + \tau; y_n, t_n) \tag{1.29}$$

$$= -\tau \frac{d^2}{dt^2} y(t_n; y_n, t_n) - \frac{\tau^2}{2} \frac{d^3}{dt^3} y(t_n + \xi_\tau; y_n, t_n)$$

Hence if M is a bound for the third derivative of y in the region under consideration, we can take h/τ times (1.29) minus h/τ^2 times (1.28) to get

$$\left| hT(\tau; y_n, t_n) + h\tau \frac{d}{d\tau} T(\tau; y_n, t_n) + \frac{hy''_n}{2} \right| \leq \frac{2}{3} h\tau M$$

where $y''_n = (d^2/dt^2)y(t_n; y_n, t_n)$. If T is a bound on $T(\tau; y_n, t_n)$ for $\tau \leq h$, then

$$|f(y(t_n + \tau; y_n, t_n), t_n + \tau) - f(z(t_n + \tau), t_n + \tau)| \leq LT\tau^2$$

and we obtain from (1.27)

$$\begin{aligned} \left| r(t_n + \tau) - \frac{R}{h} + \frac{h}{2} y''_n \right| &\leq LT\tau^2 + \frac{2}{3} h\tau M \\ &\leq \left(LT + \frac{2}{3} M \right) h^2 \end{aligned} \tag{1.30}$$

This shows that $r(t)$ is essentially h^{-1} (round-off plus truncation errors), although in this case the second derivative y''_n is evaluated at the numerical solution y_n rather than at $y(t_n)$, as was done in (1.20).

PROBLEMS

1. Are the initial value problems given below well-posed?
   (a) $y' = \sqrt{1 - y^2}$, $y(0) = 0$.
   (b) $y' = +\sqrt{y^2 - 1}$, $y(0) = 2$.

2. What step size would you use with Euler's method to integrate $y' = 2t$ from $t = 0$ to $t = 1$ in order to achieve errors of not more than the following?
   (a) 0.1.
   (b) 0.01.
   (c) 0.001.

3. Plot roughly and describe the principle features of the graph of log (error at $t = 100$) versus log (h) for the equation

   $$y' = -20(y - t^2) + 2t, \qquad y(0) = 0$$

   integrated by the Euler method with step size h. Show h in the range 0.0001 to 1. Assume that the machine used for the integration truncates (not rounds) all numbers to 10^{-6}.

4. Suppose you are asked to integrate the following equations from $t = 0$ to $t = 1$ in order to get less than 0.05 error using Euler's method. Approximately what step size would you use? Ignore round-off errors.
   (a) $dy/dt = 1 + 3t^2 - y + t^3$, $y(0) = 1$.
   (b) $dy/dt = 2t + 1000(2 + t^2) - 1000y$, $y(0) = 2$.

5. When a circle is to be continuously plotted on a computer graphical display, successive values of x and y can be computed using

$$x = r\cos\theta$$
$$y = r\sin\theta$$
$$\theta = nh \qquad n = 0, 1, \ldots$$

   This is very time consuming. Instead, we can solve the differential equation

$$\frac{dx}{d\theta} = -r\sin\theta = -y, \qquad x(0) = r$$
$$\frac{dy}{d\theta} = r\cos\theta = x, \qquad y(0) = 0$$

   If Euler's method is used, we get

$$x_{n+1} = x_n - hy_n$$
$$y_{n+1} = y_n + hx_n$$

   This is a poor method. In fact, it forms a spiral. Why? If, instead, the formula below is used, the points form a closed figure. Why? What effect would round-off errors have on your arguments?

$$x_{n+1} = x_n - hy_n$$
$$y_{n+1} = y_n + hx_{n+1}$$

6. Derive some error bounds for the following problems, each of which is to be integrated over [0, 1] with step size h and initial value $y(0) = 1$. Compare the bounds with the asymptotic error estimates for the same problems.
   (a) $y' = 2ty$.
   (b) $y' = -2ty$.
   (c) $y' = (2 + t^2 - y^2)^{1/2}$.
   (d) $y' = t^{1/2}$.
   (e) $y' = t\sqrt{y - 1 + 3t^3}$.

2 HIGHER ORDER ONE-STEP METHODS

The Euler method discussed in Chapter 1 is called a *one-step method* because it is an algorithm which prescribes the numerical technique for calculating the approximation to the solution at t_{n+1} in terms of the value at one previous step, namely at t_n. We saw that the error in Euler's method behaved like $0(h)$ as $h \longrightarrow 0$. For this reason we call it a *first order method*. If we had a method in which the error behaved like $0(h^r)$ for some $r > 0$, we would call it an *rth order method*. We note that in Euler's method, the *local truncation error* is $0(h^2)$, a power one higher than the *global error* (that is, the error over the whole range). We will show in Chapter 4 that an rth order method will have a local truncation error equal to $0(h^{r+1})$, or rather that the global error is $0(h^{-1} \times$ local truncation error), but we will leave the proofs of such theorems until that chapter, and limit our discussion in this chapter to practical aspects of one-step methods that have a higher order than Euler's method.

We first question why we should be interested in higher order methods. A detailed discussion will be delayed until Chapter 5 when we have the necessary background. A simple answer at this stage is to point out that if h is small, then h^2 is even smaller, so more accuracy can be achieved with higher order methods for small h, and as h goes to zero a higher order method will converge faster.

2.1 THE TAYLOR'S SERIES METHOD

We can view the Euler method as an approximation by the first two terms of a Taylor's series. If we can calculate the higher order derivatives of y, we can write

$$y_{n+1} = y_n + hy'_n + \frac{h^2}{2}y''_n + \cdots + \frac{h^r}{r!}y_n^{(r)} \tag{2.1}$$

We have $y'_n = f(y_n, t_n)$, so we could evaluate higher derivatives as follows. Write f for $f(y, t)$, f_t for $\partial f/\partial t$, f_y for $\partial f/\partial y$, etc. We then have for the next two derivatives

$$\begin{aligned} y'' &= f_t + f_y f \\ y''' &= f_{tt} + f_{ty}f + f_{yt}f + f_y f_t + f_{yy}f^2 + f_y^2 f \\ &= f_{tt} + 2f_{ty}f + f_y f_t + f_{yy}f^2 + f_y^2 f \end{aligned} \tag{2.2}$$

It is immediately evident that this is not a practical method unless the function f is simple enough that many of these partial derivatives vanish, but it is theoretically possible to develop as many formulas like (2.2) as necessary and to evaluate these derivatives numerically for substitution in Eq. (2.1). The local truncation error will be

$$\frac{h^{r+1}y^{(r+1)}}{(r+1)!}$$

Differentiation of reasonably complex functions $f(y, t)$ by hand is tedious and error prone. Computer techniques for symbolic differentiation† can be used, although they may lead to very complex formulas that are unnecessarily long to evaluate. In general, the Taylor's series method is not practical for computer solutions, but is more useful for developing low accuracy approximations (by hand or by desk calculator) of simple problems. The method is not useful if the function $f(y, t)$ can only be evaluated by complex subroutines which are effectively impossible to differentiate.

2.2 RICHARDSON EXTRAPOLATION TO $h = 0$

We saw that the Euler method gave an error of the form $h\delta(b) + 0(h^2)$, where $\delta(b)$ depends on the differential equation only if round-off is ignored.‡ To halve the error produced by the Euler method it is necessary to halve the step size if the $0(h^2)$ term can be neglected. Let us consider two integrations of a differential equation, one using step size h and one using step size $h/2$.

†In symbolic differentiation, the function f is represented in the computer in its algebraic form as a sequence of symbols (characters). The differentiation operation generates a new sequence of symbols which represents an algebraic form of the derivative with respect to a given variable. For a survey, see Engeli (1969).

‡From this point we will ignore round-off errors in the discussions, assuming that they are smaller than the local truncation error at each step. If this is not so, the effects of round-off can be estimated by suitably scaling the local truncation error to include the round-off error.

If the answers are $y_h(b)$ and $y_{h/2}(b)$, respectively, we can write

$$\begin{aligned} y_h(b) &= y(b) + h\delta(b) + 0(h^2) \\ y_{h/2}(b) &= y(b) + \frac{h}{2}\delta(b) + 0(h^2) \end{aligned} \tag{2.3}$$

Eliminating $\delta(b)$ from (2.3), we get

$$y(b) = 2y_{h/2}(b) - y_h(b) + 0(h^2) \tag{2.4}$$

We can use this as a better numerical approximation. This is illustrated in Table 2.1. The second column is the result of integrating $y' = -y$ by Euler's

Table 2.1 RICHARDSON EXTRAPOLATION

h	*Euler* $y_h(1)$	$2y_{h/2}(1) - y_h(1)$	*Error*	*Error*/h^2
1	0.0000000	0.5000000	0.1321206	0.1321206
$\frac{1}{2}$	0.2500000	0.3828125	0.0149331	0.0597322
$\frac{1}{4}$	0.3164063	0.3708116	0.0029321	0.0469141
$\frac{1}{8}$	0.3435089	0.3685393	0.0006599	0.0422317
$\frac{1}{16}$	0.3560741	0.3680364	0.0001569	0.0401740
$\frac{1}{32}$	0.3620552	0.3679177	0.0000382	0.0391373
$\frac{1}{64}$	0.3649865	—	—	—

method for step sizes 2^{-p}, $p = 0, 1, 2, 3, 4, 5$, and 6, as taken from Table 1.2. The third column is obtained by forming a better approximation from Eq. (2.4). We see that the new approximation is more accurate than the Euler formula, and that it gains accuracy more rapidly as h is decreased because the error is $0(h^2)$ rather than $0(h)$. This process is called the *deferred approach to the limit* and was first used by L. F. Richardson (1927). This new formula gives us a *second order method* because the error is $0(h^2)$.

2.3 SECOND ORDER RUNGE-KUTTA METHODS

Let us look at the approximation of the last section in more detail as it is applied to a single step of size h. We get

$$y_h(t_n + h) = y_n + hf(y_n, t_n)$$

with a step of size h, and

$$q_1 = y_{h/2}\left(t_n + \frac{h}{2}\right) = y_n + \frac{h}{2}f(y_n, t_n)$$

$$y_{h/2}(t_n + h) = q_1 + \frac{h}{2}f\left(q_1, t_n + \frac{h}{2}\right)$$

with two steps of size $h/2$. Therefore,

$$\begin{aligned} y(t_n + h) &= 2y_{h/2}(t_n + h) - y_h(t_n + h) + 0(h^2) \\ &= 2q_1 + hf\left(q_1, t_n + \frac{h}{2}\right) - y_n - hf(y_n, t_n) + 0(h^2) \end{aligned}$$

The actual calculations involved are

$$\begin{aligned} q_1 &= y_n + \frac{h}{2} f(y_n, t_n) \\ y_{n+1} &= y_n + hf\left(q_1, t_n + \frac{h}{2}\right) \end{aligned} \tag{2.5}$$

Thus, the form of the method is similar to Euler's method in that the value at the end of the step is obtained by adding something to the value at the beginning of the step. It is called the *midpoint method.* If we name the amount added on $h\psi(y, t, h)$, we have

$$y_{n+1} = y_n + h\psi(y_n, t_n, h) \tag{2.6}$$

where

$$\psi(y_n, t_n, h) = f\left(y_n + \frac{h}{2} f(y_n, t_n), t_n + \frac{h}{2}\right)$$

It is evident that we wish $h\psi$ to approximate

$$hy'_n + \frac{h^2}{2} y''_n + \frac{h^3}{6} y''' + \cdots$$

in order that (2.6) approximate the Taylor's series expansion about t_n. If we assume sufficient differentiability conditions on $f(y, t)$, we can estimate the agreement between (2.6) and the Taylor's series as follows:

$$\begin{aligned} q_1 &= y_n + \frac{h}{2} f(y_n, t_n) = y_n + \frac{h}{2} y'_n \\ y_{n+1} &= y_n + hf\left(y_n + \frac{h}{2} y'_n, t_n + \frac{h}{2}\right) \\ &= y_n + hf(y_n, t_n) + \frac{h^2}{2} y'_n f_y(y_n, t_n) + \frac{h^2}{2} f_t(y_n, t_n) + 0(h^3) \\ &= y_n + hy'_n + \frac{h^2}{2} (f_y y' + f_t)_n + 0(h^3) \end{aligned}$$

But

$$y'' = f_t + f_y y'$$

Therefore,

$$y_{n+1} = y_n + hy'_n + \frac{h^2}{2} y''_n + 0(h^3)$$

so that the local truncation error in (2.6) is $0(h^3)$. We can use formula (2.6) repetitively over each step rather than use the Richardson extrapolation process. This technique is illustrated by the integration of $y' = -y$, $y(0) = 1$ with step size $h = 0.25$ in Table 2.2

Table 2.2 MIDPOINT METHOD

t_n	y_n	$\frac{h}{2}f(t_n, y_n)$	$hf\left(y_n + \frac{h}{2}f(t_n, y_n), t_n + \frac{h}{2}\right)$
0.00	1	−0.125	−0.21875
0.25	0.78125	−0.0976563	−0.1708984
0.50	0.6103516	−0.0762940	−0.1335144
0.75	0.4768372	−0.0596046	−0.1043081
1.00	0.3725291	—	—

An alternative way of deriving Eq. (2.5) is to return to the Taylor's series method and ask how we can form

$$y_{n+1} = y_n + hy'_n + \frac{h^2}{2}y''_n$$

to get a second order method. If we look at $hf\left(y\left(t_n + \frac{h}{2}\right), t_n + \frac{h}{2}\right)$, we see that it is $hy'(t_n) + [h^2y''(t_n)/2] + 0(h^3)$. We do not know the value of $y(t_n + (h/2))$ but we can approximate it by the Euler method to get

$$y\left(t_n + \frac{h}{2}\right) = y(t_n) + \frac{hy'(t_n)}{2} + 0(h^2)$$

and from this we can get

$$y(t_n) + hy'(t_n) + \frac{h^2}{2}y''(t_n)$$
$$= y(t_n) + hf\left(y(t_n) + \frac{hy'(t_n)}{2}, t_n + \frac{h}{2}\right) + 0(h^3)$$

which leads to the same equation as (2.6) when y_n is substituted for $y(t_n)$.

Other approximations can be used to approximate the terms in the Taylor's series. For example, we have

$$\frac{hy'(t_n) + hy'(t_{n+1})}{2} = hy'(t_n) + \frac{h^2}{2}y''(t_n) + 0(h^3)$$

so we can write

$$y(t_{n+1}) - y(t_n) = \frac{h}{2}(f(y(t_n), t_n) + f(y(t_{n+1}), t_{n+1})) + 0(h^3) \tag{2.7}$$

If the $0(h^3)$ term is neglected and (2.7) is used to approximate the value of $y(t_{n+1})$, it is called the *trapezoidal method.* Since $y(t_{n+1})$ is not known, the right-hand side cannot be evaluated. Two approaches are possible. One is to attempt to solve the nonlinear equation (2.7) for $y(t_{n+1})$. This is called an *implicit method.* Alternatively, we can estimate $y(t_{n+1})$ by another method, such as the Euler method, and get an *explicit method.* Then we get the formula

$$\begin{aligned}\bar{y}(t_{n+1}) &= y(t_n) + hf(y(t_n), t_n) + 0(h^2)\\ y(t_{n+1}) &= y(t_n) + \frac{h}{2}(f(y(t_n), t_n) + f(\bar{y}(t_{n+1}), t_{n+1})) + 0(h^3)\end{aligned} \tag{2.8}$$

This is called the *modified trapezoidal method* or the *Heun method.* Rather than analyzing it in detail, we will look at a class of second order explicit methods. Consider

$$\begin{aligned}q_1 &= y_n + \alpha hf(y_n, t_n)\\ y_{n+1} &= y_n + \beta hf(y_n, t_n) + \gamma hf(q_1, t_n + \eta h)\end{aligned} \tag{2.9}$$

We wish to make the expansion for y_{n+1} agree as closely as possible with the Taylor's series. From (2.9),

$$\begin{aligned}y_{n+1} = y_n + \beta hy'_n + \gamma hy'_n + \alpha\gamma h^2 f_y y'_n + \gamma\eta h^2 f_t\\ + \frac{\alpha^2}{2}\gamma h^3 f_{yy}(y'_n)^2 + \alpha\gamma\eta h^3 f_{yt} y'_n + \frac{\gamma}{2}\eta^2 h^3 f_{tt} + 0(h^4)\end{aligned}$$

Matching terms with the Taylor's series about $t = t_n$, we get

$$(\beta + \gamma)hy'_n = hy'_n$$

$$\gamma(\alpha f_y y'_n + \eta f_t)h^2 = \frac{h^2}{2}y''_n = \frac{h^2}{2}(f_y y'_n + f_t)$$

Consequently,

$$\beta = 1 - \gamma \tag{2.10}$$

and

$$\alpha = \eta = \frac{1}{2\gamma}$$

The asymptotic form of the error term as it goes to zero is found by looking at the next term in the Taylor's series. The difference between the h^3 term in the method and $h^3 y'''_n/6$ is

$$\begin{aligned}&\frac{h^3}{6}(3\alpha^2\gamma f_{yy}(y'_n)^2 + 6\alpha\gamma\eta f_{yt}y'_n + 3\gamma\eta^2 f_{tt} - y'''_n)\\ &= \frac{h^3}{6}\left[\frac{3}{4\gamma}(f_{yy}(y'_n)^2 + 2f_{yt}y'_n + f_{tt}) - y'''_n\right]\end{aligned} \tag{2.11}$$

By taking $\gamma = 1$, we get the midpoint method (2.5). $\gamma = \frac{1}{2}$ gives the Heun method.† If the expansions substituted into (2.9) had contained the remainder term in the Taylor's series of $0(h^3)$, we would have gotten an expression similar to (2.11) but which could be used to bound the error by an expression of the form Mh^3. This could be used in the error bounds similar to those given

†The naming of these second order methods is not consistent in the literature. Some names used by various authors for different values of γ are shown in the table below.

in Theorem 1.2 for Euler's method. The equivalent theorem will be discussed in Chapter 4.

The methods described by (2.9) and (2.10) are called *second order Runge-Kutta methods*. The one free parameter γ can be chosen to optimize any desired feature of the process. One particular criterion for choosing γ is due to Ralston (1965), Section 5.6.3.1. Bounds are assumed for the derivatives appearing in (2.11) and γ is chosen to minimize a bound for (2.11). The assumed bound for the derivatives suggested by Lotkin (1951) is

$$\left|\frac{\partial^{i+j}}{\partial t^i\,\partial y^j} f\right| \le \frac{L^{i+j}}{M^{j-1}}, \qquad |f(y, t)| \le M \tag{2.12}$$

so that (2.11) can be bounded by

$$\begin{aligned} |\text{error}| &= \frac{h^3}{6}\left|\left[\left(\frac{3}{4\gamma} - 1\right)[f_{yy}f^2 + 2f_{yt}f + f_{tt}] - f_y^2 f - f_y f_t\right]\right| \\ &\le \frac{h^3}{6}\left[\left|\frac{3}{4\gamma} - 1\right| 4ML^2 + 2ML^2\right] \end{aligned}$$

This has a minimum of $h^3ML^2/3$ when $\gamma = \frac{3}{4}$.

2.4 EXPLICIT RUNGE-KUTTA METHODS

In the previous section we derived a class of second order methods which consisted of approximating the solution at some point $t_n + \alpha h$, and then using the value of $f(y, t)$ at this point along with its value at t_n to match terms in the Taylor's series. In this section we are going to extend this technique by approximating the solution at a number of additional points $t_n + \alpha_i h$, and then trying to match more terms in the Taylor's series to get higher order methods. Before starting on this, it is convenient to introduce some simplifying notations to handle the derivatives.

Table 2.3 COMMON SECOND ORDER RUNGE-KUTTA METHODS

Author	$\gamma = \frac{1}{2}$	$\gamma = \frac{3}{4}$	$\gamma = 1$
Ceschino and Kuntzman (1966)	Euler-Cauchy	Heun	Improved tangent
Collatz (1960)	Improved polygon	—	Improved polygon
Henrici (1962)	Heun	—	Improved polygon or modified Euler
Isaacson and Keller (1962)	Modified Euler	Heun	Euler-Cauchy

Heun (1900) discussed a class of methods, and so it appears difficult to assign his name definitely to any one of these.

We define the vector $\boldsymbol{f}$ as having two components $f^0 = 1$ and $f^1 = f(y, t)$. Thus f^0 and f^1 are the derivatives of t and y with respect to t. We can write the second derivative of y with respect to t as

$$\begin{aligned} y'' = (f^1)' &= \frac{\partial f^1}{\partial t}\frac{dt}{dt} + \frac{\partial f^1}{\partial y}\frac{dy}{dt} \\ &= \frac{\partial f^1}{\partial t} f^0 + \frac{\partial f^1}{\partial y} f^1 \end{aligned}$$

If we write z_0 for $\partial z/\partial t$ and z_1 for $\partial z/\partial y$, we get

$$y'' = f^1_0 f^0 + f^1_1 f^1 = \sum_{i=0}^{1} f^1_i f^i$$

Finally, we use the *summation convention*, which says that any repeated subscript or superscript in a multiplication term is to be summed over its range; in this case they are 0 and 1. This means that $a_i b^i$ is shorthand for $\sum_{i=0}^{1} a_i b^i$ and that $a_i b^i_j c^j$ means

$$\sum_{i=0}^{1}\sum_{j=0}^{1} a_i b^i_j c^j$$

(This convention will only be used in Chapter 2.) With this we have

$$y'' = f^1_i f^i \tag{2.13}$$

If we define $y^0 = t$, $y^1 = y$, we can write

$$(y^i)'' = f^i_j f^j$$

Note that the rule for differentiation of z with respect to t is $z_j f^j$, so we can write

$$\begin{aligned} (y^i)''' &= ((y^i)'')' \\ &= (f^i_j f^j)' \\ &= f^i_{jk} f^j f^k + f^i_j f^j_k f^k \end{aligned} \tag{2.14}$$

and

$$\begin{aligned} (y^i)^{(4)} &= (f^i_{jk} f^j f^k + f^i_j f^j_k f^k)' \\ &= f^i_{jkl} f^j f^k f^l + f^i_{jk} f^j_l f^k f^l \\ &\quad + f^i_{jk} f^j f^k_l f^l + f^i_{jl} f^j_k f^k f^l \\ &\quad + f^i_j f^j_{kl} f^k f^l + f^i_j f^j_k f^k_l f^l \end{aligned}$$

Now we note that

$$f^i_{jk} = \frac{\partial^2 f^i}{\partial y^j\,\partial y^k} = \frac{\partial^2 f^i}{\partial y^k\,\partial y^j} = f^i_{kj}$$

that is, the order of subscripts is unimportant, and that, since j, k, and l are *dummy* subscripts in the summation, we have

$$\begin{aligned} f^i_{jk} f^j f^k_l f^l &= f^i_{kj} f^k f^j_l f^l && (j \text{ and } k \text{ interchanged}) \\ &= f^i_{jk} f^j_l f^l f^k && (\text{order of differentiation changed}) \end{aligned}$$

Thus the second and third terms in $(y^i)^{(4)}$ are the same. Similarly, the fourth term is

$$f^i_{jl}f^j_k f^k f^l = f^i_{jk} f^j_l f^l f^k \qquad (l \text{ and } k \text{ interchanged})$$

Thus,

$$(y^i)^{(4)} = f^i_{jkl}f^j f^k f^l + 3f^i_{jk}f^j_l f^k f^l + f^i_j f^j_{kl} f^k f^l + f^i_j f^j_k f^k_l f^l \qquad (2.15)$$

Using this notation, we will consider the natural extension of (2.9) to one more intermediate point, or *stage*, and examine the three-stage Runge-Kutta method. Write

$$\begin{aligned} k^i_0 &= hf^i(y_n) \\ k^i_1 &= hf^i(y_n + \alpha_1 k_0) \\ k^i_2 &= hf^i(y_n + \beta_{21}k_0 + (\alpha_2 - \beta_{21})k_1) \\ &= hf^i(y_n + \alpha_2 k_0 + (\alpha_2 - \beta_{21})(k_1 - k_0)) \\ y^i_{n+1} &= y^i_n + \gamma_0 k^i_0 + \gamma_1 k^i_1 + \gamma_2 k^i_2 \end{aligned} \qquad (2.16)$$

Note that we have not given the explicit dependence of f on t since y_n can be taken as a vector containing both the dependent and independent variables as components. $f^0(y_n + \alpha k) = 1$, of course, so k^0_p is h. When $y_n = y(t_n)$, we want y_{n+1} defined by (2.16) to match as many terms in the Taylor's series for $y(t_{n+1})$ as possible. From (2.13) to (2.15),

$$\begin{aligned} y^i(t_{n+1}) = y^i + hf^i + \frac{h^2}{2}(f^i_j f^j) + \frac{h^3}{6}(f^i_{jk}f^j f^k + f^i_j f^j_k f^k) \\ + \frac{h^4}{24}(f^i_{jkl}f^j f^k f^l + 3f^i_{jk}f^j_l f^l f^k + f^i_j f^j_{kl} f^k f^l \\ + f^i_j f^j_k f^k_l f^l) + 0(h^5) \end{aligned} \qquad (2.17)$$

where everything is evaluated at $t = t_n$. From (2.16) and Taylor's series expansions, we get

$$\begin{aligned} k^i_0 &= hf^i \\ k^i_1 &= hf^i + \alpha_1 h^2 f^i_j f^j + \frac{\alpha_1^2 h^3}{2} f^i_{jk} f^j f^k + \frac{\alpha_1^3 h^4}{6} f^i_{jkl} f^j f^k f^l + 0(h^5) \\ k^i_2 &= hf^i + \alpha_2 h^2 f^i_j f^j + \frac{\alpha_2^2 h^3}{2} f^i_{jk} f^j f^k + \frac{\alpha_2^3 h^4}{6} f^i_{jkl} f^j f^k f^l \\ &\quad + (\alpha_2 - \beta_{21})hf^i_j(k^j_1 - k^j_0) + \alpha_2(\alpha_2 - \beta_{21})h^2 f^i_{jk} f^j (k^k_1 - k^k_2) \\ &\quad + 0(h^5) \\ &= hf^i + \alpha_2 h^2 f^i_j f^j + h^3\left[\frac{\alpha_2^2}{2} f^i_{jk} f^j f^k + (\alpha_2 - \beta_{21})\alpha_1 f^i_j f^j_k f^k\right] \\ &\quad + h^4\left[\frac{\alpha_2^3}{6} f^i_{jkl} f^j f^k f^l + (\alpha_2 - \beta_{21})\frac{\alpha_1^2}{2} f^i_j f^j_{kl} f^k f^l \right. \\ &\quad \left. + \alpha_2(\alpha_2 - \beta_{21})\alpha_1 f^i_{jk} f^j f^k_l f^l\right] + 0(h^5) \end{aligned}$$

Forming y_{n+1} from (2.16) and equating with (2.17), we get for terms in h, h^2, and h^3:

$$h: \quad f^i = (\gamma_0 + \gamma_1 + \gamma_2) f^i$$
$$h^2: \quad \tfrac{1}{2} f^i_j f^j = \gamma_1 \alpha_1 f^i_j f^j + \gamma_2 \alpha_2 f^i_j f^j$$
$$h^3: \quad \tfrac{1}{6}(f^i_{jk} f^j f^k + f^i_j f^j_k f^k) = \tfrac{1}{2}\gamma_1 \alpha_1^2 f^i_{jk} f^j f^k + \tfrac{1}{2}\gamma_2 \alpha_2^2 f^i_{jk} f^j f^k + \gamma_2(\alpha_2 - \beta_{21})\alpha_1 f^i_j f^j_k f^k$$

Since the partial derivatives are arbitrary, each different combination must vanish separately. Thus, we get the equations

$$1 = \gamma_0 + \gamma_1 + \gamma_2$$
$$\tfrac{1}{2} = \gamma_1 \alpha_1 + \gamma_2 \alpha_2$$
$$\tfrac{1}{6} = \tfrac{1}{2}\gamma_1 \alpha_1^2 + \tfrac{1}{2}\gamma_2 \alpha_2^2$$
$$\tfrac{1}{6} = \gamma_2(\alpha_2 - \beta_{21})\alpha_1$$

This is a system of four equations in six unknowns. It has the two-parameter family of solutions in terms of α_1 and α_2:

$$\gamma_1 = \frac{3\alpha_2 - 2}{6\alpha_1(\alpha_2 - \alpha_1)}$$
$$\gamma_2 = \frac{3\alpha_1 - 2}{6\alpha_2(\alpha_1 - \alpha_2)}$$
$$\gamma_0 = 1 - \gamma_1 - \gamma_2$$
$$\beta_{21} = \alpha_2 - \frac{1}{6\alpha_1 \gamma_2}$$

The error term $y^i_{n+1} - y^i(t_{n+1})$ is

$$\begin{aligned} -\frac{h^4}{24}[&f^i_{jkl} f^j f^k f^l (1 + 2\alpha_1\alpha_2 - 4(\alpha_1 + \alpha_2)/3) \\ &+ f^i_{jk} f^j_l f^l f^k (3 - 4\alpha_2) \\ &+ f^i_j f^j_{kl} f^k f^l (1 - 2\alpha_1) \\ &+ f^i_j f^j_k f^k_l f^l] + 0(h^5) \end{aligned} \tag{2.18}$$

We can see that no choice of the parameters α_1 and α_2 will make this last term vanish, so that the maximum order of the three-stage method is three. In fact, we can readily extend this argument to show that the maximum order of an r-stage method is r by considering the term $f^i_j f^j_k \cdots f^\mu_y f^y$ that occurs in the $(r + 1)$th derivative. Additional restrictions on the maximum order will be discussed later.

The general r-stage explicit Runge-Kutta process is given by

$$k^i_0 = hf^i(y_n)$$
$$k^i_q = hf^i\left(y_n + \sum_{j=1}^{q} \beta_{qj} k_{j-1}\right) \qquad q = 1, \ldots, r-1$$
$$y^i_{n+1} = y^i_n + \sum_{q=0}^{r-1} \gamma_q k^i_q$$

It is called an explicit method because each right-hand side of the above equations can be calculated in terms of previously calculated values only. Equations for the unknowns γ_q and β_{qj}, $1 \leq j \leq q < r$, can be derived by the method of matching terms with the Taylor's series.

2.4.1 The Classical Runge-Kutta Method

The usual Runge-Kutta method is a four-stage method of fourth order. We can expand k_0^i, k_1^i and k_2^i with $\alpha_2 - \beta_{21} = \beta_{22}$, and perform a similar expansion for k_3^i, where $k_3^i = hf^i(y_n + \beta_{31}k_0 + \beta_{32}k_1 + (\alpha_3 - \beta_{31} - \beta_{32})k_2)$. On equating coefficients up to fourth powers of h, we get the equations

$$\begin{aligned}
\gamma_0 + \gamma_1 + \gamma_2 + \gamma_3 &= 1 \\
\gamma_1\alpha_1 + \gamma_2\alpha_2 + \gamma_3\alpha_3 &= \tfrac{1}{2} \\
\gamma_1\alpha_1^2 + \gamma_2\alpha_2^2 + \gamma_3\alpha_3^2 &= \tfrac{1}{3} \\
\gamma_1\alpha_1^3 + \gamma_2\alpha_2^3 + \gamma_3\alpha_3^3 &= \tfrac{1}{4} \\
\gamma_2\alpha_1\beta_{22} + \gamma_3(\alpha_1\beta_{32} + \alpha_2\beta_{33}) &= \tfrac{1}{6} \\
\gamma_2\alpha_1\alpha_2\beta_{22} + \gamma_3\alpha_3(\alpha_1\beta_{32} + \alpha_2\beta_{33}) &= \tfrac{1}{8} \\
\gamma_2\alpha_1^2\beta_{22} + \gamma_3(\alpha_1^2\beta_{32} + \alpha_2^2\beta_{33}) &= \tfrac{1}{12} \\
\gamma_3\alpha_1\beta_{22}\beta_{33} &= \tfrac{1}{24}
\end{aligned} \tag{2.19}$$

where $\alpha_q = \sum_{j=1}^{q} \beta_{qj}$, $q = 1, 2, 3$. Ralston (1965), p. 199, gives the following two-parameter family of solutions of Eqs. (2.19) in terms of α_1 and α_2:

$$\begin{aligned}
\gamma_0 &= \frac{1}{2} + \frac{1 - 2(\alpha_1 + \alpha_2)}{12\alpha_1\alpha_2}, &\quad \gamma_1 &= \frac{2\alpha_2 - 1}{12\alpha_1(\alpha_2 - \alpha_1)(1 - \alpha_1)} \\
\gamma_2 &= \frac{1 - 2\alpha_1}{12\alpha_2(\alpha_2 - \alpha_1)(1 - \alpha_2)}, &\quad \gamma_3 &= \frac{1}{2} + \frac{2(\alpha_1 + \alpha_2) - 3}{12(1 - \alpha_1)(1 - \alpha_2)} \\
\beta_{22} &= \frac{\alpha_2(\alpha_2 - \alpha_1)}{2\alpha_1(1 - 2\alpha_1)}, &\quad \alpha_3 &= 1 \\
\beta_{32} &= \frac{(1 - \alpha_1)[\alpha_1 + \alpha_2 - 1 - (2\alpha_2 - 1)^2]}{2\alpha_1(\alpha_2 - \alpha_1)[6\alpha_1\alpha_2 - 4(\alpha_1 + \alpha_2) + 3]} \\
\beta_{33} &= \frac{(1 - 2\alpha_1)(1 - \alpha_1)(1 - \alpha_2)}{\alpha_2(\alpha_2 - \alpha_1)[6\alpha_1\alpha_2 - 4(\alpha_1 + \alpha_2) + 3]}
\end{aligned} \tag{2.20}$$

Historically, the first use of this method was with the solution corresponding to

$$\begin{aligned}
k_0 &= hf(y_n) \\
k_1 &= hf(y_n + \tfrac{1}{2}k_0) \\
k_2 &= hf(y_n + \tfrac{1}{2}k_1) \\
k_3 &= hf(y_n + k_2) \\
y_{n+1} &= y_n + \tfrac{1}{6}(k_0 + 2k_1 + 2k_2 + k_3)
\end{aligned} \tag{2.21}$$

which is obtained by setting $\alpha_1 = \alpha_2 = \frac{1}{2}$. We call this the classical Runge-Kutta formula.

The classical Runge-Kutta formula can be viewed as an attempt to extend the Simpson's quadrature rule to differential equations.

$$\int_{t_n}^{t_{n+1}} f(t)\,dt \cong \frac{h}{6}\left(f(t_n) + 4f\left(t_n + \frac{h}{2}\right) + f(t_{n+1})\right)$$

If $f(y, t)$ is a function of t only, they are identical.

2.4.2 The Ralston Runge-Kutta Method

Ralston (1965), p. 200, assumes a bound of the form given in (2.12) for the partial derivatives of $f(y, t)$, and chooses the two free parameters in (2.19) to minimize the bound on the solution error. The result is the method corresponding to

$$\alpha_1 = 0.4 \quad \text{and} \quad \alpha_2 = \tfrac{7}{8} - \tfrac{3}{16}\sqrt{5}$$

and given by

$$\begin{aligned}
k_0 &= hf(y_n) \\
k_1 &= hf(y_n + 0.4k_0) \\
k_2 &= hf(y_n + 0.29697760k_0 + 0.15875966k_1) \\
k_3 &= hf(y_n + 0.21810038k_0 - 3.05096470k_1 \\
&\qquad + 3.83286432k_2) \\
y_{n+1} &= y_n + 0.17476028k_0 - 0.55148053k_1 \\
&\qquad + 1.20553547k_2 + 0.17118478k_3
\end{aligned} \tag{2.22}$$

2.4.3 Butcher's Results on the Attainable Order of Runge-Kutta Methods

So far we have seen the following relation between order and number of function evaluations for explicit Runge-Kutta methods:

Table 2.4 MAXIMUM ORDER OF ONE- THROUGH FOUR-STAGE RUNGE-KUTTA METHODS

Number of function evaluations $= \nu$	*Maximum order* $= r(\nu)$
1	1
2	2
3	3
4	4
?	?

We have already seen that

$$r(\nu) \leq \nu$$

[See the remark following expression (2.18).] We naturally ask the question, "Do higher order methods exist?" Butcher (1965) has studied $r(\nu)$ extensively, and has shown the following relations for explicit Runge-Kutta methods:

Table 2.5 MAXIMUM ORDER OF THE RUNGE-KUTTA METHOD FOR VARIOUS NUMBERS OF STAGES

ν	$r(\nu)$
5	4
6	5
7	6
$\nu \geq 8$	$r(\nu) \leq \nu - 2$

All processes of this type are known as Runge-Kutta methods, although the original method was the classical fourth order one.

2.5 IMPLICIT RUNGE-KUTTA METHODS

The Runge-Kutta methods discussed in the previous section had the property that the right-hand sides could be explicitly calculated. If this requirement is dropped, we get the implicit Runge-Kutta methods. An r-stage implicit Runge-Kutta method has the form

$$\begin{aligned} k_q &= hf(y_n + \beta_{qj}k_j) \qquad q = 1, 2, \ldots, r \\ y_{n+1} &= y_n + \gamma_j k_j \qquad (j \text{ summed from 1 to } r) \end{aligned} \tag{2.23}$$

An example of a two-stage implicit Runge-Kutta method is given by

$$\begin{aligned} k_1 &= hf(y_n) \\ k_2 &= hf(y_n + \tfrac{1}{2}k_1 + \tfrac{1}{2}k_2) \\ y_{n+1} &= y_n + \tfrac{1}{2}k_1 + \tfrac{1}{2}k_2 \end{aligned}$$

It is, of course, the trapezoidal rule. Let us examine the general two-stage implicit Runge-Kutta method.

$$\begin{aligned} k_1 &= hf(y_n + \beta_{11}k_1 + \beta_{12}k_2) \\ k_2 &= hf(y_n + \beta_{21}k_1 + \beta_{22}k_2) \\ y_{n+1} &= y_n + \gamma_1 k_1 + \gamma_2 k_2 \end{aligned}$$

Expand by Taylor's series to get

$$\begin{aligned} k_q = {} & hf + hf_i(\beta_{q1}k_1^i + \beta_{q2}k_2^i) \\ & + \frac{h}{2}f_{ij}(\beta_{q1}k_1^i + \beta_{q2}k_2^i)(\beta_{q1}k_1^j + \beta_{q2}k_2^j) + \cdots \end{aligned}$$

Substitute for k_q on the right-hand side as many times as necessary to get

$$\begin{aligned} k_q = {} & hf + hf_i[\beta_{q1}(hf^i + hf^i_j(\beta_{11}k_1^j + \beta_{12}k_2^j) + \cdots) \\ & + \beta_{q2}(hf^i + hf^i_j(\beta_{21}k_1^j + \beta_{22}k_2^j) + \cdots)] \\ & + \frac{h}{2}f_{ij}(\beta_{q1} + \beta_{q2})hf^i(\beta_{q1} + \beta_{q2})hf^j + \cdots \\ = {} & hf + h^2 f_i f^i \alpha_q + \frac{h^3}{2} f_{ij} f^i f^j \alpha_q^2 + h^3 f_i f^i_j f^j(\beta_{q1}\alpha_1 + \beta_{q2}\alpha_2) \\ & + \frac{h^4}{6}\alpha_q^3 f_{ijk} f^i f^j f^k + h^4 f_{ij} f^i_k f^j f^k \alpha_q(\beta_{q1}\alpha_1 + \beta_{q2}\alpha_2) \\ & + \frac{h^4}{2} f_i f^i_{jk} f^j f^k(\beta_{q1}\alpha_1^2 + \beta_{q2}\alpha_2^2) \\ & + h^4 f_i f^i_j f^k(\beta_{q1}(\beta_{11}\alpha_1 + \beta_{12}\alpha_2) + \beta_{q2}(\beta_{21}\alpha_1 + \beta_{22}\alpha_2)) + 0(h^5) \end{aligned} \tag{2.24}$$

where $\alpha_1 = \beta_{11} + \beta_{12}$, $\alpha_2 = \beta_{21} + \beta_{22}$. Evaluating y_{n+1}, we get

$$\begin{aligned} y_{n+1} = {} & y_n + (\gamma_1 + \gamma_2)hf \\ & + (\gamma_1\alpha_1 + \gamma_2\alpha_2)h^2 f_i f^i \\ & + (\gamma_1\alpha_1^2 + \gamma_2\alpha_2^2)\frac{h^3}{2} f_{ij} f^i f^j \\ & + (\gamma_1(\beta_{11}\alpha_1 + \beta_{12}\alpha_2) + \gamma_2(\beta_{21}\alpha_1 + \beta_{22}\alpha_2))h^3 f_i f^i_j f^j \\ & + 0(h^4) \end{aligned} \tag{2.25}$$

From this we get equations by matching terms with a Taylor's series up to order h^3:

$$\begin{aligned} \gamma_1 + \gamma_2 &= 1 \\ \gamma_1\alpha_1 + \gamma_2\alpha_2 &= \tfrac{1}{2} \\ \gamma_1\alpha_1^2 + \gamma_2\alpha_2^2 &= \tfrac{1}{3} \\ \gamma_1(\beta_{11}\alpha_1 + \beta_{12}\alpha_2) + \gamma_2(\beta_{21}\alpha_1 + \beta_{22}\alpha_2) &= \tfrac{1}{6} \end{aligned}$$

These four equations in six unknowns have a two-parameter family of solutions. From the first three equations we get

$$\begin{aligned} \gamma_2(\alpha_2 - \alpha_1) &= \tfrac{1}{2} - \alpha_1 \\ \gamma_2(\alpha_2 - \alpha_1)(\alpha_2 + \alpha_1) &= \tfrac{1}{3} - \alpha_1^2 \end{aligned}$$

or

$$\alpha_2 = -\alpha_1 + \frac{\frac{1}{3} - \alpha_1^2}{\frac{1}{2} - \alpha_1}$$

Thus selection of $\alpha_1 \neq \frac{1}{2}$ determines α_2 and hence γ_1 and γ_2 from the first three equations. Since $\beta_{11} = \alpha_1 - \beta_{12}$ and $\beta_{21} = \alpha_2 - \beta_{22}$, the last equation is linear in β_{12} and β_{22}. Choice of one of these determines the other. Thus one solution is

$$\begin{aligned} &\alpha_1 = 0, \qquad \alpha_2 = \tfrac{2}{3}, \qquad \gamma_2 = \tfrac{3}{4}, \qquad \gamma_1 = \tfrac{1}{4}, \\ &\beta_{12} = 1, \qquad \beta_{22} = 0, \qquad \beta_{11} = -1, \qquad \beta_{21} = \tfrac{2}{3} \end{aligned}$$

to give

$$k_1 = hf(y_n - k_1 + k_2)$$
$$k_2 = hf(y_2 + \tfrac{2}{3}k_1)$$
$$y_{n+1} = y_n + \tfrac{1}{4}(k_1 + 3k_2)$$

which is a third order method. However, Butcher (1964) has investigated the r-stage implicit Runge-Kutta method and has shown that it is possible to achieve order $2r$ for such methods.† Thus, if in the above two-stage method we choose

$$\gamma_1 = \gamma_2 = \frac{1}{2}$$

$$\alpha_1 = \frac{1}{2} - \frac{\sqrt{3}}{6}$$

$$\alpha_2 = \frac{1}{2} + \frac{\sqrt{3}}{6}$$

$$\beta_{11} = \beta_{22} = \frac{1}{4}$$

$$\beta_{12} = \frac{1}{4} - \frac{\sqrt{3}}{6}$$

$$\beta_{21} = \frac{1}{4} + \frac{\sqrt{3}}{6}$$

we can see that the terms in h^4 match those in the Taylor's series for y_{n+1}. The term in h^4 in (2.25) will be

$$\frac{(\gamma_1\alpha_1^3 + \gamma_2\alpha_2^3)f_{ijk}f^if^jf^k}{6} + [\gamma_1\alpha_1(\beta_{11}\alpha_1 + \beta_{12}\alpha_2)$$
$$+ \gamma_2\alpha_2(\beta_{21}\alpha_1 + \beta_{22}\alpha_2)]f_{ij}f^i_kf^jf^k$$
$$+ \frac{[\gamma_1(\beta_{11}\alpha_1^2 + \beta_{12}\alpha_2^2) + \gamma_2(\beta_{21}\alpha_1^2 + \beta_{22}\alpha_2^2)]f_if^i_{jk}f^jf^k}{2}$$
$$+ [(\gamma_1\beta_{11} + \gamma_2\beta_{21})(\beta_{11}\alpha_1 + \beta_{12}\alpha_2)$$
$$+ (\gamma_1\beta_{21} + \gamma_2\beta_{22})(\beta_{21}\alpha_1 + \beta_{22}\alpha_2)]f_if^i_jf^j_kf^k$$
$$= \frac{[f_{ijk}f^if^jf^k + 3f_{ij}f^i_kf^jf^k + f_if^i_{jk}f^jf^k + f_if^i_jf^j_kf^k]}{24} = \frac{y^{(4)}}{24}$$

2.5.1 Practical Considerations of Implicit Runge-Kutta Methods

The explicit Runge-Kutta method is straightforward to use, whereas the implicit Runge-Kutta method requires that a simultaneous system of equations be solved at each step. These are nonlinear unless f is a linear

†These methods also have been reported independently by Ceschino and Kuntzman (1966) who also gave a proof of this result.

function of y. They can be solved by an iterative method which is guaranteed to converge for small enough h as follows: Suppose we have approximations $k_{q,(0)}$, $q = 1, 2, \ldots, r$ to the values of k_q. Define new approximations $k_{q,(m+1)}$, $m = 0, 1, 2, \ldots$ by

$$k_{q,(m+1)} = hf(y_n + \beta_{qj}k_{j,(m)}) \tag{2.26}$$

Suppose the error in our approximations to k_q is defined by

$$\epsilon_{q,(m)} = k_q - k_{q,(m)}$$

We can subtract (2.26) from (2.23) to get

$$\begin{aligned}|\epsilon_{q,(m+1)}| &= |hf(y_n + \beta_{qj}k_j) - hf(y_n + \beta_{qj}k_{j,(m)})| \\ &\leq hL|\beta_{qj}(k_j - k_{j,(m)})| \qquad \text{(by the Lipschitz condition)} \\ &\leq hL|\beta_{qj}||\epsilon_{j,(m)}|\end{aligned}$$

Thus, if $e_{(m)} = \max\limits_q |\epsilon_{q,(m)}|$, we get

$$e_{(m+1)} \leq hL \max_q \sum_{j=1}^{r} |\beta_{qj}| e_{(m)}$$

showing that $e_{(m)} \longrightarrow 0$ as $m \longrightarrow \infty$ if

$$h < \frac{1}{L \max\limits_q \sum\limits_j |\beta_{qj}|}$$

In general it is difficult to justify the additional work of the implicit methods by the increased accuracy attainable, so their utility is restricted to some special problems for which they have desirable stability characteristics.

2.6 CONVERGENCE AND STABILITY

In Chapter 4 we will prove that the Runge-Kutta and Taylor's series methods converge as $h \longrightarrow 0$ whenever their order is greater than zero. It will also be shown that convergent one-step methods are stable (that is, a small perturbation to the problem causes a bounded change as $h \longrightarrow 0$). In this section we wish to discuss the absolute stability region of the Runge-Kutta methods.

2.6.1 Stability Regions for Explicit Runge-Kutta Methods

We consider the equation $y' = \lambda y$ for complex values of λ. If we examine the fourth order explicit Runge-Kutta process for $y' = \lambda y$, we find

$$\begin{aligned}k_1 &= \lambda h(y + \alpha_1 h\lambda y) = h\lambda(1 + \alpha_1 h\lambda)y \\ k_2 &= \lambda h(1 + \beta_{21}h\lambda + \beta_{12}h\lambda(1 + \alpha_1 h\lambda))y \\ &= h\lambda(1 + \alpha_2 h\lambda + \alpha_1\beta_{12}h^2\lambda^2)y\end{aligned}$$

Similarly,

$$k_3 = h\lambda(1 + \eta_1 h\lambda + \eta_2 h^2\lambda^2 + \eta^3 h^3\lambda^3)y$$

and

$$y_{n+1} = (1 + \delta_1 h\lambda + \delta_2 h^2\lambda^2 + \delta_3 h^3\lambda^3 + \delta_4 h^4\lambda^4)y_n$$

where η_i and δ_i are combinations of the coefficients β_{ij} and γ_j. Since we know that y_{n+1} agrees with $y(t_{n+1})$ to order h^4 if y_n is exact, and since $y(t_{n+1}) = e^{\lambda h}y(t_n)$, we know that $\delta_1 = 1$, $\delta_2 = \frac{1}{2}$, $\delta_3 = \frac{1}{6}$, and $\delta_4 = \frac{1}{24}$. Thus, the growth factor for the method is

$$\sum_{n=0}^{4} \frac{(h\lambda)^n}{n!}$$

The region of absolute stability is that area in which

$$\left|1 + \mu + \frac{\mu^2}{2} + \frac{\mu^3}{6} + \frac{\mu^4}{24}\right| < 1 \tag{2.27}$$

where $\mu = h\lambda$. To find this in the complex plane, we would like to plot the locus where equality holds in (2.27). One way of getting this is to set

$$1 + \mu + \frac{\mu^2}{2} + \frac{\mu^3}{6} + \frac{\mu^4}{24} = e^{i\theta}$$

where $\mu = h\lambda$ and to determine $\mu(\theta)$ by any standard polynomial root finder. The region of absolute stability is shown in Figure 2.1.

If we consider the problem

$$y' = \lambda(y - F(t)) + F'(t), \qquad y(0) = F(0) + c_0$$

with the solution

$$y = F(t) + c_0 e^{\lambda t}$$

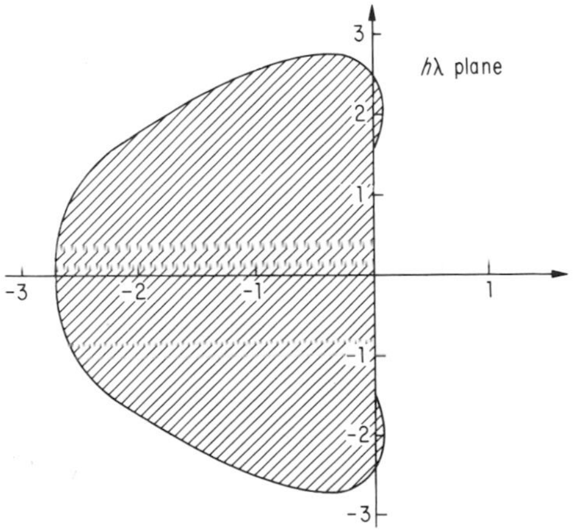

Fig. 2.1 Region of absolute stability of fourth order explicit Runge-Kutta method.

we note that because it is linear in y, we can consider the numerical solution as the sum of the results of separately integrating the inhomogeneous problem

$$y' = \lambda(y - F(t)) + F'(t), \qquad y(0) = F(0) \tag{2.28}$$

and the homogeneous problem

$$y' = \lambda y, \qquad y(0) = c_0 \tag{2.29}$$

The step size must certainly be picked small enough to accurately integrate the $F(t)$ term in (2.28). Let us suppose that $F(t)$ is smooth and remains about one in size. If $\lambda \ll 0$ (or its real part $\ll 0$), the step size will have to be small initially to accurately integrate the $c_0e^{\lambda t}$ term. Since $e^{h\lambda}$ is represented numerically by $1 + h\lambda + [(h\lambda)^2/2] + [(h\lambda)^3/6] + [(h\lambda)^4/24]$, $h\lambda$ will have to be small enough that this expression is close to $e^{h\lambda}$. After a number of steps, the $c_0e^{\lambda t}$ term will be insignificant compared to the $F(t)$ term in the answer. From then on absolute stability will be a restrictive requirement on the step size. h will have to be sufficiently small that $h\lambda$ will be in the region of absolute stability so that the numerical approximation to the $e^{\lambda t}$ term does not increase. However, this h will be much smaller than necessary to accurately integrate the $F(t)$ term. When λ is small but negative, the value of h chosen for accurate integration of $F(t)$ will put $h\lambda$ in the region of absolute stability. When $\lambda > 0$, the solution of (2.29) is growing, and h must be chosen small enough to also correctly integrate $e^{\lambda t}$. $h\lambda$ will not lie in the region of absolute stability, but now it is not important because $e^{\lambda t}$ is also growing.

Thus we see that for $\lambda < 0$ we may be interested in either accuracy or absolute stability. (The former requirement will restrict $h\lambda$ such that it is in the absolutely stable region for most methods, since if $1 + h\lambda + \cdots$ is an accurate approximation of $e^{\lambda h}$, it is necessarily less than one when $h\lambda < 0$.) If $\lambda > 0$, accuracy is the only criterion.

A q-stage explicit Runge-Kutta method will represent $e^{h\lambda}$ by $1 + h\lambda + \cdots + (h\lambda)^q/q!$, so that if it is of order q (as is possible if $q \leq 4$) the representation will be the first $(q + 1)$ terms of the exponential series. A Taylor's series method which uses up to q derivatives will behave similarly.

2.6.2 Stability Regions for Implicit Runge-Kutta Methods

Although the implicit methods have little practical importance, they have an important stability characteristic that does make them useful for some special problems to be discussed later. We will point out this property in this section for the simplest example, the trapezoidal rule. The extension to the general r-stage implicit Runge-Kutta method appears in the literature.

The trapezoidal rule is given by

$$y_{n+1} = y_n + \frac{h}{2}(y'_n + y'_{n+1})$$

If we substitute $y' = \lambda y$, we get

$$y_{n+1} = y_n + \frac{h\lambda}{2} y_n + \frac{h\lambda}{2} y_{n+1}$$

or

$$y_{n+1} = \frac{1 + (h\lambda/2)}{1 - (h\lambda/2)} y_n$$

The growth factor is thus $[1 + (h\lambda/2)]/[1 - (h\lambda/2)]$. If $h\lambda/2 = u + iv$, where u and v are real, we get

$$|y_{n+1}| = \left|\frac{(1+u)+iv}{(1-u)-iv}\right| |y_n|$$
$$= \left|\frac{1+u^2+v^2+2u}{1+u^2+v^2-2u}\right|^{1/2} |y_n|$$

so that the solution decays if $u < 0$. Dahlquist (1963) defined a method to be *A-stable if the numerical solution asymptotically approaches zero as* $n \longrightarrow \infty$ *for the differential equation* $y' = \lambda y$, *where* $\text{Re}(\lambda) < 0$. We see that the trapezoidal rule is A-stable.

The r-stage implicit Runge-Kutta method of order $2r$ is also A-stable as was pointed out by Ehle (1968). As an example, we consider the problem

$$y' = -1000(y - t^3) + 3t^2, \qquad y(0) = 0, \qquad y(1) = ?$$

using step size $h = 1, 0.1, 0.01$, and 0.001 with the trapezoidal method. The answers are shown in Table 2.6. Note that the error behaves as $0(h^2)$ in spite of the initially large value of $h(\partial f/\partial y) = -1000$.

Table 2.6 ERRORS FOR THE TRAPEZOIDAL METHOD AS A FUNCTION OF h

h	$y_h(1)$	$\epsilon_h(1) = y_n(1) - 1$	$\epsilon_h(1)/h^2$
1	1.00998000	0.99800×10^{-3}	0.998×10^{-3}
0.1	1.00000165	0.16486×10^{-5}	0.165×10^{-3}
0.01	1.00000005	0.50000×10^{-7}	0.500×10^{-3}
0.001	1.00000000	0.49996×10^{-9}	0.500×10^{-3}

PROBLEMS

1. Calculate the minimum number of the fixed step sizes that should be used to integrate the equation $y' = y$, $y(0) = 1$ from $t = 0$ to $t = 1$ if each of the following methods is used, and the requested accuracy is to be obtained.
   (a) Modified trapezoidal method with accuracy at least 0.20; 0.05.
   (b) Classical fourth order Runge-Kutta method with accuracy at least 0.01; 0.001.

2. Use the modified trapezoidal rule, Eq. (2.8), and the ordinary trapezoidal rule, Eq. (2.7), to integrate

$$y' = -1000(y - t^3) + 3t^2, \qquad y(0) = 0$$

over the interval [0, 1] using step sizes 2^{-p}, $p = 2(2)12$. (Use a digital computer!) Plot the results on log-log paper. Explain the similarities and differences.

3. If you were integrating a general equation

$$y' = f(y, t)$$

would you prefer to use the trapezoidal rule rather than the modified trapezoidal rule in view of the results of the last problem?

4. Find conditions on the value of λ such that if the midpoint rule is used to integrate

$$y' = \lambda(t^3 - y) + 3t^2, \qquad y(0) = 0$$

from $t = 0$ to $t = 1$, then the error will be $0(h^3)$. Verify by numerically integrating the problem for this value of λ and plotting the error versus h.

5. Derive the error term which is $0(h^5)$ for the general fourth order Runge-Kutta formula, and find its particular form for the classical fourth order method. What is the value of the general formula if f is assumed to be linear in y and t?

6. Calculate and draw the region of absolute stability for a third order Runge-Kutta method.

7. Discuss the region of absolute stability for the method defined by

$$y_{n+1} = y_n + \frac{h}{2}(f(y_n) + f(y_{n+1})) + \frac{h^2}{12}(f'(y_n) - f'(y_{n+1}))$$

where $f'(y)$ is to be evaluated by $f'(y) = (\partial f/\partial t) + (\partial f/\partial y)f(y)$. What is the order of this method?

3 SYSTEMS OF EQUATIONS AND EQUATIONS OF ORDER GREATER THAN ONE

The purpose of this short chapter is to discuss the manner in which the techniques of Chapters 1 and 2 can be extended in two ways: to systems of equations and to equations of higher order. The order of an equation is the order of the highest derivative appearing in it. We will assume that it can be written in the form

$$y^{(p)} = f(y, y', \ldots, y^{(p-1)}, t) \tag{3.1}$$

where f satisfies a Lipschitz condition in each of the $y^{(i)}$, and $y^{(i)}$ is a notation for d^iy/dt^i. Thus,

$$(y''')^2(2 + t^2) - y(y' + \cos t) = 0$$

is a third order equation which we would write as

$$y''' = \left[\frac{y(y' + \cos t)}{2 + t^2}\right]^{1/2} \tag{3.2}$$

but we have to require that $y(y' + \cos t)$ remain bounded away from zero in order that (3.2) satisfy a Lipschitz condition, and we have to agree on which branch of the square roots is intended so that y''' is single-valued.

A system of equations involves one independent variable t and more than one dependent variable. Thus,

$$\begin{aligned} y' &= z \\ z' &= -y \end{aligned} \tag{3.3}$$

is a system of two first order equations.

We will label the dependent variables $y^1, y^2, y^3, \ldots, y^s$ if there are s of them. The superscripts are not to be confused with the powers of y, which will be written $(y)^p$ if there is a possibility of ambiguity. The general first order system of equations takes the form

$$\begin{aligned} y^{1\prime} &= f^1(y^1, y^2, \ldots, y^s, t) \\ y^{2\prime} &= f^2(y^1, y^2, \ldots, y^s, t) \\ &\cdots \\ y^{s\prime} &= f^s(y^1, y^2, \ldots, y^s, t) \end{aligned} \tag{3.4}$$

where each of the f^i is assumed to satisfy a Lipschitz condition separately in each dependent variable and be continuous in t. Subsequent notation is simplified if we rename t as y^0, as was done in Chapter 2, and append the differential equation

$$y^{0\prime} = f^0(y^0, y^1, \ldots, y^s) = 1 \tag{3.5}$$

with the initial condition $y^0(0) = 0$. Equations (3.4) and (3.5) are now a system of $(s + 1)$ equations for the $(s + 1)$ dependent variables $y^0, \ldots, y^s$ which are independent of t, and can thus be written as

$$\mathbf{y}' = \mathbf{f}(\mathbf{y}) \tag{3.6}$$

where $\mathbf{y}$ is the column vector $[y^0, y^1, \ldots, y^s]^T$, T being the transpose operator. Two points should be stressed about this rewriting. First, if we require that f^i satisfy a Lipschitz condition in y^0, we are requiring a slightly stronger condition than continuity in t which is sufficient for the existence and convergence of all methods to be discussed. However, in practice we have the even stronger property of differentiability, so it is not a problem. Secondly, this transformation is for notational purposes only; the independent variable t should be handled separately in most numerical integrations to reduce work and round-off error. If the step size is fixed, then it is best to calculate t_n from hn, but if h varies, it may have to be calculated from $t_{n+1} = t_n + h_n$, recognizing that some round-off errors will be introduced if h is not exactly representable in the machine.

3.1 APPLICATION OF ONE-STEP TECHNIQUES TO SYSTEMS OF EQUATIONS

All of the methods discussed in Chapters 1 and 2 are directly applicable to systems of equations. Each member of the system is treated separately but simultaneously. Thus, the Euler method for

$$\begin{aligned} y' &= f(y, z) \\ z' &= g(y, z) \end{aligned} \tag{3.7}$$

is

$$\begin{aligned} y_{n+1} &= y_n + hf(y_n, z_n) \\ z_{n+1} &= z_n + hg(y_n, z_n) \end{aligned}$$

The classical four step Runge-Kutta method requires that a set of the k_q be calculated for each dependent variable. Thus, if we name y as y^1 and z as y^2 in (3.7), we could use the Runge-Kutta process given by

$$
\begin{aligned}
k_0^1 &= hf(y_n^1, y_n^2), \qquad k_0^2 = hg(y_n^1, y_n^2) \\
k_1^1 &= hf(y_n^1 + \tfrac{1}{2}k_0^1, y_n^2 + \tfrac{1}{2}k_0^2), \qquad k_1^2 = hg(y_n^1 + \tfrac{1}{2}k_0^1, y_n^2 + \tfrac{1}{2}k_0^2) \\
k_2^1 &= hf(y_n^1 + \tfrac{1}{2}k_1^1, y_n^2 + \tfrac{1}{2}k_1^2), \qquad k_2^2 = hg(y_n^1 + \tfrac{1}{2}k_1^1, y_n^2 + \tfrac{1}{2}k_1^2) \\
k_3^1 &= hf(y_n^1 + k_2^1, y_n^2 + k_2^2), \qquad k_3^2 = hg(y_n^1 + k_2^1, y_n^2 + k_2^2) \\
y_{n+1}^1 &= y_n^1 + \tfrac{1}{6}(k_0^1 + 2k_1^1 + 2k_2^1 + k_3^1) \\
y_{n+1}^2 &= y_n^2 + \tfrac{1}{6}(k_0^2 + 2k_1^2 + 2k_2^2 + k_3^2)
\end{aligned}
\tag{3.8}
$$

It will be shown in Chapter 4 that these processes have the same behavior as does the similar process for single equations. At this point we remark that the analysis in Chapter 2 which consisted of an expansion in Taylor's series can be applied directly. The notation developed there for handling the independent variable is, in fact, the same as we have used here so that the expansions by Eqs. (2.13) to (2.15) for the derivatives are valid for systems of equations where the indices i, j, etc., are allowed to take values 0 to s rather than just 0 and 1.

3.2 REDUCTION OF A HIGHER ORDER EQUATION TO A SYSTEM OF FIRST ORDER EQUATIONS

One common technique for handling equations of the form (3.1) is to transform them into an equivalent first order system. If we define the variables

$$y^i = y^{(i-1)} \qquad i = 1, 2, \ldots, p \tag{3.9}$$

we can write (3.1) as

$$(y^p)' = f(y^1, y^2, \ldots, , y^p, t) \tag{3.10a}$$

while by differentiation of (3.9) for $i = 1, \ldots, p - 1$, we get

$$(y^i)' = y^{(i)} = y^{i+1} \tag{3.10b}$$

(3.10) is a system of p first order equations which can be handled by the methods already discussed. The original problem (3.1) will have initial values specified for $y^{(i-1)}(0)$, $i = 1, 2, \ldots, p$, so (3.10) will have initial values specified for $y^i(0)$ as required.

3.3 DIRECT METHODS FOR HIGHER ORDER EQUATIONS

Rather than expand a higher order equation into a larger system, a number of authors have proposed direct methods. Some of these will be discussed in later chapters on multistep methods. Here we will outline the extension of some of the one-step methods to higher order equations such as (3.1). Whether or not it is better to handle these equations directly or to convert them to a lower order system depends on the equations. Some cases have been studied in a paper by Rutishauser (1960).

3.3.1 Taylor's Series Methods

The Taylor's series method can be extended in the obvious way as follows: We are given $y_0, y_0^{(1)}, \ldots, y_0^{(p-1)}$. Using (3.1) we can calculate $y_0^{(p)}$, and by differentiation of (3.1) a number of times we can also form $y_0^{(q)}$, $q = p + 1, \ldots, r$. We can then write

$$\begin{aligned} y_1 &= y_0 + hy_0^{(1)} + \frac{h^2}{2}y_0^{(2)} + \cdots + \frac{h^r}{r!}y_0^{(r)} \\ y_1^{(1)} &= y_0^{(1)} + hy_0^{(2)} + \frac{h^2}{2}y_0^{(3)} + \cdots + \frac{h^{r-1}}{(r-1)!}y_0^{(r)} \\ &\cdots \\ y_1^{(p-1)} &= y_0^{(p-1)} + hy_0^{(p)} + \cdots + \frac{h^{r-p+1}}{(r-p+1)!}y_0^{(r)} \end{aligned} \tag{3.11}$$

so that the values of $y, y^{(1)}, \ldots, y^{(p-1)}$ can be approximated at t_1.

3.3.2 Runge-Kutta Methods

Instead of calculating the derivatives of order greater than p explicitly, the derivative of order p can be evaluated at more than one point, as was done in the Runge-Kutta methods for first order equations. As an example, we consider the second order equation $y'' = f(y, y')$. We know y_n, y'_n. A general two-stage explicit Runge-Kutta method has the form

$$\begin{aligned} k_1 &= \frac{h^2}{2}f(y_n, y'_n) \\ k_2 &= \frac{h^2}{2}f\left(y_n + \alpha_1 hy'_n + \alpha_2 k_1, y'_n + \frac{\alpha_3 k_1}{h}\right) \\ y_{n+1} &= y_n + hy'_n + \gamma_1 k_1 + \gamma_2 k_2 \\ y'_{n+1} &= y'_n + \frac{\delta_1}{h}k_1 + \frac{\delta_2}{h}k_2 \end{aligned} \tag{3.12}$$

Expanding k_2 we get

$$\begin{aligned} k_2 = \frac{h^2}{2}\bigg[& f + (\alpha_1 hy' + \alpha_2 k_1)\frac{\partial f}{\partial y} + \alpha_1^2 \frac{h^2(y')^2}{2}\frac{\partial^2 f}{\partial y^2} \\ & + \frac{\alpha_3 k_1}{h}\frac{\partial f}{\partial y'} + \frac{\alpha_3^2 k_1^2}{2h^2}\frac{\partial^2 f}{\partial y'^2} + \alpha_1\alpha_3 k_1 \frac{\partial^2 f}{\partial y\, \partial y'}y'\bigg] + 0(h^5) \end{aligned} \tag{3.13}$$

where everything is evaluated at y_n, y'_n. We also have $y''_n = f$, $y'''_n = (\partial f/\partial y)y' + (\partial f/\partial y')y''$,

$$y_n^{(4)} = \frac{\partial^2 f}{\partial y^2}(y')^2 + 2\frac{\partial^2 f}{\partial y\, \partial y'}y'y'' + \frac{\partial f}{\partial y}y'' + \frac{\partial^2 f}{\partial y'^2}(y'')^2 + \frac{\partial f}{\partial y'}y'''$$

Substituting (3.13) into (3.12), we get

$$\begin{aligned} y_{n+1} = y_n + hy'_n + (\gamma_1 + \gamma_2)\frac{h^2}{2}y''_n \\ + \gamma_2\frac{h^3}{2}\left[\alpha_1\frac{\partial f}{\partial y}y' + \frac{\alpha_3}{2}\frac{\partial f}{\partial y'}y''\right] \\ + \gamma_2\frac{h^4}{4}\left[\alpha_2\frac{\partial f}{\partial y}y'' + \alpha_1^2\frac{\partial^2 f}{\partial y^2}(y')^2\right. \\ \left. + \alpha_1\alpha_3\frac{\partial^2 f}{\partial y\,\partial y'}y'y'' + \frac{\alpha_3^2}{4}\frac{\partial^2 f}{\partial y'^2}(y'')^2\right] + 0(h^5) \end{aligned} \tag{3.14}$$

If $\gamma_1 + \gamma_2 = 1$ and $\alpha_1\gamma_2 = \alpha_3\gamma_2/2 = \frac{1}{3}$, (3.14) will agree with the first four terms of the Taylor's expansion of $y(t_{n+1})$. The term in h^4 will not match the fifth term in the Taylor's series for any choice of the α_i and γ_i. Similarly, if $\delta_1 + \delta_2 = 2$ and $\alpha_1\delta_2 = \alpha_3\delta_2/1 = 1$, then y'_{n+1} agrees with $y'_n + hy''_n + (h^2/2)y'''_n$ to order h^2. Thus, one solution is

$$\begin{aligned} \gamma_1 = \gamma_2 = \tfrac{1}{2} \\ \alpha_1 = \alpha_2 = \tfrac{2}{3}, \qquad \alpha_3 = \tfrac{4}{3} \\ \delta_1 = \tfrac{1}{2} \\ \delta_2 = \tfrac{3}{2} \end{aligned} \tag{3.15}$$

The local truncation error is thus $0(h^4)$ in y and $0(h^3)$ in y'. The reader might therefore assume that the global errors are $0(h^3)$ and $0(h^2)$, respectively. However, y' enters into f and hence k_2, hence the global error is $0(h^2)$ for both y and y'.

Both explicit† and implicit‡ Runge-Kutta methods have been obtained for systems of higher order equations. These methods have important applications in some special problems, but a general computer program will usually only provide for first order equations. If the equations are of a restricted type, then these methods can be more practical. The system

$$\mathbf{y}'' = \mathbf{f}(\mathbf{y})$$

with first derivatives absent frequently occurs in celestial mechanics. In the simple case $y'' = f(y)$, $y^{(3)} = f_y y'$, $y^{(4)} = f_y y'' + f_{yy}(y')^2$, and $f_{y'} = 0$. Consequently, (3.14) is replaced by

$$\begin{aligned} y_{n+1} = y_n + hy'_n + \frac{h^2}{2}y''_n(\gamma_1 + \gamma_2) \\ + \frac{h^3}{2}\gamma_2\alpha_1 f_y y' + \frac{h^4}{4}\gamma_2[\alpha_2 f_y y'' + \alpha_1^2 f_{yy}(y')^2] + 0(h^5) \end{aligned} \tag{3.16}$$

while y'_{n+1} is given by

$$\begin{aligned} y'_{n+1} = y'_n + \frac{h}{2}y''_n(\delta_1 + \delta_2) + \frac{h^2}{2}\delta_2\alpha_1 f_y y' \\ + \frac{h^3}{4}\delta_2[\alpha_2 f_y y'' + \alpha_1^2 f_{yy}(y')^2] + 0(h^4) \end{aligned} \tag{3.17}$$

†Zurmuhl (1968) and Henrici (1962), Section 4.2.3.
‡Cooper (1967).

The first four terms of both (3.16) and (3.17) can be made to agree with the Taylor's series terms by choosing

$$\delta_2\alpha_2 = \tfrac{2}{3}, \qquad \delta_2\alpha_1^2 = \tfrac{2}{3}$$
$$\delta_1 + \delta_2 = 2, \qquad \delta_2\alpha_1 = 1$$
$$\gamma_2\alpha_1 = \tfrac{1}{3}, \qquad \gamma_1 + \gamma_2 = 1$$

which gives the *Nyström formula for special second order equations:*

$$k_1 = \frac{h^2 f(y_n, t_n)}{2}, \qquad k_2 = \frac{h^2 f[y_n + (2hy'_n/3) + (4k_1/9)]}{2}$$
$$y_{n+1} = y_n + hy'_n + \frac{1}{2}(k_1 + k_2), \qquad y'_{n+1} = y'_n + \frac{k_1 + 3k_2}{2h}$$

This has a local truncation error of $0(h^4)$ in both y and y', giving a global error of $0(h^3)$ in each component.

PROBLEMS

1. Consider methods of the form

$$y'_{n+1} = y'_n + \alpha_1 hf_n + \alpha_2 hf_{n+1}$$
$$y_{n+1} = y_n + hy'_n + \beta_1 h^2 f_n + \beta_2 h^2 f_{n+1}$$

for the second order equation

$$y'' = f(y, y', t)$$

f_{n+1} is taken to mean $f(y_{n+1}, y'_{n+1}, t_{n+1})$. What is the highest order that can be achieved globally? Is this order changed if $\partial f/\partial y' \equiv 0$? Give the coefficients for an example of this method achieving the orders you quote.

2. Verify that the method

$$\mathbf{y}_{n+1} = \mathbf{y}_n + \tfrac{1}{2}h[\mathbf{f}(\mathbf{y}_n) + \mathbf{f}(\mathbf{y}_{n+1})] + \tfrac{1}{12}h^2[\mathbf{f}'(\mathbf{y}_n) - \mathbf{f}'(\mathbf{y}_{n+1})]$$

is of fourth order for a system of equations of the first order.

3. What is the highest order method of the Runge-Kutta type that you can find for the special third order equation

$$y''' = f(y)$$

using two stages? The two-stage method has the form

$$k_0 = \frac{h^3 f(y_n)}{6}$$
$$k_1 = h^3 f\left(y_n + \alpha hy'_n + \alpha^2\frac{h^2}{2}y''_n + \alpha^3 k_0\right)$$
$$y_{n+1} = y_n + hy'_n + \frac{h^2}{2}y''_n + \beta_1 k_0 + \beta_2 k_1$$
$$y'_{n+1} = y'_n + hy''_n + \gamma_1\frac{k_0}{h} + \gamma_2\frac{k_1}{h}$$

$$y''_{n+1} = y''_n + \delta_1 \frac{k_0}{h^2} + \delta_2 \frac{k_1}{h^2}$$

4. What is the most general form of a third order equation for which the two-stage Runge-Kutta method described in your answer to Problem 3 will continue to give this higher order?

4 CONVERGENCE, ERROR BOUNDS, AND ERROR ESTIMATES FOR ONE-STEP METHODS

In this chapter we are going to ask the questions:

1. Under what conditions do methods converge?
2. What is the order of convergence?
3. What type of error bounds can be obtained?
4. What can be said about the asymptotic form of the error as $h \longrightarrow 0$ (as in Theorem 1.4 for Euler's method)?

We have previously examined the simplest one-step method in Chapter 1; now we would like to investigate the following one-step methods:

1. Higher order methods for single equations of the first order;
2. Methods for systems of equations of the first order;
3. Methods for systems of equations of order ≥ 1.

Each class is a special case of the subsequent class, so we will prove results for class (3) and discuss applications of these theorems to simpler cases from the other classes of problems.

Although we will be talking about direct methods for systems of higher order equations, we will almost always use the notational reduction of Section 3.2 to talk about a larger first order system. For example, if we had the pair of equations

$$y'' = f(y, y', z, z', z'', t)$$
$$z''' = g(y, y', z, z', z'', t)$$

we would write

$$y^1 = y, \qquad y^2 = y', \qquad y^3 = z, \qquad y^4 = z', \qquad y^5 = z''$$

so that we could express this as the system

$$\mathbf{y}' = \mathbf{f}(\mathbf{y}, t)$$

where

$$f^1 = y^2$$
$$f^2 = f(y^1, y^2, y^3, y^4, y^5, t)$$
$$f^3 = y^4$$
$$f^4 = y^5$$
$$f^5 = g(y^1, y^2, y^3, y^4, y^5, t)$$

4.1 VECTOR AND MATRIX NORMS

When we discussed the error in Euler's method for a single equation, we put a bound on its absolute value. When we deal with a system of equations, there will be an error in each member of the system so there will be a vector of errors. It will be necessary to derive bounds for these errors which will ensure that they are all small, so it is convenient to have a single quantity which is a measure of the size of all of the errors. It should have all the properties of the absolute value operator as far as possible so that the theorems for a single equation can be simply extended. We call such a quantity a *norm* of a vector **a**, written $||\mathbf{a}||$. An example of a norm is $||\mathbf{a}|| = \max_i |a^i|$ where the a^i are the components of **a**. This is called the max-norm. We see that if $||\mathbf{a}|| \leq e$, then each $|a^i| \leq e$. Thus, if we bound $||\mathbf{a}||$ we bound each component. It is possible to define many different norms. Formally they must satisfy the property that $||\mathbf{a}||$ is a nonnegative real number such that

$$||\mathbf{a}|| = 0 \Longleftrightarrow \mathbf{a} = \mathbf{0} \tag{4.1}$$

(This guarantees that if the norm converges to zero, so do the components of the vector.)

$$||\mathbf{a} + \mathbf{b}|| \leq ||\mathbf{a}|| + ||\mathbf{b}|| \tag{4.2}$$

(This is the triangle inequality which allows us to expand complex expressions and bound the individual parts separately.)

and

$$||\lambda\mathbf{a}|| = |\lambda|\,||\mathbf{a}|| \tag{4.3}$$

for any scalar λ.

A vector may be premultiplied by a matrix to obtain another vector. A norm for a matrix, written $||A||$, is called *consistent* with a vector norm if

$$||A\mathbf{a}|| \leq ||A||\,||\mathbf{a}|| \tag{4.4}$$

for all matrices A and vectors **a**. A matrix norm must also satisfy properties (4.1), (4.2), (4.3), and

$$||AB|| \leq ||A||\,||B|| \tag{4.5}$$

It can be seen that if $||A||$ is defined by $\sup_{||\mathbf{a}||\neq 0} ||A\mathbf{a}||/||\mathbf{a}||$, it satisfies (4.1) through (4.4). A consistent matrix norm for the max-norm given above is

$$||A|| = \max_i \sum_j |A^i_j|$$

Equation (4.4) can be verified by observing that

$$\begin{aligned} ||A\mathbf{a}|| &= \max_i |\sum_j A^i_j a^j| \\ &\leq \max_i \sum_j |A^i_j| \max_k |a^k| \\ &= ||A|| \, ||\mathbf{a}|| \end{aligned}$$

while (4.5) follows from

$$\begin{aligned} ||AB|| &= \max_i \sum_k |\sum_j A^i_j B^j_k| \\ &\leq \max_i \sum_j [|A^i_j| \sum_k |B^j_k|] \\ &\leq \max_i \sum_j |A^i_j| \max_l \sum_k |B^l_k| \\ &= ||A|| \, ||B|| \end{aligned}$$

Another example of a norm is

$$||\mathbf{a}|| = \sum_i |a^i| \tag{4.6}$$

with the consistent matrix norm

$$||A|| = \max_j \sum_i |A^i_j| \tag{4.7}$$

Frequently we refer to the L_p norm. By this is meant

$$||a||_p = (\sum_i |a^i|^p)^{1/p}$$

For $p = 1$, this is $\sum_i |a^i|$ discussed above. For $p = \infty$ (that is, the limit as $p \to \infty$), we get the max-norm. $p = 2$ corresponds to the familiar Euclidean norm of distance

$$||a||_2 = \sqrt{\sum_i |a^i|^2}$$

In subsequent work almost any norm could be used, although the max-norm is usually the simplest. A different norm may lead to different error bounds, with the max-norm frequently yielding the sharpest.

4.2 EXISTENCE AND THE LIPSCHITZ CONDITION

A direct analog of Theorem 1.1 provides an existence proof for systems of equations of any order. If these equations are written as

$$\mathbf{y}' = \mathbf{f}(\mathbf{y}, t)$$

we can state

THEOREM 4.1

If $\mathbf{f}(\mathbf{y}, t)$ *is a continuous function of t and satisfies the Lipschitz condition in* $\mathbf{y}$ *in the region* $0 \leq t \leq b$, $-\infty \leq y^i \leq \infty$, *then there exists a unique differentiable function* $\mathbf{y}(t)$ *such that*

$$\mathbf{y}(0) = \mathbf{y}_0$$

$$\frac{d\mathbf{y}(t)}{dt} = \mathbf{f}(\mathbf{y}(t), t)$$

This result can also be found in any standard text on differential equations.

By the Lipschitz condition for a system we mean that there exists a constant L such that

$$||\mathbf{f}(\mathbf{y}, t) - \mathbf{f}(\mathbf{y}^*, t)|| \leq L||\mathbf{y} - \mathbf{y}^*|| \tag{4.8}$$

If $\mathbf{f}$ satisfies a Lipschitz condition in this sense, it also satisfies one separately in each of its dependent variables and vice versa.

In practice, we will only need the Lipschitz condition in a finite region within which $\mathbf{y}$ can be shown to lie (by bounds on the size of $||\mathbf{f}||$). If, in that region, $\mathbf{f}$ has continuous derivatives with respect to $\mathbf{y}$, then it satisfies a Lipschitz condition since by the mean value theorem

$$f^i(\mathbf{y}, t) - f^i(\mathbf{y}^*, t) = \sum_j \frac{\partial f^i}{\partial y^j}(\xi^j)(y^j - y^{j*})$$

where $\{\xi^j\}$ is a set of points in the region. If

$$K^i_j = \max_{\text{region}} \left| \frac{\partial f^i}{\partial y^j} \right|$$

then

$$|f^i(\mathbf{y}, t) - f^i(\mathbf{y}^*, t)| \leq \sum_j K^i_j |y^j - y^{j*}|$$

If we define the Lipschitz constant L to be the norm of the matrix $K = \{K^i_j\}$, we get (4.8).

4.3 CONVERGENCE AND STABILITY

The common basis of the methods described in earlier chapters was that each prescribed an amount which was to be added onto $\mathbf{y}_n$ to get $\mathbf{y}_{n+1}$. Formally, we define a one-step method by:

DEFINITION 4.1

A one-step method for approximating the solution of a differential equation is a method which can be written in the form

$$\mathbf{y}_{n+1} = \mathbf{y}_n + h\boldsymbol{\psi}(\mathbf{y}_n, t_n, h) \tag{4.9}$$

where the increment function $\boldsymbol{\psi}$ is determined by $\mathbf{f}$ and is a function of $\mathbf{y}_n$, t_n and h only.

Convergence is defined for one-step methods by:

DEFINITION 4.2

The one-step method (4.9) is convergent if $\mathbf{y}_n \longrightarrow \mathbf{y}(t)$ for all $0 \leq t \leq b$ as $n \longrightarrow \infty$ and $\mathbf{y}_0 \longrightarrow \mathbf{y}(0)$ with $h = t/n$ for any differential equation $\mathbf{y}' = \mathbf{f}(\mathbf{y})$ which satisfies a Lipschitz condition.

We note that errors are permitted in the starting value $\mathbf{y}_0$ since in practice we cannot represent $\mathbf{y}(0)$ exactly in finite precision. Convergence assures us that the true solution can be approximated arbitrarily closely by making h smaller and using greater precision.

Stability is concerned with the effect of perturbations on the numerical solution.

DEFINITION 4.3

A one-step method is stable if for each differential equation satisfying a Lipschitz condition there exist positive constants h_0 and K such that the difference between two different numerical solutions $\mathbf{y}_n$ and $\tilde{\mathbf{y}}_n$ each satisfying (4.9) is such that

$$\|\mathbf{y}_n - \tilde{\mathbf{y}}_n\| \leq K\|\mathbf{y}_0 - \tilde{\mathbf{y}}_0\|$$

for all $0 \leq h \leq h_0$.

Stability is nearly automatic for one-step methods as the following theorem indicates.

THEOREM 4.2

If $\boldsymbol{\psi}(\mathbf{y}, t, h)$ *satisfies a Lipschitz condition in* $\mathbf{y}$, *then the method given by* (4.9) *is stable.*

The proof is left as a simple exercise.

We assume that $\mathbf{f}$ satisfies the conditions of Theorem 4.1 throughout. We can then see that for all of the methods discussed earlier, the $\boldsymbol{\psi}$ will also satisfy these conditions for $0 \leq h \leq h_0$.

In the case of the midpoint rule, for example,

$$\boldsymbol{\psi}(\mathbf{y}, t, h) = \mathbf{f}\left(\mathbf{y} + \frac{h}{2}\mathbf{f}(\mathbf{y}, t), t + \frac{h}{2}\right)$$

which is continuous in t and $\mathbf{y}$ if $\mathbf{f}$ is, and

$$\begin{aligned}
&\| \boldsymbol{\psi}(\mathbf{y}, t, h) - \boldsymbol{\psi}(\mathbf{y}^*, t, h) \| \\
&= \left\| \mathbf{f}\left(\mathbf{y} + \frac{h}{2}\mathbf{f}(\mathbf{y}, t), t + \frac{h}{2}\right) - \mathbf{f}\left(\mathbf{y}^* + \frac{h}{2}\mathbf{f}(\mathbf{y}^*, t), t + \frac{h}{2}\right) \right\| \\
&\leq L \left\| \mathbf{y} + \frac{h}{2}\mathbf{f}(\mathbf{y}, t) - \mathbf{y}^* - \frac{h}{2}\mathbf{f}(\mathbf{y}^*, t) \right\| \\
&\leq L \| \mathbf{y} - \mathbf{y}^* \| + L\frac{h}{2} \| \mathbf{f}(\mathbf{y}, t) - \mathbf{f}(\mathbf{y}^*, t) \| \\
&\leq L\left(1 + \frac{Lh}{2}\right) \| \mathbf{y} - \mathbf{y}^* \|
\end{aligned}$$

Thus $\boldsymbol{\psi}$ satisfies a Lipschitz condition in $\mathbf{y}$ for $0 \leq h \leq h_0$. Also note that $\boldsymbol{\psi}$ is continuous in h if $\mathbf{f}$ is continuous in $\mathbf{y}$ and t. In the case of the Euler method, $\boldsymbol{\psi} = \mathbf{f}(\mathbf{y}, t)$, so it is not surprising that we can show:

THEOREM 4.3

If $\boldsymbol{\psi}(\mathbf{y}, t, h)$ *is continuous in* $\mathbf{y}$, t, h *for* $0 \leq t \leq b$, $0 \leq h \leq h_0$, *and all* $\mathbf{y}$, *and if it satisfies a Lipschitz condition in* $\mathbf{y}$ *in that region, a necessary and sufficient condition for convergence is that*

$$\boldsymbol{\psi}(\mathbf{y}(t), t, 0) = \mathbf{f}(\mathbf{y}(t), t) \tag{4.10}$$

Equation (4.10) is called the condition of *consistency*. Since, by suitable choice of initial conditions, $\mathbf{y}(t)$ can take on any value for a given t, (4.10) will hold for any $\mathbf{y}$ in the form

$$\boldsymbol{\psi}(\mathbf{y}, t, 0) = \mathbf{f}(\mathbf{y}, t)$$

Proof: Let $\boldsymbol{\psi}(\mathbf{y}, \mathbf{t}, 0) = \mathbf{g}(\mathbf{y}, t)$

Since $\mathbf{g}$ satisfies the conditions of Theorem 4.1, the differential equation

$$\mathbf{z}' = \mathbf{g}(\mathbf{z}, t), \qquad \mathbf{z}_0 = \mathbf{y}_0 \tag{4.11}$$

has a unique differentiable solution. We will show that the numerical solution given by (4.9) converges to $\mathbf{z}(t)$, and hence that $\mathbf{f} = \mathbf{g}$ is a necessary and sufficient condition. The numerical solution satisfies

$$\mathbf{y}_{n+1} = \mathbf{y}_n + h\boldsymbol{\psi}(\mathbf{y}_n, t_n, h) \tag{4.12}$$

By the mean value theorem,

$$z^i(t_{n+1}) = z^i(t_n) + hg^i(\mathbf{z}(t_n + \theta^i h), t_n + \theta^i h) \qquad \text{for } 0 < \theta^i < 1$$

Subtracting this from (4.12) and setting $\mathbf{e}_n = \mathbf{y}_n - \mathbf{z}(t_n)$, we get

$$\begin{aligned}
e^i_{n+1} = e^i_n + h[&\psi^i(\mathbf{y}_n, t_n, h) - \psi^i(\mathbf{z}(t_n), t_n, h) \\
&+ \psi^i(\mathbf{z}(t_n), t_n, h) - \psi^i(\mathbf{z}(t_n), t_n, 0) \\
&+ \psi^i(\mathbf{z}(t_n), t_n, 0) - g^i(\mathbf{z}(t_n + \theta^i h), t_n + \theta^i h)]
\end{aligned} \tag{4.13}$$

As in Theorem 1.3, we can continue without additional hypotheses, but if we also assume that $\boldsymbol{\psi}$ satisfies a Lipschitz condition in t and h, as will happen in practice, we get the following bounds

$$\|\boldsymbol{\psi}(\mathbf{y}_n, t_n, h) - \boldsymbol{\psi}(\mathbf{z}(t_n), t_n, h)\| \leq L\|\mathbf{y}_n - \mathbf{z}(t_n)\| = L\|\mathbf{e}_n\|$$
$$\|\boldsymbol{\psi}(\mathbf{z}(t_n), t_n, h) - \boldsymbol{\psi}(\mathbf{z}(t_n), t_n, 0)\| \leq L_1 h$$

and

$$\begin{aligned} &|\{\psi^i(\mathbf{z}(t_n), t_n, 0) - g^i(\mathbf{z}(t_n + \theta^i h), t_n + \theta^i h)\}| \\ &= |\{g^i(\mathbf{z}(t_n), t_n) - g^i(\mathbf{z}(t_n + \theta^i h), t_n + \theta^i h)\}| \\ &\leq L|z'(t_n + \xi\theta^i h)|\theta^i h + L_3\theta^i h \leq L_4 h \end{aligned}$$

Hence the norm of the last line in (4.13) can be bounded by $L_2 h$. Substituting these in (4.13), we get

$$\begin{aligned} \|e_{n+1}\| &\leq \|e_n\| + hL\|e_n\| + h^2(L_1 + L_2) \\ &= (1 + hL)\|e_n\| + h^2(L_1 + L_2) \end{aligned} \tag{4.14}$$

This is a difference equation of the type in Lemma 1.1 from which we have

$$\|e_N\| \leq (L_1 + L_2)h\frac{e^{Lb} - 1}{L} + e^{Lb}\|e_0\|$$

This converges to zero as h and $\|e_0\| \longrightarrow 0$, so the numerical solution converges to the solution of (4.11). Sufficiency of the condition $\mathbf{g}(\mathbf{y}, t) = \mathbf{f}(\mathbf{y}, t)$ follows immediately. If, on the other hand, we have convergence, then $\mathbf{z}(t)$, the solution of (4.11), is identical to $\mathbf{y}(t)$, the solution of $\mathbf{y}'(t) = \mathbf{f}(\mathbf{y}(t), t)$. Suppose also that $\mathbf{f}$ and $\mathbf{g}$ differ at some point $(\mathbf{y}_a, t_a)$. If we consider the initial value problem starting from $(\mathbf{y}_a, t_a)$ we have

$$\mathbf{y}'(t_a) = \mathbf{f}(\mathbf{y}(t_a), t_a) \neq \mathbf{g}(\mathbf{y}(t_a), t_a) = \mathbf{g}(\mathbf{z}(t_a), t_a) = \mathbf{z}'(t_a)$$

leading to a contradiction.

Examples of the Application of Theorem 4.3

A. Fourth order classical Runge-Kutta methods applied to the system of first order equations $\mathbf{y}' = \mathbf{f}(\mathbf{y})$.

We are given that $\mathbf{f}$ satisfies a Lipschitz condition. Therefore, $\mathbf{k}_0(\mathbf{y}) = h\mathbf{f}(\mathbf{y})$ satisfies

$$\|\mathbf{k}_0(\mathbf{y}) - \mathbf{k}_0(\mathbf{y}^*)\| \leq hL\|\mathbf{y} - \mathbf{y}^*\|$$

$\mathbf{k}_1(\mathbf{y}) = h\mathbf{f}(\mathbf{y} + \frac{1}{2}\mathbf{k}_0(\mathbf{y}))$ satisfies

$$\begin{aligned} \|\mathbf{k}_1(\mathbf{y}) - \mathbf{k}_1(\mathbf{y}^*)\| &\leq hL\|\mathbf{y} - \mathbf{y}^* + \tfrac{1}{2}\mathbf{k}_0(\mathbf{y}) - \tfrac{1}{2}\mathbf{k}_0(\mathbf{y}^*)\| \\ &\leq hL(1 + \tfrac{1}{2}hL)\|\mathbf{y} - \mathbf{y}^*\| \end{aligned}$$

$\mathbf{k}_2(\mathbf{y}) = h\mathbf{f}(\mathbf{y} + \frac{1}{2}\mathbf{k}_1(\mathbf{y}))$ satisfies

$$\begin{aligned} \|\mathbf{k}_2(\mathbf{y}) - \mathbf{k}_2(\mathbf{y}^*)\| &\leq hL|\mathbf{y} - \mathbf{y}^* + \tfrac{1}{2}\mathbf{k}_1(\mathbf{y}) - \tfrac{1}{2}\mathbf{k}_1(\mathbf{y}^*)\| \\ &\leq hL(1 + \tfrac{1}{2}hL + \tfrac{1}{4}(hL)^2)\|\mathbf{y} - \mathbf{y}^*\| \end{aligned}$$

and $\mathbf{k}_3(\mathbf{y}) = h\mathbf{f}(\mathbf{y} + \mathbf{k}_2(\mathbf{y}))$ satisfies

$$\|\mathbf{k}_3(\mathbf{y}) - \mathbf{k}_3(\mathbf{y}^*)\| \leq hL\,|\mathbf{y} - \mathbf{y}^* + \mathbf{k}_2(\mathbf{y}) - \mathbf{k}_2(\mathbf{y}^*)\|$$
$$\leq hL(1 + hL + \tfrac{1}{2}(hL)^2 + \tfrac{1}{4}(hL)^3)\,\|\mathbf{y} - \mathbf{y}^*\|$$

Therefore,

$$\boldsymbol{\psi}(\mathbf{y}, t, h) = \frac{1}{6h}(\mathbf{k}_0 + 2\mathbf{k}_1 + 2\mathbf{k}_2 + \mathbf{k}_3)$$

satisfies

$$\|\boldsymbol{\psi}(\mathbf{y}, t, h) - \boldsymbol{\psi}(\mathbf{y}^*, t, h)\|$$
$$\leq \frac{L}{6}\Big(1 + 2 + hL + 2 + hL + \frac{1}{2}(hL)^2$$
$$+ 1 + hL + \frac{1}{2}(hL)^2 + \frac{1}{4}(hL)^3\Big)\,\|\mathbf{y} - \mathbf{y}^*\|$$
$$= L\Big(1 + \frac{1}{2}hL + \frac{1}{6}(hL)^2 + \frac{1}{24}(hL)^3\Big)\,\|\mathbf{y} - \mathbf{y}^*\|$$

Hence $\boldsymbol{\psi}$ satisfies a Lipschitz condition in $\mathbf{y}$. It can also be seen to be continuous in h so we can conclude that the classical fourth order Runge-Kutta method converges for a system of equations.

B. A second order Taylor's series method applied to the second order equation

$$y'' = f(y, y', t)$$

where f, $\partial f/\partial y$ and $\partial f/\partial t$ satisfy Lipschitz conditions in t, y, and y'.

$$\begin{aligned} y_{n+1} &= y_n + hy'_n + \frac{h^2}{2}f_n + \frac{h^3}{6}\frac{df_n}{dt} \\ y'_{n+1} &= y'_n + hf_n + \frac{h^2}{2}\frac{df_n}{dt} \end{aligned} \tag{4.15}$$

If the vector $\mathbf{y}$ is $[y, y']^T$, then

$$\boldsymbol{\psi}(\mathbf{y}, t, h) = \left[y'_n + \frac{h}{2}f_n + \frac{h^2}{6}\frac{df_n}{dt}, f_n + \frac{h}{2}\frac{df_n}{dt}\right]^T$$

Evidently $\boldsymbol{\psi}$ satisfies a Lipschitz condition in t, y, and y' and $\boldsymbol{\psi}(\mathbf{y}, t, 0) = [y'_n, f_n]^T$ so the method converges. (In fact, we do not need a Lipschitz condition on df/dt, only boundedness. The proof of Theorem 4.3 can be modified for that case.)

4.4 ERROR BOUNDS AND ORDER OF CONVERGENCE

Theorem 4.3 gave conditions for convergence, but did not specify the rate at which the answers converged. To examine this we need to define the *local truncation error* $\mathbf{d}_n(h)$. It is given by

$$\mathbf{d}_n(h) = h\boldsymbol{\psi}(\mathbf{y}(t_n), t_n, h) - (\mathbf{y}(t_{n+1}) - \mathbf{y}(t_n)) \tag{4.16}$$

Thus it is the amount by which the solution of the differential equation fails to satisfy the equation used in the numerical method. We have seen that $\mathbf{d}_n(h)$ can be expressed as a power series in h with coefficients which are polynomials in the derivatives of the solution. By means of a remainder form of the Taylor's series expansion used to derive $\mathbf{d}_n(h)$, bounds can be obtained. If we have such a bound in the form

$$\|\mathbf{d}_n(h)\| \leq Dh^{r+1} \tag{4.17}$$

for all $0 \leq h \leq h_0$ and all t and $\mathbf{y}$ considered, we can show that the global error is of order r. Thus the method is said to have order r.

THEOREM 4.4

If $\boldsymbol{\psi}$ *satisfies the conditions of Theorem 4.3 and if* $\mathbf{d}_n(h)$ *defined by* (4.16) *satisfies* (4.17), *then the error is bounded by*

$$\|\mathbf{y}_n - \mathbf{y}(t_n)\| \leq Dh^r \frac{e^{Lb} - 1}{L} + e^{Lb}\|\mathbf{y}_0 - \mathbf{y}(t_0)\| \tag{4.18}$$

Proof: If we write $\mathbf{e}_n = \mathbf{y}_n - \mathbf{y}(t_n)$ and subtract $\mathbf{y}(t_{n+1})$ from both sides of (4.12), we get

$$\begin{aligned} \mathbf{e}_{n+1} &= \mathbf{y}_n - \mathbf{y}(t_n) + [h\boldsymbol{\psi}(\mathbf{y}_n, t_n, h) - (\mathbf{y}(t_{n+1}) - \mathbf{y}(t_n))] \\ &= \mathbf{e}_n + h[\boldsymbol{\psi}(\mathbf{y}_n, t_n, h) - \boldsymbol{\psi}(\mathbf{y}(t_n), t_n, h)] + \mathbf{d}_n(h) \end{aligned} \tag{4.19}$$

Therefore,

$$\begin{aligned} \|\mathbf{e}_{n+1}\| &\leq \|\mathbf{e}_n\| + hL\|\mathbf{e}_n\| + h^{r+1}D \\ &= (1 + hL)\|\mathbf{e}_n\| + h^{r+1}D \end{aligned}$$

From Lemma 1.1 we obtain (4.18) directly.

We see that the global order of the error is one lower than the order of the local truncation error.

Example of the Application of Theorem 4.4

Consider the method defined by the equations in (4.15). If $y(t)$ has a continuous fourth derivative, we can express $\mathbf{d}_n(h)$ as

$$\left[-\frac{h^4}{24}y^{(4)}(\xi_n), -\frac{h^3}{6}y^{(4)}(\xi_n')\right]^T$$

Consequently, $\|\mathbf{d}_n(h)\| \leq h^3 D$ for $h \leq h_0$, where D is proportional to $\max|y^{(4)}|$. The global rate of convergence is quadratic, even though the local truncation error in y_n is $0(h^4)$. We can observe this in the example

$$y'' = y, \qquad y(0) = 1, \qquad y'(0) = 1$$

integrated over $t = 0(h)1$, $h = 2^{-m}$, $m = 0, 1, 2, \ldots, 6$ as shown in Table 4.1. The error in both y and y' is behaving as $0(h^2)$.

Table 4.1 INTEGRATION OF $y'' = y$

h	$y_h(1)$	$e = y_h(1) -2.71828$	$\frac{e}{h^2}$	$y_h'(1)$	$e' = y_h'(1) -2.71828$	$\frac{e'}{h^2}$
1	2.6666667	−0.0516152	−0.0516	2.5000000	−0.2182818	−0.2183
$\frac{1}{2}$	2.6979167	−0.0203652	−0.0815	2.6510417	−0.0672402	−0.2690
$\frac{1}{4}$	2.7118655	−0.0064163	−0.1027	2.6997631	−0.0185187	−0.2963
$\frac{1}{8}$	2.7164846	−0.0017973	−0.1150	2.7134331	−0.0048487	−0.3103
$\frac{1}{16}$	2.7178066	−0.0004752	−0.1217	2.7170421	−0.0012398	−0.3174
$\frac{1}{32}$	2.7181597	−0.0001222	−0.1251	2.7179684	−0.0003134	−0.3209
$\frac{1}{64}$	2.7182509	−0.0000310	−0.1269	2.7182031	−0.0000788	−0.3227

Theorem 4.4 also takes care of putting an error bound on the accumulated round-off error. This is handled by including the round-off in the definition of $h\boldsymbol{\psi}$, so that the (4.16) definition of $\mathbf{d}_n(h)$ includes the round-off and truncation errors. As long as a bound of the form of (4.17) can be imposed, Theorem 4.4 can be applied.

4.5 ASYMPTOTIC ERROR ESTIMATES

In the case of the Euler method we saw that the error bounds are not very realistic estimates of the size of the error. The same is true in the general case. In the example at the end of the last section, we have

$$\mathbf{d}_n(h) = \left[-\frac{h^4}{24}y^{(4)}(\xi_n), \ -\frac{h^3}{6}y^{(4)}(\xi_n')\right]^T$$

Using the max-norm, we get

$$\|\mathbf{d}_n(h)\| \le h^3 \max_{0 \le t \le 1} \left|\frac{y^{(4)}t}{6}\right|$$

for $0 \le t \le 1$, $0 \le h \le 1$. Since $y(t) = e^t$, $D = e/6$.

$$\frac{\partial \boldsymbol{\psi}}{\partial(y, y')} = \begin{bmatrix} \frac{h}{2} & 1 + \frac{h^2}{6} \\ 1 & \frac{h}{2} \end{bmatrix}$$

so $L = 1 + (h/2) + (h^2/6) \le \frac{5}{3}$ for $h < 1$.

Using these figures, the error bound given by (4.18) is

$$h^2 \frac{e}{10}(e^{5/3} - 1) \cong 1.167h^2$$

From Table 4.1 we see that the error is roughly asymptotic to $0.128h^2$ and $0.325h^2$ in y and y', so the bound is larger by a factor of more than three. In complex problems the bound may be much larger.

Therefore, we again look for error estimates of an asymptotic form as $h \longrightarrow 0$. If we can write the local truncation error in the form

$$\begin{aligned} h\boldsymbol{\psi}(\mathbf{y}(t), t, h) - (\mathbf{y}(t+h) - \mathbf{y}(t)) &= \mathbf{d}_n(h) \\ &= h^{r+1}\boldsymbol{\phi}(\mathbf{y}, t) + 0(h^{r+2}) \end{aligned} \tag{4.20}$$

[which we can if $\boldsymbol{\psi}$ and $\mathbf{f}$ have continuous $(r + 1)$th derivatives], then $\boldsymbol{\phi}$ is called the *principal error function.* We know from Theorem 4.4 that the error is $0(h^r)$, so we look for an error expression of the form

$$\mathbf{e}_n = h^r\boldsymbol{\delta}(t_n) + 0(h^{r+1})$$

In order to examine the $\boldsymbol{\delta}(t)$ term we start by substituting

$$\mathbf{e}_n = h^r\boldsymbol{\delta}_n$$

into (4.19) to get

$$\begin{aligned} \boldsymbol{\delta}_{n+1} = \boldsymbol{\delta}_n &+ h^{1-r}[\boldsymbol{\psi}(\mathbf{y}(t_n) + h^r\boldsymbol{\delta}_n, t_n, h) - \boldsymbol{\psi}(\mathbf{y}(t_n), t_n, h)] \\ &+ h\boldsymbol{\phi}(\mathbf{y}(t_n), t_n) + 0(h^2) \end{aligned} \tag{4.21}$$

If we assume that $\boldsymbol{\psi}$ is twice continuously differentiable with respect to its arguments $\mathbf{y}$ and h and that these derivatives are bounded in the region in which we are working (continuity of the second derivatives implies this), we can write

$$\begin{aligned} \boldsymbol{\psi}(\mathbf{y}(t_n) + h^r\boldsymbol{\delta}_n, t_n, h) = \boldsymbol{\psi}(\mathbf{y}(t_n), t_n, h) &+ \boldsymbol{\psi}_{\mathbf{y}}(\mathbf{y}(t_n), t_n, h)h^r\boldsymbol{\delta}_n \\ &+ \tfrac{1}{2}\boldsymbol{\psi}_{\mathbf{yy}}(\mathbf{y}(t_n) + \boldsymbol{\xi}, t_n, h)h^{2r}\boldsymbol{\delta}_n\boldsymbol{\delta}_n \end{aligned}$$

where $\boldsymbol{\psi}_{\mathbf{y}}$ is a matrix whose components are $\partial\psi^i/\partial y^j$, $\boldsymbol{\psi}_{\mathbf{yy}}$ is a third order tensor† whose components are $\partial^2\psi^i/\partial y^j\,\partial y^k$, and $\boldsymbol{\xi}$ is a vector each of whose components is less than those of $\mathbf{e}_n$. Since we have assumed that second derivatives are bounded, and we have already shown (Theorem 4.4) that $\|\boldsymbol{\delta}_n\|$ is bounded, the last term can be written as $\mathbf{k}_1h^{2r}$ where $\|\mathbf{k}_1\|$ is bounded. We can re-express the second term by the mean value theorem as

$$\boldsymbol{\psi}_{\mathbf{y}}(\mathbf{y}(t_n), t_n, h)h^r\boldsymbol{\delta}_n = \boldsymbol{\psi}_{\mathbf{y}}(\mathbf{y}(t_n), t_n, 0)h^r\boldsymbol{\delta}_n + \boldsymbol{\psi}_{\mathbf{y}h}(\mathbf{y}(t_n), t_n, \xi')h^{r+1}\boldsymbol{\delta}_n$$

Again by the boundedness of second derivatives, the last term is of the form $\mathbf{k}_2h^{r+1}$, where $\|\mathbf{k}_2\|$ is bounded. Finally note that $\boldsymbol{\psi}(\mathbf{y}, t, 0) = \mathbf{f}(\mathbf{y}, t)$, so $\boldsymbol{\psi}_{\mathbf{y}}(\mathbf{y}, t, 0) = \mathbf{f}_{\mathbf{y}}(\mathbf{y}, t)$. Substituting into (4.21), we get

$$\boldsymbol{\delta}_{n+1} = \boldsymbol{\delta}_n + h[\mathbf{f}_{\mathbf{y}}(\mathbf{y}(t_n), t_n)\boldsymbol{\delta}_n + \boldsymbol{\phi}(\mathbf{y}(t_n), t_n) + h\mathbf{k}_2 + h^r\mathbf{k}_1] \tag{4.22}$$

If we view (4.22) as a numerical method for solving the differential equation

$$\boldsymbol{\delta}'(t) = \mathbf{f}_{\mathbf{y}}(\mathbf{y}(t), t)\boldsymbol{\delta}(t) + \boldsymbol{\phi}(\mathbf{y}(t), t), \qquad \boldsymbol{\delta}(0) = \frac{\mathbf{e}_0}{h^r} \tag{4.23}$$

we see that the increment for the method satisfies the conditions of Theorem 4.3, so $\boldsymbol{\delta}_n$ converges to $\boldsymbol{\delta}(t_n)$ and it also satisfies the conditions of Theorem 4.4 with $r = 1$ (in that Theorem) with

†A simplified view of a tensor is that it is a generalized matrix. For our purposes we just note that $\boldsymbol{\psi}_{\mathbf{yy}}\boldsymbol{\delta\delta}$ is a notation for a vector with components $\sum_{jk} (\partial^2\psi^i/\partial y^j\,\partial y^k)\,\delta^j\delta^k$, or in our earlier notation, $\psi^i_{jk}\,\delta^j\delta^k$.

$$D = \max_t \tfrac{1}{2}\|\boldsymbol{\delta}''(t)\| + h_0^{r-1}\max\|\mathbf{k}_1\| + \max\|\mathbf{k}_2\|$$

for $0 \le h \le h_0$. Consequently, we have shown the following:

THEOREM 4.5

If the truncation error can be expressed as in (4.20) *and* $\boldsymbol{\psi}$ *has continuous second derivatives, then the error satisfies*

$$\mathbf{e}_n = h^r\boldsymbol{\delta}(t_n) + 0(h^{r+1}) \tag{4.24}$$

where $\boldsymbol{\delta}(t)$ *is the solution of* (4.23).

This theorem is seldom used as a means of directly estimating the error since the evaluation of $\mathbf{f}_\mathbf{y}$ and $\boldsymbol{\phi}$ are too complex for most equations and methods of interest. However, it is used to justify indirect error estimates such as extrapolation to the limit. As with the Euler method for single equations, we can compute with two different step sizes h and qh to get an estimate of $\mathbf{y}(b)$. Call these values $\mathbf{y}_h(b)$ and $\mathbf{y}_{qh}(b)$. From (4.24)

$$\mathbf{y}_h(b) = \mathbf{y}(b) + h^r\boldsymbol{\delta}(b) + 0(h^{r+1})$$
$$\mathbf{y}_{qh}(b) = \mathbf{y}(b) + q^rh^r\boldsymbol{\delta}(b) + 0(h^{r+1})$$

or

$$\boldsymbol{\delta}(b) = h^{-r}\frac{\mathbf{y}_h(b) - \mathbf{y}_{qh}(b)}{1 - q^r} + 0(h)$$

Alternatively, we can get a better approximation to $\mathbf{y}(b)$ by

$$\mathbf{y}(b) = \frac{\mathbf{y}_{qh}(b) - q^r\mathbf{y}_h(b)}{1 - q^r} + 0(h^{r+1})$$

This is shown in Table 4.2 for $q = \frac{1}{2}$ applied to the results of Table 4.1. We see that we now have $0(h^3)$ convergence.

Example of the Application of Theorem 4.5

Again we consider the second order Taylor's series method used to produce Table 4.1. We have

$$\mathbf{d}_n(h) = h^3\left[0, -\frac{e^{t_n}}{6}\right]^T + 0(h^4)$$

so

$$\boldsymbol{\phi}(\mathbf{y}, t) = \left[0, -\frac{e^t}{6}\right]^T$$

The differential equation is $y'' = y$, which we have written as

$$\mathbf{y}' = \begin{bmatrix} y \\ y' \end{bmatrix}' = \begin{bmatrix} y' \\ y \end{bmatrix} = \mathbf{f}(\mathbf{y}, t)$$

so

$$\mathbf{f}_\mathbf{y} = \begin{bmatrix} 0 & 1 \\ 1 & 0 \end{bmatrix}$$

Table 4.2 EXTRAPOLATION OF THE RESULTS OF $y'' = y$ BY A SECOND ORDER TAYLOR'S METHOD

h	$y_h(1)$	$\dfrac{4y_{h/2}(1) - y_h(1)}{3}$	*Error*	*Error*/h^3
1	2.6666667	2.7083333	−0.00994850	−0.0796
$\frac{1}{2}$	2.6979167	2.7165152	−0.00176668	−0.1131
$\frac{1}{4}$	2.7118655	2.7180242	−0.00025760	−0.1319
$\frac{1}{8}$	2.7164846	2.7182473	−0.00003455	−0.1415
$\frac{1}{16}$	2.7178066	2.7182774	−0.00000446	−0.1463
$\frac{1}{32}$	2.7181597	2.7182813	−0.00000057	−0.1487
$\frac{1}{64}$	2.7182509	2.7182818	−0.00000007	−0.1498
$\frac{1}{128}$	2.7182740	—	—	—
h	$y'_h(1)$	$\dfrac{4y'_{h/2}(1) - y'_n(1)}{3}$	*Error*	*Error*/h^3
1	2.5000000	2.7013889	−0.01689294	−0.1351
$\frac{1}{2}$	2.6510417	2.7160036	−0.00227827	−0.1458
$\frac{1}{4}$	2.6997631	2.7179898	−0.00029206	−0.1495
$\frac{1}{8}$	2.7134331	2.7182451	−0.00003678	−0.1506
$\frac{1}{16}$	2.7170421	2.7182772	−0.00000461	−0.1509
$\frac{1}{32}$	2.7179684	2.7182813	−0.00000058	−0.1510
$\frac{1}{64}$	2.7182031	2.7182818	−0.00000007	−0.1510
$\frac{1}{128}$	2.7182621	—	—	—

Consequently, $\boldsymbol{\delta}(t)$ satisfies

$$\begin{bmatrix} \delta^1(t) \\ \delta^2(t) \end{bmatrix}' = \begin{bmatrix} 0 & 1 \\ 1 & 0 \end{bmatrix} \begin{bmatrix} \delta^1(t) \\ \delta^2(t) \end{bmatrix} - \begin{bmatrix} 0 \\ \dfrac{e^t}{6} \end{bmatrix}$$

which has the solution

$$\delta^1(t) = -\tfrac{1}{24}(2te^t + e^{-t} - e^t)$$
$$\delta^2(t) = -\tfrac{1}{24}(2te^t + e^t - e^{-t})$$

Consequently,

$$\delta^1(1) = -\tfrac{1}{24}(e + e^{-1}) \cong -0.1286$$
$$\delta^2(1) = -\tfrac{1}{24}(3e - e^{-1}) \cong -0.3245$$

which agree well with the errors in Table 4.1.

4.5.1 The Perturbation Due to the Numerical Approximation

The numerical solution $\mathbf{y}_n$ can be expressed as $\mathbf{z}(t_n)$, where $\mathbf{z}$ is the solution of a perturbed differential equation. The analysis is analogous to that in

Section 1.3.5. Define the local error for the solution $\mathbf{y}(t_n + \tau; \mathbf{y}_n, t_n)$ to be

$$\begin{aligned}\mathbf{d}(\tau; \mathbf{y}_n, t_n) &= \tau^{r+1}\mathbf{T}(\tau; \mathbf{y}_n, t_n)\\ &= \mathbf{y}_n + \tau\boldsymbol{\psi}(\mathbf{y}_n, t_n, \tau) - \mathbf{y}(t_n + \tau; \mathbf{y}_n, t_n)\end{aligned}$$

and let $\mathbf{z}(t)$ be defined in $[t_n, t_{n+1}]$ by

$$\mathbf{z}(t_n + \tau) = \frac{\tau\mathbf{R}}{h} + h^r\tau\mathbf{T}(\tau; \mathbf{y}_n, t_n) + \mathbf{y}(t_n + \tau; \mathbf{y}_n, t_n)$$

The residual is then given by

$$\begin{aligned}\mathbf{r}(t_n + \tau) &= \mathbf{z}'(t_n + \tau) - \mathbf{f}(\mathbf{z}(t_n + \tau), t_n + \tau)\\ &= \frac{\mathbf{R}}{h} + h^r\mathbf{T}(\tau; \mathbf{y}_n, t_n) + h^r\tau\frac{d}{d\tau}\mathbf{T}(\tau; \mathbf{y}_n, t_n)\\ &\quad + \mathbf{f}(\mathbf{y}(t_n + \tau; \mathbf{y}_n, t_n), t_n + \tau) - \mathbf{f}(\mathbf{z}(t_n + \tau), t_n + \tau)\end{aligned} \tag{4.25}$$

For any given method we can bound $\|\mathbf{r}(t)\|$ or get a close approximation to it. Thus for the rth order Taylor's series method for a system of first order equations,

$$\mathbf{d}(\tau; \mathbf{y}_n, t_n) = -\frac{\tau^{r+1}}{(r+1)!}\frac{d^{r+1}}{dt^{r+1}}\mathbf{y}(t_n; \mathbf{y}_n, t_n) - \frac{\tau^{r+2}}{(r+2)!}\frac{d^{r+2}}{dt^{r+2}}\mathbf{y}(\xi_\tau; \mathbf{y}_n, t_n)$$

while

$$\begin{aligned}\frac{d}{d\tau}\mathbf{d}(\tau; \mathbf{y}_n, t_n) &= (r+1)\tau^r\mathbf{T}(\tau; \mathbf{y}_n, t_n) + \tau^{r+1}\frac{d}{d\tau}\mathbf{T}(\tau; \mathbf{y}_n, t_n)\\ &= -\frac{\tau^r}{r!}\frac{d^{r+1}}{dt^{r+1}}\mathbf{y}(t_n; \mathbf{y}_n, t_n) - \frac{\tau^{r+1}}{(r+1)!}\frac{d^{r+2}}{dt^{r+2}}\mathbf{y}(\tilde{\xi}_\tau; \mathbf{y}_n, t_n)\end{aligned}$$

If the $(r + 2)$th derivatives are bounded by M, $\mathbf{T}(\tau; \mathbf{y}_n, t_n)$ is bounded by T, and the Lipschitz constant for f is L, we get

$$\left\|\mathbf{r}(t_n + \tau) - \frac{\mathbf{R}}{h} + \frac{h^r\mathbf{y}_n^{(r+1)}}{(r+1)!}\right\| \le \left(LT + \frac{2r+2}{(r+2)!}M\right)h^{r+1} \tag{4.26}$$

$y_n^{(r+1)}$ is evaluated for the solution of the differential equation passing through the computed value y_n which is, in practice, the only one we can evaluate. If we write the local truncation error for a general rth order one step method as

$$h^{r+1}\boldsymbol{\phi}(\mathbf{y}, t) + h^{r+2}\tilde{\mathbf{T}}(h, \mathbf{y}, t)$$

and assume that $\tilde{\mathbf{T}}$ has a continuous bounded derivative with respect to h, we can similarly show that the residual satisfies

$$\left\|\mathbf{r}(t_n + \tau) - \frac{\mathbf{R}}{h} - h^r\boldsymbol{\phi}(\mathbf{y}_n, t_n)\right\| = 0(h^{r+1})$$

This can be used to either bound or estimate the errors. The result of Theorem 4.5 can thus be obtained in which $\boldsymbol{\phi}$ is evaluated at $\mathbf{z}(t)$ and t along the computed solution.

4.6 GENERAL APPLICATION OF ERROR BOUND AND ESTIMATE THEOREMS

Theorems 4.4 and 4.5 were stated in terms of conditions on the increment function $\boldsymbol{\psi}$. These conditions involved both the existence, continuity, and boundedness of derivatives, and the asymptotic form of the local truncation error. In practice we know something about the differential equation $\mathbf{f}$ and have to relate this to $\boldsymbol{\psi}$. In this section we want to examine this relation in order to see what are sufficient conditions on $\mathbf{f}$ for a given method to have the desired behavior.

First let us note that although we stated the existence Theorem 4.1 in terms of an unbounded region of $\mathbf{y}$- space, we can always work with a bounded closed region for $0 \leq t \leq b$ and $0 \leq h \leq h_0$, so that continuity of a derivative is sufficient for its boundedness. We required a continuity condition on $\mathbf{f}$, so it is bounded. Let us suppose, in addition, that the first partial derivatives of $\mathbf{f}$ exist and are continuous. From the differential equation

$$\mathbf{y}'' = \frac{d}{dt}(\mathbf{f}(\mathbf{y}, t)) = \frac{\partial \mathbf{f}}{\partial t} + \frac{\partial \mathbf{f}}{\partial \mathbf{y}}\mathbf{y}'$$

From the existence theorem, $\mathbf{y}'$ exists and is continuous, hence $\mathbf{y}''$ exists and is continuous. It is obvious that we can repeat this and say:

If $\mathbf{f}(\mathbf{y}, t)$ *is* q *times continuously differentiable with respect to all of its arguments, then* $\mathbf{y}^{(q+1)}$ *exists and is continuous.*

In many applications $\mathbf{f}(\mathbf{y}, t)$ is analytic in the region of interest, but there are important applications where certain derivatives may be discontinuous, in which case the solution will have discontinuities in its derivatives. In that case it makes no sense to use methods which require the continuity of those derivatives in order to apply Theorems 4.4 and 4.5. If the order of the highest continuous derivative is q, and the order of the method is $r \geq q - 1$, then Theorem 4.4 proves convergence of order $q - 1$. However, Theorem 4.5 is not applicable.

The truncation error is defined by (4.16). If we wish to get a bound of the form of (4.17), we can expand $\mathbf{d}_n(h)$ as a power series in h using Taylor's series with remainder terms containing h^{r+1}. For this we require that $\boldsymbol{\psi}$ have a continuous h-derivative of order r, and that $\mathbf{y}$ be $(r + 1)$ times continuously differentiable. If we wish to derive the asymptotic form (4.20), we can expand by Taylor's series with remainder terms that stop at h^{r+2}. For this we will need a continuous $(r + 1)$th derivative of $\boldsymbol{\psi}$ with respect to h and a continuous $(r + 2)$th derivative of $\mathbf{y}$. Then we can write

$$\begin{aligned}\boldsymbol{\phi}(\mathbf{y}, t) &= \frac{1}{r!}\frac{\partial^r}{\partial h^r}[h^r\boldsymbol{\phi}(\mathbf{y}, t) + 0(h^{r+1})]_{h=0} = \frac{1}{r!}\frac{\partial^r}{\partial h^r}\left[\frac{\mathbf{d}_n(h)}{h}\right]_{h=0} \\ &= \frac{1}{r!}\frac{\partial^r \boldsymbol{\psi}}{\partial h^r}(\mathbf{y}, t, h)\bigg|_{h=0} - \frac{1}{(r+1)!}\frac{d^r}{dt^r}\mathbf{f}(\mathbf{y}, t)\end{aligned} \tag{4.27}$$

With these comments we can examine the two types of one-step methods discussed in earlier chapters.

4.6.1 Taylor's Series Methods

In this case we defined the rth order method by

$$\boldsymbol{\psi}(\mathbf{y}, t, h) = \sum_{q=0}^{r-1} \frac{h^q}{(q+1)!} \frac{d^q}{dt^q} \mathbf{f}(\mathbf{y}, t)$$

If $\mathbf{f}$ has continuous rth order derivatives, $\boldsymbol{\psi}$ obviously satisfies the conditions of Theorem 4.4 with

$$D = \frac{1}{(r+1)!} \max_{t,\mathbf{y}} \left\| \frac{d^r}{dt^r} \mathbf{f}(\mathbf{y}, t) \right\|$$

If $\mathbf{f}$ is $(r+1)$ times continuously differentiable, then we can write

$$\boldsymbol{\phi}(\mathbf{y}, t) = -\frac{1}{(r+1)!} \mathbf{y}^{(r+1)}$$

$\boldsymbol{\psi}$ is twice continuously differentiable with respect to all its arguments since $r \geq 1$, so Theorem 4.5 holds.

4.6.2 Runge-Kutta Methods

Explicit Runge-Kutta methods consist of successive substitutions into the functions $\mathbf{f}$. Consequently, $\boldsymbol{\psi}$ is as many times differentiable as $\mathbf{f}$. If $\mathbf{f}$ is r times continuously differentiable, there exists a D such that Theorem 4.4 holds. If $\mathbf{f}$ is $(r+1)$ times continuously differentiable, then Theorem 4.5 holds. However, the evaluation of $\boldsymbol{\phi}$ is a nontrivial problem in general.

In the case of the midpoint rule we can apply Eq. (4.27) to get

$$\boldsymbol{\phi} = \frac{1}{2} \frac{\partial^2}{\partial h^2} \boldsymbol{\psi}(\mathbf{y}, t, h)_{h=0} - \frac{1}{6} \frac{d^2}{dt^2} \mathbf{f}$$

where

$$\psi^i = f^i\left(\mathbf{y} + \frac{h}{2}\mathbf{f}(\mathbf{y})\right)$$

Therefore,

$$\frac{\partial \psi^i}{\partial h} = \frac{1}{2} f^i_j\left(\mathbf{y} + \frac{h}{2}\mathbf{f}(\mathbf{y})\right) f^j$$

and

$$\frac{1}{2} \frac{\partial^2 \psi^i}{\partial h^2}\bigg|_{h=0} = \frac{1}{8} f^i_{jk} f^j f^k$$

whence

$$\phi^i = -\tfrac{1}{24} f^i_{jk} f^j f^k - \tfrac{1}{6} f^i_j f^j_k f^k$$

agreeing with Eq. (2.11).

The implicit Runge-Kutta method has the same behavior for small enough h. $\boldsymbol{\psi}$ depends on the solution of the nonlinear equations (2.23). It was shown in Section 2.5.2 that the method of successive substitution given by Eqs.

(2.26) converged with $h \leq h_0$ for some h_0. Within this region the solution can be written as a power series in h by such substitutions as were done in Eqs. (2.24). The coefficient of h^r in $\boldsymbol{\psi}$ will contain up to rth order derivatives of $\mathbf{f}$, hence $\boldsymbol{\psi}$ is continuously differentiable with respect to h up to the same order as $\mathbf{f}$ is differentiable with respect to $\mathbf{y}$ and t. Differentiation of the relations defining $\boldsymbol{\psi}$ with respect to $\mathbf{y}$ and t will show that $\boldsymbol{\psi}$ has continuous second derivatives with respect to all of its arguments for small enough h if $\mathbf{f}$ has continuous second derivatives.

4.6.3 The Need for Continuous Derivatives

It is reasonable to ask if continuous derivatives of order r or $r + 1$ are needed in order that the results of Theorems 4.4 and 4.5 hold. The answer is that it depends. Many problems will have simple discontinuities in some derivative at isolated points. If these points are never in the interior of an integration step, then the discontinuities can be ignored. (However, if these derivatives have to be evaluated, care must be taken to use the appropriate value when integrating over the intervals on the left and the right of the discontinuities.) On the other hand, some discontinuities cannot be ignored. Consider $y' = 2.5t^{3/2}$, $t \epsilon [0, 1]$, $y(0) = 0$. Table 4.3 shows the result of applying the trapezoidal rule and classical Runge-Kutta method to this. They are second and fourth order methods respectively, but we can see that the Runge-Kutta error is certainly not $0(h^4)$ because the third derivative of the solution misbehaves at $t = 0$.

Table 4.3 INTEGRATION OF $y' = 2.5t^{3/2}$ FOR t FROM 0 TO 1

h	*Trapezoidal*	*Error*/h^2	*Fourth order Runge-Kutta*	*Error*/h^4	*Error*/$h^{5/2}$
1	1.250000	0.2500	1.00592232	0.00592	0.005922
$\frac{1}{2}$	1.066942	0.2678	1.00107979	0.01728	0.006108
$\frac{1}{4}$	1.017545	0.2807	1.00019312	0.04944	0.006180
$\frac{1}{8}$	1.004531	0.2900	1.00003428	0.14043	0.006206
$\frac{1}{16}$	1.001159	0.2966	1.00000607	0.39778	0.006215
$\frac{1}{32}$	1.000294	0.3012	1.00000107	1.12568	0.006219
$\frac{1}{64}$	1.000074	0.3045	1.00000019	3.18462	0.006220

4.7 VARIABLE STEP SIZE

Theorems 4.3 to 4.5 were based on the interval being divided into a number of equal length intervals. In practice the step length is adjusted over the interval to take account of the changing behavior of the solution. Strategies for choosing the step size will be discussed in the next chapter. We will complete the discussion of the underlying theory of one-step methods by showing that the results of this chapter are also true for variable step sizes.

The theorems discussed the convergence as a function of $h \longrightarrow 0$. We will now interpret h as the maximum step size and assume that there is a function $\theta(t)$ such that $0 < \Delta \leq \theta(t) \leq 1$ for $t \epsilon [0, b]$ and such that the step from t_n to t_{n+1} is given by

$$\begin{aligned} h_n &= h\theta(t_n) \\ t_{n+1} &= t_n + h_n \end{aligned} \tag{4.28}$$

We see that if $h > 0$, a finite number of steps will cover the interval $[0, b]$ since $h_n \geq \Delta h > 0$. We can then show that Theorems 4.3 and 4.4 remain true, while Theorem 4.5 is modified to define the amplified asymptotic error by

$$\boldsymbol{\delta}'(t) = \frac{\partial \mathbf{f}}{\partial \mathbf{y}} \boldsymbol{\delta}(t) + \theta^r(t)\boldsymbol{\phi}(\mathbf{y}(t), t) \tag{4.29}$$

Proof of Theorem 4.3: We note that throughout the earlier proof h must be replaced by $\theta(t_n)h$. Thus, Eq. (4.14) is modified to read

$$\|\mathbf{e}_{n+1}\| \leq (1 + \theta(t_n)hL)\,\|\mathbf{e}_n\| + \theta^2(t_n)h^2(L_1 + L_2) \tag{4.30}$$

Since $|\theta(t_n)| \leq 1$, the last term can be bounded by $kh^2\theta(t_n)$. We now show that (4.30) implies that

$$\|\mathbf{e}_n\| \leq \frac{\prod_{i=0}^{n-1} [1 + hL\theta(t_i)](L\|\mathbf{e}_0\| + kh) - kh}{L} \tag{4.31}$$

It is true for $n = 0$, so by induction we look at $\|\mathbf{e}_{n+1}\|$. From (4.30) we get

$$\begin{aligned} \|\mathbf{e}_{n+1}\| &\leq (1 + \theta(t_n)hL)\frac{\prod_{0}^{n-1} [1 + \theta(t_i)hL](L\|\mathbf{e}_0\| + kh) - kh}{L} + kh^2\theta(t_n) \\ &= \frac{\prod_{0}^{n} [1 + \theta(t_i)hL](L\|\mathbf{e}_0\| + kh) - kh}{L} \end{aligned}$$

which proves (4.31) true for all n. Finally, we note that $1 + \theta(t_n)hL \leq e^{\theta(t_n)hL}$ so that

$$\begin{aligned} \prod_{0}^{n-1} (1 + \theta(t_i)hL) &\leq e^{\sum_0^{n-1} \theta(t_i)hL} \\ &= e^{t_n L} \\ &\leq e^{bL} \end{aligned}$$

Hence the numerical solution converges to $\mathbf{z}(t)$ as in the earlier proof and the result follows.

Proof of Theorem 4.4: Again we can continue to the point

$$\|\mathbf{e}_{n+1}\| \leq \|\mathbf{e}_n\|(1 + \theta(t_n)hL) + \theta^{r+1}(t_n)h^{r+1}D$$

By direct analogy with (4.31) we get

$$\|\mathbf{e}_n\| \le \frac{\prod_0^{n-1} [1 + hL\theta(t_i)](L\|\mathbf{e}_0\| + Dh^r) - Dh^r}{L}$$

$$\le Dh^r \frac{e^{bL} - 1}{L} + e^{bL}\|\mathbf{e}_0\|$$

Proof of Theorem 4.5: The proof follows the earlier proof up to Eq. (4.22), which now reads

$$\boldsymbol{\delta}_{n+1} = \boldsymbol{\delta}_n + \theta(t_n)h[\mathbf{f}_\mathbf{y}(\mathbf{y}(t_n), t_n)\boldsymbol{\delta}_n + \theta^r(t_n)\boldsymbol{\phi}(\mathbf{y}(t_n), t_n) + \theta(t_n)h\mathbf{k}_2 + \theta^r(t_n)h^r\mathbf{k}_1]$$

This can be viewed as a variable step size Euler method which satisfies the requirement of Theorems 4.3 and 4.4 for solving the differential equation (4.29). (These have just been proved for variable step sizes.) The result follows immediately.

PROBLEMS

1. Show that if $\mathbf{f}(\mathbf{y})$ satisfies a Lipschitz condition separately in each of its variables, then

$$\|\mathbf{f}(\mathbf{y}) - \mathbf{f}(\mathbf{y}^*)\| \le L\|\mathbf{y} - \mathbf{y}^*\|$$

for any norm and vice versa.

2. Prove that the method defined by Eqs. (4.15) converges if f satisfies Lipschitz conditions in y, y', and t, and df/dt is only continuous in y, y', and t.

3. Apply Theorems 4.3, 4.4, and 4.5 to the differential equations

$$y' = -2y + z + e^t$$
$$z' = 3y$$
$$y(0) = 1.5, \qquad z(0) = -.25$$

using Heun's method. Compare your results with the computed solution at $t = 1$ for $h = 2^{-m}$, $m = 0, 1, \ldots, 6$.

4. What fixed step size would you use to integrate

$$y'' - y' + 2y = e^t$$
$$y(0) = y'(0) = 2$$

over $t \epsilon [0, 1]$ if the error in y is to be less than 10^{-8} using the second order Taylor's series method? What is the error in y' approximately for this step size?

5. Justify the statement in the second paragraph of Section 4.6 to the effect that we can work with a closed bounded region. (You may assume the validity of the existence theorem and the existence of a Lipschitz constant. It is necessary to show that there exists a finite region which contains the differential equation solution and all approximations we compute for $0 \le h \le h_0$.)

6. Show that the increment function $\boldsymbol{\psi}$ for an implicit Runge-Kutta method has continuous second derivatives with respect to $\mathbf{y}$, h, and t if $\mathbf{f}$ is twice continuously differentiable and h is sufficiently small.

7. If we assume that the best choice of step is such that the size of the local truncation error is a constant times the step size, what function $\theta(t)$ should be used to specify a variable step size for the differentiable equation

$$y' = (e^{-t} + \sin t - y) + \cos t, \qquad y(0) = 1$$

using the midpoint method? What is the asymptotic form of the error under these conditions?

8. Explain the behavior in the last column of Table 4.3.

5 THE CHOICE OF STEP SIZE AND ORDER

We have discussed one-step methods of a variety of types and orders. In choosing a method for a particular problem, the analyst must select between various classes of methods, then select the order and step size of the method. Any one of these parameters could either be held constant throughout the integration or varied according to the nature of the problem. If a parameter is to be varied it may be chosen automatically by the program or may be specified by the analyst prior to the integration. (The latter might be useful if an equation is to be integrated many times with only small variations in the initial data and end conditions.) In this chapter we will discuss the choice of step size and order. The choice of methods will be discussed in a later chapter after other classes of methods have been discussed.

The objective of the analyst must be to achieve a result close to the desired goal with a minimum of effort—the effort being both his and that of the computer. We will only examine the problem of minimizing the numerical work, but it should never be forgotten that a large expenditure of human effort to save a few seconds of computer time is not economic, and that short well-behaved problems should be integrated by the most readily available, reliable programs. The desired goal will be taken to be the solution of the problem with an error at the end of the interval within a specified tolerance. We will call this the end point problem. In some problems the solution may be needed at several points. These can be treated as a sequence of end point problems.

In a small class of problems it is possible to guarantee that the error is bounded by some number. These are problems in which we can get reasonable a priori bounds on the derivatives appearing in the error and can bound the

Lipschitz constant. In general, however, we can only hope to approximate the error and to try and minimize the work for this approximation. There are two problems to be investigated, the choice of an optimum step and order combination, and the estimation of the derivatives that appear in the error formulas (from the computed results). We will be concerned with techniques that estimate these quantities during the calculation of the solution since these are the techniques needed in automatic programs.

One other area that will not be examined in this book is the a posteriori computation of error estimates or bounds for the given solution. This could be done by evaluating either (4.18) or (4.24) during the integration. The former requires a computational procedure to find the maximum value of a number of partial derivatives. For this the functions f must be known algebraically. The latter requires that double the number of equations be integrated. However (4.23) is sometimes solved to low accuracy when the original differential equations are being integrated to very high accuracy in order to check that the errors do not appear to grow rapidly. Another approach to getting rigorous error bounds is by means of *interval analysis*. See Moore (1966), p. 90. In this approach intervals are constructed which are guaranteed to contain the solution. It requires that the function $\mathbf{f}(\mathbf{y}, t)$ be replaced with an interval function $\mathbf{F}(\mathbf{Y}, T)$ such that if $\mathbf{y}$ and t exist in the intervals $\mathbf{Y}$ and T, then $\mathbf{f}(\mathbf{y}, t)$ exists in the interval $\mathbf{F}(\mathbf{Y}, T)$. Finding such an interval function is not always simple and the method tends to produce pessimistic error bounds (that is, large intervals). Currently, therefore, it is only suitable for a small class of problems.

5.1 THE CHOICE OF ORDER

Let us suppose that we wish to integrate a problem over a fixed range $[0, b]$ with a fixed order method. Let us further assume that we have a number of methods in our bag of tricks with orders $r = 1, 2, \ldots$, etc., and have previously been told the best choice of step size for each method by means of h_r and $\theta_r(t)$. By "best" we mean that the amount of computational work is minimized for the given error tolerance if h_n is chosen by

$$h_n = h_r\theta_r(t_n) \tag{5.1}$$

in the rth order method. From the extension of Theorem 4.5 to variable step sizes we have

$$\mathbf{e}_N = \boldsymbol{\delta}_r(t_N)h_r^r + 0(h_r^{r+1}) \tag{5.2}$$

where $\boldsymbol{\delta}_r(t_N)$ is the solution of (4.29). The size of h_r obviously depends on the error tolerance E permitted. We will assume for now that $\theta_r(t)$ is independent of E. [This will be shown to be true to $0(h)$ in the next section.] We neglect the $0(h_r^{r+1})$ term in (5.2) and consider this as an equation for determining h_r

by means of

$$E \geq \|\mathbf{e}_N\| \cong \|\boldsymbol{\delta}_r(t_N)\| h_r^r \tag{5.3}$$

Our objective is to determine the best order method. Suppose that the amount of work per step is k_r in the rth order method. (Frequently, this can be taken to be the number of times that the derivatives f must be evaluated. Thus, for the Runge-Kutta method $k_r = r$, $r \leq 4$.) The total amount of work over the interval [0, b] is therefore Nk_r, where $t_N = b$. The number of steps N is $0(1/h_r)$† so the total work W_r is approximately L_r/h_r, where the L_r are constants for a given problem. Figure 5.1 shows the error $\|\mathbf{e}_N\|$ from (5.3) plotted against the inverse of work $W_r^{-1} = h_r/L_r$ for different order methods. The points where the curves cross each other depend on the values of $\|\delta_r(t_N)\|$ and L_r, so they vary from equation to equation. However we can see two important features in Figure 5.1. For sufficiently small $\|\mathbf{e}_N\|$, the largest (hence best) value of W_r^{-1} will occur for the highest order method. Similarly, for sufficiently large values of $\|\mathbf{e}_N\|$, the best W_r^{-1} will occur for $r = 1$. The latter may be modified because for large h it may not be reasonable to neglect the $0(h^{r+1})$ term in (5.2). However, experience with practical problems has indicated that low accuracy problems are best tackled using low order methods.

Figure 5.1 can also be used to dispel some common fallacies. For example,

†This can be shown as follows. Define $n(t, h_r)$ to be the number of steps needed to reach t from 0 using the rule in (5.1) if t is one of the mesh points t_n. Define $n(t, h_r)$ to be linear between mesh points. We have from (5.1)

$$n(t_q + \theta_r(t_q)h_r, h_r) = n(t_{q+1}, h_r) = q + 1 = n(t_q, h_r) + 1$$

or

$$\frac{n(t_q + \theta_r(t_q)h_r, h_r) - n(t_q, h_r)}{h_r\theta_r(t_q)} = \frac{1}{h_r\theta_r(t_q)}$$

Therefore,

$$\frac{\partial n}{\partial t}(t, h_r) = \frac{1}{h_r\theta_r(t_q)}$$

where t_q is such that $t \in (t_q, t_{q+1})$. Consequently,

$$\begin{aligned} N = n(b, h_r) &= \int_0^b \frac{\partial n}{\partial t}(t, h_r)\,dt \\ &= \int_0^b \frac{dt}{h_r\theta_r(t_q)} \\ &= \frac{1}{h_r}\int_0^b \frac{dt}{\theta_r(t)} + \frac{1}{h_r}\int_0^b \left\{\frac{1}{\theta_r(t_q)} - \frac{1}{\theta_r(t)}\right\} dt \end{aligned}$$

Under the assumption that θ_r has a continuous first derivative and is bounded away from 0 by Δ, the last term has the bound

$$\frac{1}{h_r}\int_0^b \frac{\max|\theta'|\max|t_q - t|}{\Delta^2}\,dt = \frac{\max|\theta'|}{\Delta^2}\,b$$

Therefore,

$$N = \frac{1}{h_r}\int_0^b \frac{dt}{\theta_r(t)} + 0(1) \qquad \text{as } h_r \longrightarrow 0 \tag{5.4}$$

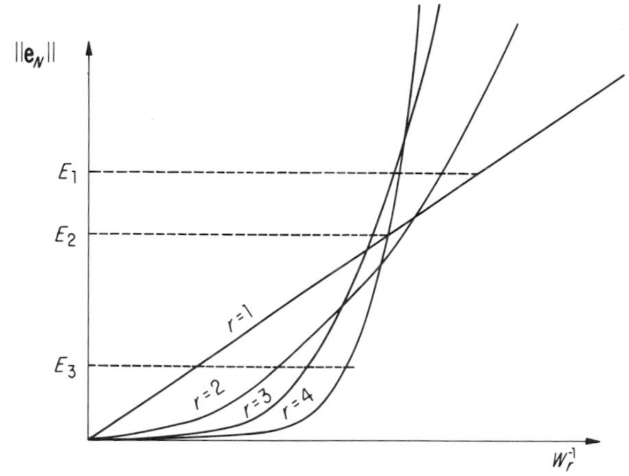

Fig. 5.1 Error versus the inverse of work for various order methods.

it may not be true that for a given step size (or W_r^{-1}) more accuracy can be obtained by going to a higher order method. In the figure shown, error E_1 is most economically obtained using a first order method, E_2 with a second order method, and E_3 with a fourth order method.

We have assumed that the order was to be held constant over the whole interval. Such is not the case, but we can apply the same considerations to any subinterval in which the order is constant, down to the point where the subinterval consists of one step only, from which we can see that it is better to use the order method that gives the smallest amount of work per unit step. This statement depends on the assumption that a smaller error in a given step gives a smaller error in the answer. This may not be true if later errors cancel out early errors because of sign changes in the truncation error.

To examine the choice of order at each step we can start with Eq. (4.19). We will assume the $||\mathbf{e}_n|| \leq E$ for all n and that as $E \longrightarrow 0$, $\mathbf{T}(h, t_n)$ defined as $\mathbf{d}_n(h)/h$ is such that $||\mathbf{T}_n(h)|| = 0(E)$. We will also assume continuous bounded second derivatives for $\boldsymbol{\psi}$ so we can write

$$\mathbf{e}_{n+1} = \mathbf{e}_n + h_n[\boldsymbol{\psi}_\mathbf{y}(\mathbf{y}(t_n), t_n, 0)\mathbf{e}_n + \mathbf{T}(h_n, t_n)] + 0(E^2h_n + Eh_n^2)$$

Since $\boldsymbol{\psi}(\mathbf{y}, t, 0) = \mathbf{f}(\mathbf{y}, t)$ and $h_n = h\theta(t_n) \leq h$, this is the solution by Euler's method of

$$\mathbf{e}'(t) = \mathbf{f}_\mathbf{y}\mathbf{e}(t) + \mathbf{T}(h\theta(t), t), \qquad \mathbf{e}(0) = 0 \tag{5.5}$$

with an additional error of $0(E^2 + Eh)$. Since the error in the Euler solution is $0(h||\mathbf{e}''(t)||) = 0(hE)$, $\mathbf{e}_n = \mathbf{e}(t_n) + 0(E^2 + hE)$.

The solution of (5.5) is given by

$$\mathbf{e}(t) = \int_0^t G(\tau, t)\mathbf{T}(\theta(\tau)h, \tau)\, d\tau \tag{5.6}$$

where $G(\tau, t)$ is a matrix which satisfies the homogeneous version of (5.5), namely

$$\frac{\partial G}{\partial t}(\tau, t) = \mathbf{f}_y G(\tau, t), \qquad G(\tau, \tau) = I \tag{5.7}$$

Note that G is independent of θ, h, and $\mathbf{T}$. If we assume that all components of $G(\tau, t)$ and $\mathbf{T}(\theta(\tau)h, \tau)$ have the same sign throughout the interval, it is then evident that a change in the order of the method that reduces $\mathbf{T}$ will reduce the total error $\|\mathbf{e}(t_N)\|$.

It will also be useful to rewrite (5.6) as

$$\begin{aligned} \mathbf{e}(t_N) &= \sum_{n=0}^{N-1} \int_{t_n}^{t_{n+1}} G(\tau, t_N)\mathbf{T}(\theta(\tau)h, \tau)\, d\tau \\ &= \sum_{n=0}^{N-1} G(t_n, t_N)\mathbf{T}(\theta(t_n)h, t_n)h\theta(t_n) + 0(Eh) \\ &= \sum_{n=0}^{N-1} G(t_n, t_N)\mathbf{d}_n(\theta(t_n)h) + 0(Eh) \end{aligned} \tag{5.8}$$

from which we see that $G(t_n, t_N)$ is the amplification applied to an error committed at the nth step as the integration proceeds to t_N.

5.2 CHOICE OF STEP SIZE

We previously assumed that the step size function $\theta(t)$ was given. Now we assume that we know the best order to use at each point in the interval so that $\mathbf{T}(\theta(t)h, t)$ is known when $\theta(t)$ is known. We wish to minimize the amount of work subject to a bound of E on $\|\mathbf{e}_N\|$. Suppose that the work per step at the point t due to the step size being used is $k(t)$. The total amount of work is seen to be

$$W = \frac{1}{h}\int_0^b \frac{k(t)\, dt}{\theta(t)} + 0(1) \tag{5.9}$$

by application of (5.4) to each subinterval with fixed order. From (5.6)

$$\|\mathbf{e}_N\| \le \int_0^b \|G(t, b)\|\, \|\mathbf{T}(\theta(t)h, t)\|\, dt$$

We will therefore try and minimize (5.9) subject to the side condition

$$E = \int_0^b \|G(t, b)\|\, \|\mathbf{T}(\theta(t)h, t)\|\, dt \tag{5.10}$$

[In the case of a single equation with constant signs in the integrand the norms are not needed and we can be more certain that we get the true minimum. See Greenspan, *et al.* (1965) and Morrison (1962) for single equations.]

We use the standard techniques from the calculus of variations and Lagrange undetermined multipliers† to see that the solution to this problem is given by

$$||G(t, b)|| \frac{\partial}{\partial\theta} ||\mathbf{T}(\theta(t)h, t)|| = \frac{\lambda k(t)}{\theta^2(t)} \tag{5.11}$$

The scaled truncation error will be approximated by the first nonvanishing term in its Taylor's series expansion, so we have

$$\left\|\frac{\mathbf{d}_n(\theta h)}{\theta h}\right\| = ||\mathbf{T}(\theta h, t_n)|| = \theta^{r(t)}h^{r(t)} ||\boldsymbol{\phi}(\mathbf{y}, t_n)|| + 0(h^{r+1})$$

where $r(t)$ is the order used at the point t. Thus (5.11) implies

$$r(t)\theta^{r(t)-1}h^{r(t)} ||G(t, b)|| \, ||\boldsymbol{\phi}(\mathbf{y}, t)|| \cong \frac{\lambda k(t)}{\theta^2(t)}$$

or

$$||G(t_n, b)|| \, ||\mathbf{d}_n(\theta(t_n)h)|| \cong \frac{h\lambda k(t_n)}{r(t_n)} \tag{5.14}$$

In view of (5.8) this implies that the choice of step size which minimizes work subject to satisfying (5.10) is obtained when the contribution of the error at the nth step to the error at the end of the interval is equal to $h\lambda k/r$. The constant λ must be picked to satisfy (5.10).

For a method whose order is fixed at r, (5.14) implies that

$$\theta^{r+1}(t_n) \cong \lambda h^{-r} \frac{k}{r} \frac{1}{||G(t_n, b)|| \quad ||\phi(y, t_n)||}$$

If θ is normalized so that its maximum value is 1, we can see that θ is independent of the desired error bound E which is obtained by adjusting h.

Unfortunately this prescription for choosing θ and r is not practical since the function $G(t, b)$ is not usually known *a priori*. Until the integration has been completed, it is not possible to calculate the effect that the errors will have at the end of the interval of integration. In the next section we will discuss some practical approaches that do not give the optimum method but do attempt to overcome this problem. First, let us apply these results to some examples.

†We are looking for a function $\tilde{\theta}(t)$ to minimize (5.9). Any other function $\theta(t) = \epsilon\eta(t) + \tilde{\theta}(t)$ will therefore have a minimum at $\epsilon = 0$. Hence

$$\frac{dW}{d\epsilon} = \int_0^b \frac{k(t)}{\tilde{\theta}^2(t)}\eta(t)\,dt - 0 \tag{5.12}$$

However (5.10) must continue to hold for $\epsilon \neq 0$, so $\eta(t)$ must be such that

$$0 = \frac{dE}{d\epsilon} = \int_0^b ||G(t, b)|| \frac{\partial}{\partial\theta} ||\mathbf{T}(\tilde{\theta}(t)h, t)|| \, \eta(t)\,dt \tag{5.13}$$

(5.13) will imply (5.12) for any $\eta(t)$ satisfying (5.13) if one integrand is a constant times the other. This gives Eq. (5.11). A more general discussion can be found in Courant (1950), pp. 188 and 498.

EXAMPLE 1

$$\mathbf{y}' = \mathbf{f}(t)$$

Since $\mathbf{f}$ is independent of $\mathbf{y}$, $G(\mathbf{b}, t) = I$. Therefore (5.14) requires that

$$\|\mathbf{d}(\theta(t)h)\| = \frac{h\lambda k(t)}{r(t)} \tag{5.15}$$

$k(t)$ may depend on the order $r(t)$. If we are considering using a Runge-Kutta method for $r \leq 4$, then $k(t) = r(t)$. Thus, the best strategy is to commit the same size error at each step, and to choose the order from step to step to make the work per unit step as small as possible.

EXAMPLE 2

$$y' = 3t^2$$

This is a special case of Example 1. Suppose we are considering either the trapezoidal rule or Euler's rule ($r = 2$ or 1). The amount of work in either method is similar, so take $k(t) = 1$. For Euler's method (5.15) implies that

$$\|d(\theta(t)h)\| = 3t\theta^2(t)h^2 = \lambda h$$

while for the trapezoidal method

$$\|d(\theta(t)h)\| = \frac{1}{2}\theta^3(t)h^3 = \frac{\lambda h}{2}$$

Since the work is proportional to $\int dt/\theta$, we wish to maximize θ, so we choose

$$\theta = \max\left\{\left[\frac{\lambda}{3ht}\right]^{1/2}, \left[\frac{\lambda}{h^2}\right]^{1/3}\right\} \tag{5.16}$$

and either the Euler rule or the trapezoidal rule according to which term is the maximum. We can take $h = 1$ since it is only a scale for θ. Consequently, for $t \leq t_0$, where $t_0^3 = \lambda/27$, the Euler method should be used. For larger t, the trapezoidal rule is better. The total error in the interval $[a, b]$ will be approximately

$$\int_a^b \frac{1}{\theta(t)}\lambda\, dt$$

with

$$\begin{aligned}
\theta(t) = \left[\frac{\lambda}{3t}\right]^{1/2} &= 2(b^{3/2} - a^{3/2})\sqrt{\frac{\lambda}{3}} \qquad \text{if } b \leq t_0 \\
&= 2(t_0^{3/2} - a^{3/2})\sqrt{\frac{\lambda}{3}} + (b - t_0)\lambda^{2/3} \qquad \text{if } a < t_0 < b \\
&= (b - a)\lambda^{2/3} \qquad \text{if } t_0 \leq a
\end{aligned}$$

If the allowed error is large, a large λ can be used, so t_0 would be large and the Euler method would be best. As the error is reduced, λ

must be decreased, and it becomes advantageous to use the trapezoidal rule toward the end of the interval. For small enough errors, the trapezoidal rule should be used over the whole interval.

It can be seen that even in this simple example, choosing λ to obtain a particular error bound is not simple.

EXAMPLE 3

$$y' = \alpha y, \; y(0) = 1$$

For this problem $G(\tau, t) = e^{\alpha(t-\tau)}$, so from (5.14) $\theta(t)$ should be chosen so that

$$e^{\alpha(b-t_n)} \| d_n(\theta(t_n)h) \| \cong \frac{h\lambda k(t_n)}{r(t)}$$

If a constant order is used, the right-hand side is constant, so we see that the error d_n should be proportional to $e^{\alpha t}$, which is the solution. Thus a constant relative error is the best strategy.

5.3 THE PRACTICAL CONTROL OF ERRORS

In the previous sections we developed the prescription for the optimum choice of step size and order by using asymptotic approximations, but we saw that these prescriptions were not too practical. In a typical situation the problem is to integrate a set of equations producing a given number of significant digits of accuracy in the final answer. Nothing is known initially about $G(\tau, t)$ or even $\partial \mathbf{f}/\partial \mathbf{y}$.

To simplify the problem initially, let us suppose that $\| G(\tau, t) \| < 1$ so that it is known that errors are not magnified. The total error is then bounded by the sum of the errors at each step. The optimum solution for this bound is to set these errors to $\lambda k(t)/r(t)$. If a value of λ is known, each step can be chosen by this formula such that when the end of the interval of integration has been reached, the bound on the error committed has been achieved in an optimum way! Unfortunately, this is not what we want. Our problem is that we do not know what values $r(t)$ will take at later stages, nor how many steps will be used, yet these determine λ.

The problem of $r(t)$ can be disposed of by saying that $k(t) = r(t)$. This is a reasonable approximation for many methods. Now we must set the error at each step equal to λ, although we should still choose the order that gives the least work per unit step using the best approximation to $k(t)$ that we have. With this prescription, the total error is simply $N\lambda$, but N is still unknown. A way around this is to control the error in each step to be less than $\lambda h \theta$, that is, to make the error per unit step constant while minimizing the work per unit step. Then the total error in an interval of length b is λb so λ can be taken as E/b. The resulting choice is not the optimum, but is not too different and leads to an answer in which the error is of the right size. [We must remember that terms like $0(E^2h + Eh^2)$ were ignored in the derivation of the error so our prescription will not guarantee the desired accuracy.]

What if $G(\tau, t)$ is not the identity? In general we can say nothing, but for a reasonably large class of problems we can justify an approach similar to the above method. If we consider linear equations of the form

$$\mathbf{y}' = A\mathbf{y} + \mathbf{g}(t) \tag{5.17}$$

we have $G(\tau, t) = \exp\{A(t - \tau)\}$. The solution of (5.17) is

$$\mathbf{y}(t) = \int_0^t G(\tau, t)\mathbf{g}(\tau)\, d\tau + G(0, t)\mathbf{y}(0) \tag{5.18}$$

If the components of G and $\mathbf{g}$ are such that there is little cancellation and $\mathbf{y}(0)$ is such that the behavior of $\mathbf{y}(t)$ parallels the behavior of the second term, we can use the solution to control the error. That is, we will control the error to be equal to $Eh\theta \,||\,\mathbf{y}(t)\,||/b$ at each step. The discussion in Section 5.2 indicated that the errors should be such that their contributions to the final error are equal. Thus $||\, G(t_n, t_N)\mathbf{d}_n(\theta(t_n)h)\,|| = ||\, G(t_{n+1}, t_N)\mathbf{d}_{n+1}(\theta(t_{n+1})h)\,||$. For this problem

$$G(a, c) = G(a, b)G(b, c) = G(b, c)G(a, b)$$

so we can require instead that

$$||\, G(t_n, t_{n+1})\mathbf{d}_n(\theta(t_n)h)\,|| = ||\,\mathbf{d}_{n+1}(\theta(t_{n+1})h)\,|| \tag{5.19}$$

That is, the contribution of the error committed at the nth step to the total error at the $(n + 1)$th step is of the same size as the error committed at the $(n + 1)$th step. From the assumption that the behavior of $y(t)$ in (5.18) parallels the last term, we see that

$$||\, G(t_n, t_{n+1})\,|| \cong \frac{||\,\mathbf{y}(t_{n+1})\,||}{||\,\mathbf{y}(t_n)\,||}$$

Thus (5.19) is approximately satisfied if we make the truncation error proportional to $||\,\mathbf{y}(t)\,||$. By introducing the additional factor $Eh\theta/b$, we will have a total error approximately equal to E relative to the final answer. Thus, if E is 10^{-k}, we hope to get k significant digits. We can see this by substituting in (5.8) to get

$$\begin{aligned}
||\,\mathbf{e}(b)\,|| &\leq \sum_{n=0}^{N-1} ||\, G(t_n, b)\,|| \frac{Eh\theta(t_n)\,||\,\mathbf{y}(t_n)\,||}{b} + 0(Eh) \\
&\cong \sum_{n=0}^{N-1} ||\, G(t_n, b)\mathbf{y}(t_n)\,|| \frac{Eh\theta(t_n)}{b} \\
&\cong \sum_{n=0}^{N-1} ||\,\mathbf{y}(b)\,|| \frac{Eh\theta(t_n)}{b} \\
&= ||\,\mathbf{y}(b)\,||\, E
\end{aligned}$$

It must be emphasized that these are very crude approximations and do not give any guarantee of accuracy. It is simple to construct examples in which the actual error is arbitrarily smaller or larger than E. Consider, for example,

$$y' = \lambda(y - e^t) + e^t, \qquad y(0) = 1$$

This has the solution $y(t) = e^t$, so errors will be controlled relative to e^t. However, $G(\tau, t) = e^{\lambda(t-\tau)}$ and errors should be controlled relative to that. If $\lambda \gg 0$, then the actual error at t_N will be much greater than E, whereas if $\lambda < 0$, the error will be less.

5.4 ESTIMATION OF THE LOCAL TRUNCATION ERROR

The criteria developed for step and order control have assumed a knowledge of the size of the local truncation error. In a general purpose program it is necessary to estimate the local truncation error from a knowledge of the numerical solution only, and to use this information to control the step and order selection. If a Taylor's series method is used, it is possible to evaluate the next few derivatives in order to decide which order to use. (However, note that the work of the highest order method considered will have already been done in calculating these derivatives.) The Taylor's series methods are not usually practical, which leaves the Runge-Kutta methods. Their error term involves a combination of partial derivatives of f which it is not practical to compute. There are, however, two commonly used techniques for error control with the Runge-Kutta methods.

5.4.1 Step Doubling

This technique is applicable to all methods but is usually restricted to the Runge-Kutta method because there are better techniques for other methods. Each basic step of size h is done twice, once as two steps of size $h/2$ and once as one step of size h. Since the error has the form $h^{r+1}\boldsymbol{\phi}(y, t) + 0(h^{r+2})$, the two results can be compared to estimate $\|\boldsymbol{\phi}\|$. If the result of one step of size h is $\mathbf{y}_2$, while the result of two steps of size $h/2$ is $\mathbf{y}_1$, we have

$$\|\mathbf{y}_2 - \mathbf{y}_1\| = h^{r+1}(1 - 2^{-r})\|\boldsymbol{\phi}(t)\| + 0(h^{r+2})$$

This can only be used to estimate the error for the order of method being used. No techniques are currently available for selecting the order in Runge-Kutta methods.

The program on pp. 83–84 illustrates a typical fourth order Runge-Kutta subroutine for performing one double step of the integration at a time. The errors in each component are computed and divided by the numbers YMAX(I). These can be set initially to give more weight to some components than others, and are updated to contain the maximum value of Y(I) that has been computed thus far. The worst-case scaled error is restricted to be less than EPS, an input parameter. The value of h for the next step is computed to make h nearly as large as necessary to make the next scaled error equal to EPS. The factor 0.99 in the statement following statement no. 6 serves to keep H slightly smaller so that if the truncation error on the next step is slightly larger, the H recommended will be small enough.

If the scaled error is greater than EPS, the step is rejected and repeated with the newly recommended H subject to a minimum step size of HMIN (a parameter to the subroutine). This program keeps the relative error approximately constant in the max-norm. If the error is to be made propor-

Table 5.1 FOURTH ORDER RUNGE-KUTTA RESULTS

k	*Relative error*	*No. function evaluations*	*No. steps*
3	0.67500D-02	187	17
4	0.21449D-02	286	26
5	0.38661D-03	440	40
6	0.62225D-04	704	64
7	0.98484D-05	1100	100
8	0.15670D-05	1738	158
9	0.24796D-06	2728	248
10	0.39216D-07	4367	397
11	0.62065D-08	6798	618

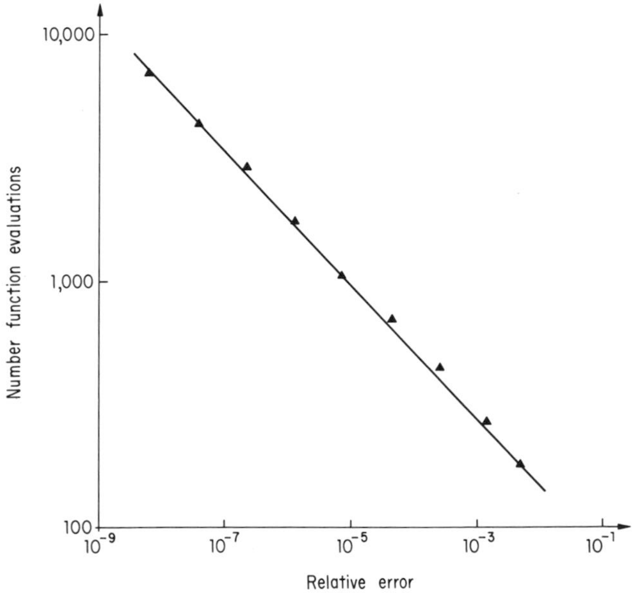

Fig. 5.2 Fourth order Runge-Kutta method results.

tional to hE, all references to EPS should be replaced with H*EPS and the exponent -0.2 in the statement following statement no. 6 should be changed to -0.25.

The result of integrating $y' = -y$ from $y(0) = 1$ to $t = 20$ is shown in Table 5.1 for different relative errors E. EPS was set to 10^{-k}, while YMAX(1) was set to Y(1) prior to each step. The graph of actual relative error at $t = 20$ versus number of function evaluations is shown in Figure 5.2.

```
      SUBROUTINE DIFSUB (N,T,Y,DY,H,HMIN,EPS,YMAX,ERROR,KFLAG,
     1                   JSTART)
      IMPLICIT REAL *8 (A-H,O-Z)
C**********************************************************************
C*   THE PARAMETERS TO THIS INTEGRATION SUBROUTINE HAVE               *
C*   THE FOLLOWING MEANINGS..                                         *
C*     N        THE NUMBER OF FIRST ORDER DIFFERENTIAL EQUATIONS      *
C*     T        THE INDEPENDENT VARIABLE                              *
C*     Y        THE DEPENDENT VARIABLES, UP TO 10 ARE ALLOWED.        *
C*     DY       AN ARRAY OF 10 LOCATIONS WHICH WILL CONTAIN THE       *
C*                VALUES OF THE DERIVATIVES AT THE START OF THE INTERVAL.*
C*     H        THE STEP SIZE THAT SHOULD BE ATTEMPTED.  IT MAY BE    *
C*                INCREASED OR DECREASED BY THE SUBROUTINE.           *
C*     HMIN     THE MINIMUM STEP SIZE THAT SHOULD BE ALLOWED ON THIS  *
C*                STEP.                                               *
C*     EPS      THE ERROR TEST CONSTANT.  THE ESTIMATED ERRORS ARE    *
C*                REQUIRED TO BE LESS THAN EPS*YMAX IN EACH COMPONENT. *
C*                IF YMAX IS ORIGINALLY SET TO +1 IN EACH COMPONENT,  *
C*                THE ERROR TEST WILL BE RELATIVE FOR THOSE COMPONENTS *
C*                GREATER THAN 1 AND ABSOLUTE FOR THE OTHERS.         *
C*     YMAX     THE MAXIMUM VALUES OF THE DEPENDENT VARIABLES ARE     *
C*                SAVED IN THIS ARRAY. IT SHOULD BE SET TO +1 BEFORE  *
C*                THE FIRST ENTRY. (SEE THE DESCRIPTION OF EPS).      *
C*     ERROR    THE ESTIMATED SINGLE STEP ERROR IN EACH COMPONENT     *
C*     KFLAG    A COMPLETION CODE WITH THE FOLLOWING MEANINGS..       *
C*                        -1    THE STEP WAS TAKEN WITH H = HMIN      *
C*                              BUT THE REQUESTED ERROR WAS NOT ACHIEVED. *
C*                        +1    THE STEP WAS SUCCESSFUL.              *
C*     JSTART AN INITIALIZATION INDICATOR WITH THE MEANING..          *
C*                        -1    REPEAT THE LAST STEP, RESTORING THE   *
C*                              VALUES OF Y AND YMAX THAT WERE USED   *
C*                              LAST TIME.                            *
C*                        +1    TAKE A NEW STEP.                      *
C**********************************************************************
      DIMENSION Y(10),DY(10),YMAX(10),YSAVE(10),Y1(10),Y2(10),Y3(10)
     1          ,ERROR(10),DYN(10),YMAXSV(10)
      IF(JSTART.LT.0) GO TO 2
C**********************************************************************
C* SAVE THE VALUES OF Y AND YMAX IN CASE A RESTART IS NECESSARY.      *
C**********************************************************************
      DO 1 I = 1,N
      YSAVE(I) = Y(I)
    1 YMAXSV(I) = YMAX(I)
C**********************************************************************
C*   CALCULATE THE INITIAL DERIVATIVES.                               *
C**********************************************************************
      CALL DIFFUN(T,Y,DYN)
      GO TO 4
C**********************************************************************
C*   RESTORE THE INITIAL VALUES OF Y AND YMAX FOR A RESTART.          *
C**********************************************************************
    2 DO 3 I = 1,N
      Y(I) = YSAVE(I)
    3 YMAX(I) = YMAXSV(I)
    4 KFLAG = 1
C**********************************************************************
C*   SAVE THE FINAL VALUE OF T AND CALCULATE THE HALF STEP.           *
```

```
C***********************************************************************
 5    A = H + T
      HHALF = H*0.5D0
C***********************************************************************
C*   PERFORM ONE FULL RUNGE KUTTA STEP                                 *
C***********************************************************************
      CALL RK1(N,T,YSAVE,DYN,H,Y1)
C***********************************************************************
C*   NOW PERFORM TWO HALF INTERVAL RUNGE KUTTA STEPS                   *
C***********************************************************************
      CALL RK1(N,T,YSAVE,DYN,HHALF,Y2)
       THALF = T + HHALF
      CALL DIFFUN(THALF,Y2,DY)
      CALL RK1(N,THALF,Y2,DY,HHALF,Y3)
      ERRMAX = 0
C***********************************************************************
C*   CALCULATE THE NEW MAX Y'S, THE ERRORS AND THE MAX                 *
C*   RELATIVE ERRORS.                                                  *
C***********************************************************************
      DO 6 I = 1,N
        YMAX(I) = DMAX1(YMAX(I),DABS(Y1(I)),DABS(Y2(I)),DABS(Y3(I)))
        ERROR(I) = DABS((Y3(I) - Y1(I))/15.0D0)
        ERRMAX = DMAX1(ERRMAX,ERROR(I)/(EPS*YMAX(I)))
C***********************************************************************
C*   CALCULATE THE IMPROVED VALUE OF Y BY ELIMINATING THE              *
C*   ESTIMATED ERROR.                                                  *
C***********************************************************************
        Y(I) = (16.0D0*Y3(I) -Y1(I))/15.0D0
 6      CONTINUE
      IF (ERRMAX.EQ.0) H = H*2.0D0
      IF (ERRMAX.GT.0) H = H*ERRMAX**(-0.2)*0.99
      IF (ERRMAX.GT.1.0D0) GO TO 8
      KFLAG = 1
 7    T = A
      RETURN
 8    IF (H.GT.HMIN) GO TO 5
      IF (KFLAG.LT.0) GO TO 7
      H = HMIN
      KFLAG = -1
      GO TO 5
      END
C***********************************************************************
C*   THIS SUBROUTINE PERFORMS ONE RUNGE KUTTA STEP.                    *
C*.     ARGUMENTS ARE..                                                *
C*      N    -  NUMBER OF EQUATIONS.                                   *
C*      T    -  INITIAL VALUE OF INDEPENDENT VARIABLE.                 *
C*      Y    -  INITIAL VALUE OF DEPENDENT VARIABLES.                  *
C*      DY   -  INITIAL VALUE OF DERIVATIVES.                          *
C*      H    -  STEP SIZE                                              *
C*      Y1   -  THE ANSWER IS RETURNED HERE.                           *
C***********************************************************************
      SUBROUTINE RK1(N,T,Y,DY,H,Y1)
      IMPLICIT REAL*8 (A-H,O-Z)
      DIMENSION Y(10),DY(10),Y1(10),Y2(10),Y3(10),DY1(10)
      HHALF = H*0.5D0
      DO 1 I = 1,N
 1      Y2(I) = Y(I) + HHALF*DY(I)
      CALL DIFFUN(T + HHALF,Y2,DY1)
      DO 2 I = 1,N
        Y3(I) = Y(I) + HHALF*DY1(I)
 2      Y2(I) = Y2(I) + 2*Y3(I)
      CALL DIFFUN(T + HHALF,Y3,DY1)
      DO 3 I = 1,N
        Y3(I) = Y(I) + H*DY1(I)
 3      Y2(I) = Y2(I) + Y3(I)
      CALL DIFFUN(T + H,Y3,DY1)
      DO 4 I = 1,N
 4      Y1(I) = (Y2(I) - Y(I) + HHALF*DY1(I))/3.0D0
      RETURN
      END
                                                              0127 CARDS
```

5.4.2 The Runge-Kutta-Merson Method

This method is a fourth order Runge-Kutta process that simultaneously gives an approximation to the single step error. The cost of this is one additional function evaluation. Thus, two steps will take 10 function evaluations against 11 for the step doubling process described in the previous section. The equations are given by Merson (1957).

$$\begin{aligned}
\eta_0 &= y_n, \qquad k_0 = hf(\eta_0) \\
\eta_1 &= \eta_0 + \frac{k_0}{3}, \qquad k_1 = hf(\eta_1) \\
\eta_2 &= \eta_0 + \frac{k_0 + k_1}{6}, \qquad k_2 = hf(\eta_2) \\
\eta_3 &= \eta_0 + \frac{k_0 + 3k_2}{8}, \qquad k_3 = hf(\eta_3) \\
\eta_4 &= \eta_0 + \frac{k_0 - 3k_2 + 4k_3}{2}, \qquad k_4 = hf(\eta_4) \\
y_{n+1} = \eta_5 &= \eta_0 + \frac{k_0 + 4k_3 + k_4}{6}
\end{aligned} \tag{5.20}$$

It is easy to verify that

$$\begin{aligned}
\eta_1 &= y\left(t_n + \frac{h}{3}\right) + 0(h^2) \\
\eta_2 &= y\left(t_n + \frac{h}{3}\right) + 0(h^3) \\
\eta_3 &= y\left(t_n + \frac{h}{2}\right) + 0(h^4) \\
\eta_4 &= y(t_n + h) + 0(h^4)
\end{aligned}$$

and finally

$$\eta_5 = y_{n+1} = y(t_{n+1}) + 0(h^5)$$

by consideration of quadrature formulas only. For example, η_2 is obtained by a trapezoidal rule with an additional error of $0(h^3)$ from k_1, while η_5 is obtained by Simpson's rule with additional errors of $0(h^5)$ from k_3 and k_4. If we examine the $0(h^4)$ error in η_4, we will see that it contains only the terms $f_{ijk}f^if^jf^k$, $f_if^i_{jk}f^jf^k$, and $f_{ij}f^i_kf^jf^k$. If all second partial derivatives of f are zero, that is, if $f(y, t)$ has the form $Ay + bt$, then $\eta_4 = y(t_{n+1}) + 0(h^5)$. In that case Merson showed that

$$\eta_4 = y(t_{n+1}) - \tfrac{1}{120}h^5y^{(5)} + 0(h^6) \tag{5.21}$$

and

$$\eta_5 = y(t_{n+1}) - \tfrac{1}{720}h^5y^{(5)} + 0(h^6) \tag{5.22}$$

Thus the difference $\eta_5 - \eta_4$ is an indication of the local error. This method has been used successfully in a number of automatic step selection methods.

PROBLEMS

1. Discuss the best choice of step and order for the equations
   (a) $y' = y$, $y(0) = 1$.
   (b) $y' = -y$, $y(0) = 1$.
   using a Taylor's series method. Show that a constant order method is optimum.
2. Express the error term in Euler's method for $y' = 3t^2$ **exactly** and use this to determine the optimum step size distribution.
3. Integrate the differential equation in Problem 2 using the rule you derive from $y(0.1) = 10^{-3}$ to $t = 1$. Also use the rules proposed in Sections 5.2 and 5.3, that is, single step truncation error constant and single step truncation error proportional to step length. Compare the graphs of number of steps versus accuracy in the three cases.
4. Verify the last statement in the penultimate paragraph of Section 5.4.1.
5. Verify Eqs. (5.21) and (5.22).
6. Repeat the integration shown in Table 5.1 with the local error bounded by EPS*H* | YMAX(I)| instead of EPS* |YMAX(I)|. Plot the results and compare them with the graph in Figure 5.2.
7. Integrate

   $$y' = -1000(y - e^t) + e^t, \qquad y(0) = 1$$

   by the trapezoidal rule over [0, 1] using a step control determined by
   (a) Local error relative to the solution held constant.
   (b) Equation (5.14).
   Repeat the integrations for several different desired error bounds and plot the error against the number of function evaluations.

6 EXTRAPOLATION METHODS

Earlier we saw that the Richardson extrapolation process could be used to improve the order of a numerical solution by trying to eliminate the error term. It is natural to consider the solution for a step of size h at some fixed point t to be of the form

$$y(t, h) = y(t) + \sum_{i=1}^{m} \tau_i(t)h^i + 0(h^{m+1}) \tag{6.1}$$

and to attempt to eliminate $\tau_1, \ldots, \tau_m$ by evaluating $y(t, h)$ for $h = h_0$, $h_1, \ldots, h_m$. The conditions under which an expansion of the form (6.1) exists have been investigated by Gragg (1963) and Stetter (1965). If such a series exists, then we can consider approximating it by some function $R_m(t, h)$ with $(m + 1)$ unknowns determined by the requirement that

$$R_m(t, h_j) = y(t, h_j) \qquad j = 0, 1, \ldots, m \tag{6.2}$$

in order to approximate $y(t)$ by $R_m(t, 0)$.

6.1 POLYNOMIAL EXTRAPOLATION

If, for example, $R_m(t, h)$ is a polynomial of degree m in h, then if (6.1) holds $R_m(t, 0)$ would $y(t) + 0(h^{m+1})$ as h goes to 0, where h is a measure of the way the h_i go to zero simultaneously in the sense that $h_i = \alpha_i h$ with fixed α_i, $i = 0, 1, \ldots$. This is simply polynomial interpolation. Polynomial extrapolation is performed by integrating the differential equation over the interval $(0, t)$ using step size h_i to get the results $y(t, h_i)$ for $i = 0, 1, \ldots, m$ where $h_0 > h_1 > \cdots > h_m > 0$. The polynomial in h of degree m that passes through these results is then evaluated at $h = 0$.

The simplest way to do this is often the Aitken interpolation process. [See Hildebrand (1956), p. 49.] In this process, define $R^i_m(h)$ as the unique polynomial of degree m in h passing through (h_j, y_j), $j = i, i+1, \ldots, i+m$. Thus $R^i_0(h) = y(t, h_i)$. It is evident that the function

$$P(h) = \alpha(h)R^i_{m-1}(h) + (1 - \alpha(h))R^{i+1}_{m-1}(h)$$

takes the value $y(t, h_j)$ at $h = h_j$ for $j = i+1, \ldots, i+m-1$. We can also make this valid for $j = i$ and $j = i+m$ by requiring that $\alpha(h_i) = 1$ and $\alpha(h_{i+m}) = 0$. If $\alpha(h)$ is linear in h, the degree of $P(h)$ is m, then $P(h)$ is $R^i_m(h)$, the unique polynomial of degree m with the required values. Consequently, we have

$$\alpha(h) = \frac{h - h_{i+m}}{h_i - h_{i+m}}$$

to get

$$R^i_m(h) = \frac{1}{h_i - h_{i+m}}[(h - h_{i+m})R^i_{m-1}(h) + (h_i - h)R^{i+1}_{m-1}(h)] \qquad (6.3)$$

This provides a rule for constructing a triangle of approximations to $R^i_m(h)$. In this case we wish to know the values of $R^i_m = R^i_m(0)$. They can be constructed as shown in Table 6.1, where two values are used to form the next approximation (as shown in the table) by the relation

$$R^i_m = \frac{h_i R^{i+1}_{m-1} - h_{i+m}R^i_{m-1}}{h_i - h_{i+m}} = R^{i+1}_{m-1} + \frac{R^{i+1}_{m-1} - R^i_{m-1}}{(h_i/h_{i+m}) - 1} \qquad (6.4)$$

If this technique were applied to the results of a method of order one, such as the Euler method, then the values R^i_m would form better approximations to the solution as i and/or m increase. In fact, it can be shown that the errors in R^i_m would be bounded by something of the form $M_m h_i^m$ if suitable differentiability conditions hold. Thus $R^i_m \longrightarrow y(t, 0)$ as $i \longrightarrow \infty$ such that $h_i \longrightarrow 0$. Since the M_m will depend on the derivatives of the solution and on the method, nothing can be said about convergence as $m \longrightarrow \infty$ for fixed i without further restrictions which would be impossible to verify in general. However, if the approximations do not appear to be converging as m increases for a fixed i,

Table 6.1 AITKEN INTERPOLATION

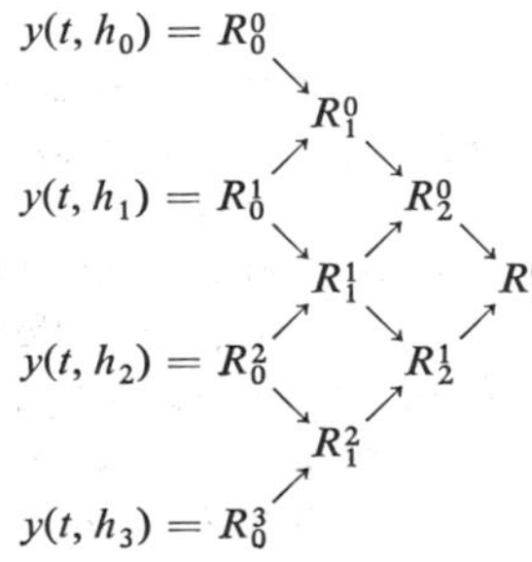

the intial step size can effectively be reduced by holding m constant and increasing i.

6.1.1 Example of Polynomial Extrapolation

An example of this process is shown in Table 6.2 below. The equation $y' = -y, y(0) = 1$ was once again integrated with step sizes $h = 2^{-i}$, $i = 0, 1, \ldots, 9$. The corresponding errors are shown in Table 6.3. Each column converges at an increasing power of h. Thus the fourth column of answers converges as h^4. This is shown in Table 6.4, which consists of the errors in Table 6.3 divided by h_i^m.

6.1.2 Round-Off Effects

We can see that this type of method has the potential for being of arbitrarily large order (by making m larger) and for reducing the step without voiding the results of work at larger steps (by increasing i for fixed m). In practice, these procedures are limited by rounding errors and instabilities. Increasing m or i increases the number of calculations in the basic integration by the Euler method at the rate of 2^{m+i}. This means that if worst-case round-off errors occur at each step, the answers will lose one bit of precision for each increase in i or m. One way of reducing this problem somewhat is to use a different strategy for selecting the smaller step sizes. In the previous example we halved the step for each additional line of extrapolations. Instead, suppose we approximately reduce the step by an amount $1/q$ $(1 < q < 2)$ each time. The worst-case rounding error in R_0^i will be proportional to q^i since this is approximately the number of steps. From (6.4) we see that the error in the next column in R_1^i could be as large as

$$q^{i+1} + \frac{q^{i+1} + q^i}{q - 1} = q^{i+1}\left(\frac{q + q^{-1}}{q - 1}\right)$$

At the next step, the error in R_2^i could be as much as

$$\frac{q + q^{-1}}{q - 1}\left[q^{i+2} + \frac{q^{i+2} + q^{i+1}}{q^2 - 1}\right] = q^{i+2}\frac{q + q^{-1}}{q - 1} \cdot \frac{q^2 + q^{-1}}{q^2 - 1}$$

and the error in R_m^i could be as large as

$$q^{i+m}\frac{q + q^{-1}}{q - 1} \cdot \frac{q^2 + q^{-1}}{q^2 - 1} \cdots \frac{q^m + q^{-1}}{q^m - 1} \tag{6.5}$$

It is true that by making q close to 1, the effect of the q^{i+m} term can be reduced, but the second part of expression (6.5) increases rapidly as $q \longrightarrow 1$.

6.1.3 Stability

The stability of all of the methods to be discussed in this chapter is still an open question. Experience has shown that equations with very negative values of $\partial f/\partial y$ cause severe instability problems, until the basic step size h_0

Table 6.2 R_m^i FOR $y' = -y$, $y(0) = 1$

i \ m	0	1	2	3	4	5	6
0	0.0	0.500000000	0.343750000	0.370105743	0.367774922	0.367881947	0.367879410
1	0.250000000	0.382812500	0.366811275	0.367920598	0.367878603	0.367879450	0.367879441
2	0.316406250	0.370811582	0.367781933	0.367881228	0.367879424	0.367879441	0.367879441
3	0.343608916	0.368539345	0.367868816	0.367879536	0.367879441	0.367879441	0.367879441
4	0.356074130	0.368036448	0.367878196	0.367879447	0.367879441	0.367879441	
5	0.362055289	0.367917759	0.367879290	0.367879442	0.367879441		
6	0.364986524	0.367888908	0.367879423	0.367879441			
7	0.366437716	0.367881794	0.367879439				
8	0.367159755	0.367880028					
9	0.367519891						

Table 6.3 ERRORS IN R_m^i

i \ m	0	1	2	3	4	5	6
0	−0.367879441	0.132120559	−0.024129441	0.002226302	−0.000104519	0.000002506	−0.000000031
1	−0.117879441	0.014933059	−0.001068166	0.000041157	−0.000000838	0.000000009	−0.0
2	−0.051473191	0.002932140	−0.000097508	0.000001786	−0.000000018	0.0	−0.0
3	−0.024270525	0.000659904	−0.000010625	0.000000095	−0.0	0.0	−0.0
4	−0.011805311	0.000157007	−0.000001245	0.000000006	−0.0	0.0	
5	−0.005824152	0.000038318	−0.000000151	0.0	−0.0		
6	−0.002892917	0.000009466	−0.000000019	0.0			
7	−0.001441725	0.000002353	−0.000000002				
8	−0.000719686	0.000000586					
9	−0.000359550						

Table 6.4 ERRORS IN R_m^i/h_i^m

i \ m	0	1	2	3	4	5	6
0	−0.367879441	0.132120559	−0.024129441	0.002226302	−0.000104519	0.000002506	−0.000000031
1	−0.235758882	0.059732235	−0.008545326	0.000658514	−0.000026827	0.000000572	−0.000000006
2	−0.205892765	0.046914247	−0.006240528	0.000457311	−0.000017967	0.000000372	−0.000000004
3	−0.194164203	0.042233851	−0.005440234	0.000389934	−0.000015080	0.000000309	−0.000000007
4	−0.188884972	0.040193764	−0.005099041	0.000361659	−0.000013883	0.000000254	
5	−0.186372861	0.039237693	−0.004940815	0.000348643	−0.000013391		
6	−0.185146683	0.038774492	−0.004864550	0.000342366			
7	−0.184540832	0.038546466	−0.004827103				
8	−0.184239688	0.038433331					
9	−0.184089557						

is reduced to the point where $|h_0\, \partial f/\partial y|$ is close to 1, but it is an area which needs a lot of attention.

6.1.4 Higher Order Methods

The basic method used in the earlier polynomial interpolation was the Euler method, a first order method. In order to achieve a fourth order method, for example, it is necessary to perform three extrapolations. If we use step sizes in the sequence $h, h/2, h/4, h/8, \ldots$, a total of 12 evaluations of the derivative function f are needed, compared with the four of a Runge-Kutta fourth order method. By using the sequence $h, h/2, h/3, h/4$ (which is the minimum since the steps must be different and must be submultiples of the basic step h), we can reduce the number of function evaluations to seven, although the rounding errors may be increased as discussed in the previous subsection. An alternative approach is to start with a higher order method. If an rth order method were used, for example, the approximations R^i_m would have $0(h^{r+m})$ errors when appropriate extrapolation formula were used. If a numerical integration formula is available for which

$$y_h(t) = y(t) + \sum_{i=1}^{\infty} \tau_i h^{ri}$$

that is, the integral is a function of h^r, an extrapolation formula can increase the order by r at each stage so that R^i_m would be in error by $0(h^{r(m+1)})$. One such formula is known for $r = 2$. It has the following form: Define

$$t_n = nh$$

$$\eta(0, h) = y(0) \tag{6.6a}$$

$$\eta(t_1, h) = y(0) + hf(y_0, 0) \tag{6.6b}$$

$$\begin{aligned} \eta(t_{n+1}, h) = \eta(t_{n-1}, h) + 2hf(\eta(t_n, h), t_n) \\ n = 1, 2, \ldots, N-1 \qquad \text{where } t_N = t \end{aligned} \tag{6.6c}$$

$$y(t, h) = \tfrac{1}{2}[\eta(t_{N-1}, h) + \eta(t_N, h) + hf(\eta(t_N, h), t_N)] \tag{6.7}$$

Gragg (1965) has shown that $y(t, h)$ defined in this way is an even power series in h provided that the number of steps is always even or always odd and provided that $y(t)$ is sufficiently differentiable. Bulirsch and Stoer(1966) recommend using an all-even sequence by starting with $h_0 = H/2$, where H is the width of the basic interval to be covered.

A new initial value problem is done over each interval of length H starting from $t = 0$. The extrapolation process is applied to the results $y(H, h_i)$ of the first interval to find $y(H)$ to the desired accuracy. Starting from there, the extrapolation process is performed over the interval $[H, 2H]$ to find $y(2H)$ and so on.

An example of this process is shown in Tables 6.5 and 6.6. The sequence of h values used is $H/2, H/4, H/6, H/8, H/12, H/16, \ldots$. The extrapolation

Table 6.5 ERRORS IN R_m^i FOR POLYNOMIAL EXTRAPOLATION

i	h_i m	0	1	2	3	4	5
0	$\frac{1}{2}$	7 120 559	1 912 226	60 374	953	7	0
1	$\frac{1}{4}$	3 214 309	266 135	4 667	33	0	
2	$\frac{1}{6}$	1 576 434	70 034	548	2		
3	$\frac{1}{8}$	917 384	17 919	79			
4	$\frac{1}{12}$	417 681	4 539				
5	$\frac{1}{16}$	236 932					

Table 6.6 NUMBER OF FUNCTION EVALUATIONS IN TABLE 6.3/TABLE 6.5

i m	0	1	2	3	4	5
0	$\frac{1}{3}$	$\frac{2}{7}$	$\frac{5}{13}$	$\frac{12}{21}$	$\frac{27}{33}$	$\frac{58}{49}$
1	$\frac{2}{5}$	$\frac{5}{11}$	$\frac{12}{19}$	$\frac{27}{31}$	$\frac{58}{47}$	
2	$\frac{4}{7}$	$\frac{11}{15}$	$\frac{26}{27}$	$\frac{57}{43}$		
3	$\frac{8}{9}$	$\frac{23}{21}$	$\frac{54}{37}$			
4	$\frac{16}{13}$	$\frac{47}{29}$				
5	$\frac{32}{17}$					

formula used is similar to (6.4) with h_i/h_{i+m} replaced by $(h_i/h_{i+m})^2$. Once again $y' = -y$ was integrated from 0 to 1 with $y(0) = 1$ and $H = 1$. The errors in R_m^i times 10^9 are shown in Table 6.5. They can be seen to be considerably less than the corresponding figures in Table 6.3. The number of function evaluations for the two tables is not simply related, so Table 6.6 shows the number of function evaluations for Table 6.3 divided by the number of function evaluations for Table 6.5. As expected, the work required in Table 6.5 for high accuracy is less.

6.2 RATIONAL FUNCTION EXTRAPOLATION

For many applications, rational function interpolation is more accurate than polynomial interpolation. A rational function is one which is the quotient of two polynomials. We will restrict ourselves to rational functions of the form

$$R_m(h) = \frac{P_m(h)}{Q_m(h)} = \frac{p_0 + p_1 h + \cdots + p_\mu h^\mu}{q_0 + q_1 h + \cdots + q_\nu h^\nu} \tag{6.8}$$

where $\mu = [m/2]$ (integer part of $m/2$) and $\nu = m - \mu = [(m + 1)/2]$. (These are called the *diagonal rational polynomials.*) If (6.8) is used in place of the polynomial approximation of the previous section, we get the Bulirsch-Stoer (1966) algorithm. In this method the second order integration rule described by (6.6) is used to integrate the equations for the sequence of steps $H/2$, $H/4$, $H/6$, $H/8$, $H/12, \ldots$ to get results which we call $R_0^0, R_0^1, R_0^2, \ldots$, etc. These numbers are then combined in a manner similar to that of Eq. (6.4) for polynomials in order to achieve the results of a rational interpolation. The arithmetic relation used in place of Eq. (6.4) is considerably more complex, so it is important to put it in a form which causes as few rounding errors as possible.

As each new value of $y(t, h_m) = R_0^m$ is calculated, values of R_k^{m-k}, $k = 1, 2, \ldots, m$ can be calculated. The process can be stopped when two successive approximations R_k^{m-k} and R_k^{m-k+1} are very close. Their difference is used as an indication of the amount of error still present. If it is necessary to continue until a small h_m is used, it is an indication that the original value of H was too large. Bulirsch and Stoer's proposal to control this was to continue the process until m was 6. If it had already converged within the error criteria, then the result could be accepted, but if it had converged for $m < 6$, the basic step H could be increased in order to save work. (However, note that the step may have to be limited in order that the results are obtained at desired printing points.) If the method had not converged by the time that m was 6, then successive values of R_6^{m-6} are calculated until convergence occurs. If m gets too large, it is necessary to reduce the basic step H because round-off accumulation in the smaller substeps h_i may be voiding the extrapolation hypothesis, while stability problems in the large substeps may prevent convergence of the extrapolation. A typical program might continue to R_6^{10} and, if the differences are not yet small, reduce the basic step H and start again.

First we will examine rational function interpolation and derive some recurrence relations. Suppose that we have formulas of the type (6.8) such that $R_m(h_i) = y_i$, $i = 0, 1, \ldots, m$ for $m = 0, 1, \ldots, n$. We wish to construct a similar rational function for $m = n + 1$. Since

$$\frac{P_m(h_i)}{Q_m(h_i)} = y_i \qquad i \leq m \leq n \tag{6.9}$$

the test function $T_m(h, y) = P_m(h) - yQ_m(h)$ has zeros on the points (h_i, y_i), $i \leq m \leq n$. Consider the function

$$T_{n+1}(h, y) = \alpha T_n(h, y) + \beta(h)T_{n-1}(h, y) \tag{6.10}$$

This certainly has zeros on the points (h_i, y_i), $i \leq n - 1$. If we choose $\beta(h_n) = 0$, then $T_{n+1}(h_n, y_n) = 0$. We now wish to arrange that $T_{n+1}(h_{n+1}, y_{n+1}) = 0$. This requires that

$$\alpha T_n(h_{n+1}, y_{n+1}) + \beta(h_{n+1})T_{n-1}(h_{n+1}, y_{n+1}) = 0 \tag{6.11}$$

Note that T_n and T_{n-1} are polynomials in h of degree $[(n + 1)/2]$ and $[n/2]$, and are linear in y. We would like T_{n+1} to be a polynomial of degree $[(n+2)/2]$ in h and linear in y so that we can express it as $P_{n+1}(h) + yQ_{n+1}(h)$, where $P_{n+1}(h)$ has degree $[(n+1)/2]$ and $Q_{n+1}(h)$ has degree $[(n+2)/2]$. This will be true if in the defining relation (6.10) α is constant and β is a linear function of h only. Thus we can choose $\beta(h) = (h - h_n)$ and α to satisfy (6.11) as

$$\alpha = (h_n - h_{n+1})\frac{T_{n-1}(h_{n+1}, y_{n+1})}{T_n(h_{n+1}, y_{n+1})}$$

Substituting in (6.10), we get

$$\begin{aligned} P_{n+1}(h) - yQ_{n+1}(h) &= T_{n+1}(h, y) \\ &= (h_n - h_{n+1})\frac{T_{n-1}(h_{n+1}, y_{n+1})}{T_n(h_{n+1}, y_{n+1})}T_n(h, y) + (h - h_n)T_{n-1}(h, y) \\ &= (h_n - h_{n+1})\frac{T_{n-1}(h_{n+1}, y_{n+1})}{T_n(h_{n+1}, y_{n+1})}P_n(h) + (h - h_n)P_{n-1}(h) \\ &\quad - y\left[(h_n - h_{n+1})\frac{T_{n-1}(h_{n+1}, y_{n+1})}{T_n(h_{n+1}, y_{n+1})}Q_n(h) + (h - h_n)Q_{n-1}(h)\right] \end{aligned} \tag{6.12}$$

(6.12) provides a recurrence relation for P_{n+1} and Q_{n+1} in terms of P_n, Q_n, P_{n-1}, Q_{n-1}, y_{n+1}, h_{n+1}, and h_n by equating terms independent of y and linear in y. This relation is not convenient for direct use since it requires that the two polynomials in h be calculated at each step. We notice that, in the case of polynomial interpolation, formula (6.4) gave a relation for the new value of $R^i_m(0) = R^i_m$. We would like to get a similar relation for the $P_n(0)/Q_n(0)$. It has been shown [Stoer (1961) and Bulirsch and Stoer (1964)] that a scheme can be derived for rational interpolation. If we define $R^i_m(h)$ as the rational approximation which agrees with $y(x, h)$ at $h = h_i, h_{i+1}, \ldots, h_{i+m}$, where $h_i > h_{i+1} > \cdots > h_{i+m}$ and $R^i_m(0) = R^i_m$, then the R's can be obtained by the formula

$$\begin{aligned} R^i_{-1} &= 0 \qquad \text{(This is used to get the recurrence started.)} \\ R^i_0 &= y(t, h_i) \\ R^i_m &= R^{i+1}_{m-1} + \frac{R^{i+1}_{m-1} - R^i_{m-1}}{\left(\dfrac{h_i}{h_{i+m}}\right)^2\left[1 - \dfrac{R^{i+1}_{m-1} - R^i_{m-1}}{R^{i+1}_{m-1} - R^{i+1}_{m-2}}\right] - 1} \qquad m \geq 1 \end{aligned} \tag{6.13}$$

(These formulas assume that the approximation is a function in h^2, not h.) This formula involves calculating the differences of numbers that are getting closer and closer to the answer R^0_∞, so less floating point error is obtained by calculating the differences directly. Defining

$$D^i_m = R^i_m - R^{i+1}_{m-1}$$

$$C^i_m = R^i_m - R^i_{m-1}$$

and

$$W^i_m = R^i_m - R^{i-1}_m$$

we can convert (6.13) to

$$D^i_m = \frac{C^{i+1}_{m-1} \cdot W^{i+1}_{m-1}}{\left(\frac{h_i}{h_{i+m}}\right)^2 D^i_{m-1} - C^{i+1}_{m-1}} \qquad m \geq 1$$

$$C^i_m = \frac{\left(\frac{h_i}{h_{i+m}}\right)^2 D^i_{m-1} \cdot W^{i+1}_{m-1}}{\left(\frac{h_i}{h_{i+m}}\right)^2 D^i_{m-1} - C^{i+1}_{m-1}} \qquad m \geq 1$$

and

$$W^i_m = C^i_m - D^{i-1}_m$$

with

$$C^i_0 = D^i_0 = y(t, h_i)$$
$$W^i_0 = y(t, h_i) - y(t, h_{i-1})$$

It is not necessary to store more than a single array containing the D^i_{m-1}, $i = m, m - 1, \ldots, 0$ during the calculation. This is illustrated in the following 360/FORTRAN program which performs integration by one step of size H using either rational function or polynomial extrapolation. This program is based on the FORTRAN version by Clark (1966) of the Bulirsch and Stoer Algol program (1966).

```
      SUBROUTINE DIFSUB (N,T,Y,DY,H,HMIN,EPS,MF,YMAX,ERROR,KFLAG,
     1                  JSTART,MAXORD,MAXPTS)
      IMPLICIT REAL*8 (A-H,O-Z)
C***********************************************************************
C*    THE PARAMETERS TO THIS INTEGRATION SUBROUTINE HAVE               *
C*    THE FOLLOWING MEANINGS..                                         *
C*      N        THE NUMBER OF FIRST ORDER DIFFERENTIAL EQUATIONS      *
C*      T        THE INDEPENDENT VARIABLE                              *
C*      Y        THE DEPENDENT VARIABLES, UP TO 10 ARE ALLOWED.        *
C*      DY       AN ARRAY OF 10 LOCATIONS WHICH WILL CONTAIN THE       *
C*                 VALUES OF THE DERIVATIVES ON EXIT.                  *
C*      H        THE STEP SIZE THAT SHOULD BE ATTEMPTED.  IT MAY BE    *
C*                 INCREASED OR DECREASED BY THE SUBROUTINE.           *
C*      HMIN     THE MINIMUM STEP SIZE THAT SHOULD BE ALLOWED ON THIS  *
C*                 STEP.                                               *
C*      EPS      THE ERROR TEST CONSTANT.  THE ESTIMATED ERRORS ARE    *
C*                 REQUIRED TO BE LESS THAN EPS*YMAX IN EACH COMPONENT. *
C*                 IF YMAX IS ORIGINALLY SET TO +1 IN EACH COMPONENT,  *
C*                 THE ERROR TEST WILL BE RELATIVE FOR THOSE COMPONENTS *
C*                 GREATER THAN 1 AND ABSOLUTE FOR THE OTHERS.         *
C*      MF       THE METHOD INDICATOR.  THE FOLLOWING ARE ALLOWED..    *
C*                        0    BULIRSCH-STOER RATIONAL EXTRAPOLATION   *
C*                        1    POLYNOMIAL EXTRAPOLATION                *
C*      YMAX     THE MAXIMUM VALUES OF THE DEPENDENT VARIABLES ARE     *
C*                 SAVED IN THIS ARRAY. IT SHOULD BE SET TO +1 BEFORE  *
C*                 THE FIRST ENTRY. (SEE THE DESCRIPTION OF EPS.)      *
C*      ERROR    THE ESTIMATED SINGLE STEP ERROR IN EACH COMPONENT     *
C*      KFLAG    A COMPLETION CODE WITH THE FOLLOWING MEANINGS..       *
C*                       -1    THE STEP WAS TAKEN WITH H = HMIN        *
C*                             BUT THE REQUESTED ERROR WAS NOT ACHIEVED. *
C*                       +1    THE STEP WAS SUCCESSFUL.                *
C*      JSTART   AN INITIALIZATION INDICATOR WITH THE MEANING..        *
C*                       -1    REPEAT THE LAST STEP, RESTORING THE     *
C*                             VALUES OF Y AND YMAX THAT WERE USED     *
C*                             LAST TIME.                              *
C*                       +1    TAKE A NEW STEP.                        *
C*      MAXORD   THE MAXIMUM ORDER OF EXTRAPOLATION ALLOWED. IT MUST   *
C*                 BE LESS THAN 11.                                    *
C*      MAXPTS   THE MAXIMUM NUMBER OF DIFFERENT SUB STEP SIZES USED   *
C*                 IN THE EXTRAPOLATION PROCESS.                       *
C***********************************************************************
```

```
      DIMENSION Y(10),DY(10),YMAX(10),YSAVE(10),YNM1(10),YN(10),DYN(10)
     1          ,YMAXSV(10),QUOT(11,2),EXTRAP(10,11),YNM1HV(10,12)
     2          ,YNHV(10,12),YMAXHV(10,12),ERROR(10)
C***********************************************************************
C*    THE ARRAYS ARE USED FOR THE FOLLOWING DATA..                     *
C*     YSAVE    THE INITIAL VALUES OF Y ARE SAVED FOR A RESTART        *
C*     YNM1     Y(N-1), THE PREVIOUS VALUE OF Y IN THE MIDPOINT METHOD *
C*     YN       Y(N), THE CURRENT VALUE OF Y IN THE MIDPOINT INTEGRATION *
C*     DYN      THE INITIAL VALUE OF THE DERIVATIVE OF Y.              *
C*     YMAXSV   THE SAVED VALUES OF YMAX AT THE INITIAL POINT.         *
C*     QUOT     THE QUOTIENTS (H(I)/H(I+M))**2 USED IN THE EXTRAPOLATION.*
C*     EXTRAP   THE MOST RECENT EXTRAPOLATED VALUES OF Y IN THE CASE   *
C*               OF POLYNOMIAL EXTRAPOLATION, OR OF THE DIFFERENCES IN *
C*               THE CASE OF RATIONAL FUNCTION EXTRAPOLATION.          *
C*     YNM1HV   THE VALUES OF YNM1 AT THE MIDPOINT OF THE BASIC INTERVAL *
C*               IF THE NUMBER OF SUB STEPS IS DIVISIBLE BY 4. THIS    *
C*               INFORMATION IS USED TO AVOID REDOING THE INTEGRATION IN *
C*               CASE THE STEP IS HALVED.                              *
C*     YNHV     THE SIMILAR VALUES OF YN                               *
C*     YMAXHV   AND THE SAME FOR YMAX                                  *
C*     ERROR    THE ESTIMATES OF THE SINGLE STEP ERRORS ARE SAVED HERE. *
C***********************************************************************
      DATA QUOT/1.,2.25,4.,9.,16.,36.,64.,144.,256.,576.,1024.,
     1      1.,1.7777777777777777,4.,7.1111111111111111,
     2      16.,28.4444444444444444,64.,113.7777777777777,
     3       256.,455.1111111111111,1024./
      DATA FMAX/10000000./
C***********************************************************************
C*    FMAX IS A NUMBER SMALLER THAN THE FIRST INTEGER THAT CANNOT BE   *
C*    REPRESENTED EXACTLY IN FLOATING POINT.                           *
C***********************************************************************
      IF(JSTART.LT.0) GO TO 2
C***********************************************************************
C* SAVE THE VALUES OF Y AND YMAX IN CASE A RESTART IS NECESSARY.       *
C***********************************************************************
      DO 1 I = 1,N
      YSAVE(I) = Y(I)
 1    YMAXSV(I) = YMAX(I)
      CALL DIFFUN(T,Y,DYN)
      GO TO 4
C***********************************************************************
C*    RESTORE THE VALUES OF Y AND YMAX FOR A RESTART.                  *
C***********************************************************************
 2    DO 3 I = 1,N
      Y(I) = YSAVE(I)
 3    YMAX(I) = YMAXSV(I)
 4    CONTINUE
C***********************************************************************
C*    THE FOLLOWING COUNTERS AND SWITCHES ARE USED..                   *
C*     J        IS THE COUNT THROUGH THE DIFFERENT SUB STEPS G USED.   *
C*     JODD     IS 1 IF J IS ODD, 2 IF J IS EVEN                       *
C*     JHVSV    IS THE NUMBER OF SUBSTEP SIZES FOR WHICH HALF WAY      *
C*               INFORMATION HAS BEEN SAVED.                           *
C*     JHVSV1   THE VALUE OF JHVSV FROM THE PREVIOUS CYCLE             *
C*     M        THE NUMBER OF PAIRS OF SUB STEPS WHICH MAKE UP THE STEP H*
C*               M TAKES THE SEQUENCE 1,2,3,4,6,8,12,16, ETC.          *
C*     MNEXT    THE NEXT VALUE OF M                                    *
C*     MTWO     THE NEXT BUT ONE VALUE OF M.                           *
C*     QUOTSV   THE LAST VALUE OF QUOT IS IRREGULAR DUE TO THE FACT THAT *
C*               THE SEQUENCE OF THE MULTIPLES 9/4,16/9 (ODD) OR       *
C*               16/9,9/4 (EVEN UNTIL THE FINAL MULTIPLE OF 4. HOWEVER, *
C*               (H(0)/H(M))**2 IS ALWAYS M**2. THE REGULAR VALUE OF   *
C*               QUOT IS SAVED IN QUOTSV, AND REPLACED BY M**2.        *
C*     KONV     IS SET TO +1 INITIALLY, AND RESET TO -1 IF THE ERROR   *
C*               TEST FAILS.                                           *
C***********************************************************************
 5    JHVSV1 = 0
      KFLAG = 1
 6    JHVSV = 0
 7    A = H + T
      JODD = 1
      M = 1
      MNEXT = 2
      MTWO = 3
      DO 23 J = 1,MAXPTS
        QUOTSV = QUOT(J,JODD)
        QUOT(J,JODD) = M*M
        KONV = 1
        IF (J.LE.(MAXORD/2)) KONV = -1
```

```
      IF (J.LE.(MAXORD+1))  GO TO 8
      L = MAXORD + 1
      HCHNGE = .7071068D0*HCHNGE
      GO TO 9
 8    L = J
      HCHNGE = 1.0D0 + (MAXORD + 1 - J)/6.0D0
 9    B = H/M
      G = B*0.5D0
      IF (J.GT.JHVSV1) GO TO 11
C*********************************************************************
C*   THE VALUES OF THE MIDPOINT INTEGRATION WERE SAVED AT THE        *
C*   HALF WAY POINT IN THE PREVIOUS INTEGRATION. USE THEM.           *
C*********************************************************************
      DO 10 I = 1,N
        YN(I) = YNHV(I,J)
        YNM1(I) = YNM1HV(I,J)
 10     YMAX(I) = YMAXHV(I,J)
      GO TO 16
C*********************************************************************
C*   INTEGRATE OVER THE RANGE H BY 2*M STEPS OF A MIDPOINT METHOD.   *
C*********************************************************************
 11   DO 12 I = 1,N
        YNM1(I) = YSAVE(I)
        YN(I) = YSAVE(I) + G*DYN(I)
 12     YMAX(I) = YMAXSV(I)
      M2 = M + M
      TU = T
      DO 15 K = 2,M2
        TU = TU + G
        CALL DIFFUN(TU,YN,DY)
        DO 13 I = 1,N
          U = YNM1(I) + B*DY(I)
          YNM1(I) = YN(I)
          YN(I) = U
          U = DABS(U)
 13       IF (U.GT.YMAX(I)) YMAX(I) = U
        IF ((K.NE.M).OR.(JHVSV1.NE.0).OR.(K.EQ.3)) GO TO 15
        JHVSV = JHVSV + 1
        DO 14 I = 1,N
          YNHV(I,JHVSV) = YN(I)
          YNM1HV(I,JHVSV) = YNM1(I)
 14       YMAXHV(I,JHVSV) = YMAX(I)
 15     CONTINUE
 16   CALL DIFFUN(A,YN,DY)
      DO 22 I = 1,N
        V = EXTRAP(I,1)
C*********************************************************************
C*   CALCULATE THE FINAL VALUE TO BE USED IN THE EXTRAPOLATION PROCESS *
C*********************************************************************
        TA = (YN(I) + YNM1(I) + G*DY(I))*0.5D0
        C = TA
C*********************************************************************
C*   INSERT THE INTEGRAL AS THE FIRST EXTRAPOLATED VALUE.            *
C*********************************************************************
        EXTRAP(I,1) = TA
        IF (L.LT.2) GO TO 21
        IF (DABS(V)*FMAX.LT.DABS(C)) GO TO 27
        IF (MF.GT.0) GO TO 19
C*********************************************************************
C*   PERFORM THE EXTRAPOLATION BY RATIONAL FUNCTIONS ON THE          *
C*   SECOND AND SUBSEQUENT INTEGRALS.                                *
C*********************************************************************
        DO 18 K = 2,L
          B1 = QUOT(K,JODD)*V
          B = B1 - C
          U = V
          IF (B.EQ.0) GO TO 17
          B = (C - V)/B
          U = C*B
          C = B1*B
 17       V = EXTRAP(I,K)
          EXTRAP(I,K) = U
          TA = TA + U
 18       CONTINUE
        GO TO 21
C*********************************************************************
C*   PERFORM THE EXTRAPOLATION BY POLYNOMIALS ON THE                 *
C*   SECOND AND SUBSEQUENT INTEGRALS.                                *
C*********************************************************************
```

```
19        DO 20 K = 2,L
            TA = TA + (TA - V)/(QUOT(K,JODD) - 1.0D0)
            V = EXTRAP(I,K)
            EXTRAP(I,K) = TA
20          CONTINUE
          GO TO 21
21        U = DABS(TA)
          IF (U.GT.YMAX(I)) YMAX(I) = U
          ERROR(I) = DABS(Y(I) - TA)
          Y(I) = TA
          IF (ERROR(I).GT.EPS*YMAX(I)) KONV = -1
22        CONTINUE
        QUOT(J,JODD) = QUOTSV
        IF (KONV.GT.0) GO TO 25
        JODD = 3-JODD
        M = MNEXT
        MNEXT = MTWO
        MTWO = M + M
23      CONTINUE
      JHVSV1 = JHVSV
24    IF (DABS(H).LE.HMIN) GO TO 26
      H = H*0.5D0
      IF (DABS(H).GE.HMIN) GO TO 6
      H = DSIGN(HMIN,H)
      GO TO 5
25    H = H*HCHNGE
      T = A
      RETURN
26    KFLAG = -1
      GO TO 25
27    QUOT(J,JODD) = QUOTSV
      GO TO 24
      END
                                                        0232 CARDS
```

The errors times 10^9 in the R^i_m from using this program to integrate $y' = -y$ from 0 to 1 with $H = 1$ are shown in Table 6.7. They are slightly smaller than those in Table 6.5. Their computation required the same number of function evaluations, but slightly more arithmetic in the extrapolation process. The advantage of Bulirsch-Stoer is also shown in Figure 6.1. This

Table 6.7 ERRORS IN THE BULIRSCH-STOER METHOD

i	h_i m	0	1	2	3	4	5
0	$\frac{1}{2}$	7 120 559	1 930 247	−287 188	32	0	0
1	$\frac{1}{4}$	3 214 309	276 508	−9 062	1	0	
2	$\frac{1}{6}$	1 576 434	13 481	−907	0		
3	$\frac{1}{8}$	917 384	8 893	−122			
4	$\frac{1}{12}$	417 681	4 800				
5	$\frac{1}{16}$	236 932					

is taken from Clark (1966). It shows the errors in three problems for a variety of requested errors versus the number of function evaluations. In each case MAXORD was set to 6 and MAXPTS to 10. The negative exponential problem was $y' = -y, y(0) = 1$, integrated to $t = 20$ with the value of YMAX reset to the current value of y before each step. The error relative

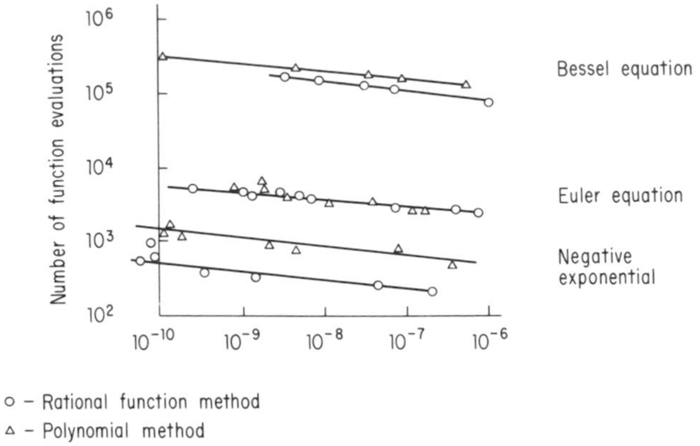

Fig. 6.1 Comparison of rational function and polynomial extrapolation.

to exp (—20) is plotted. The Euler equation is the system

$$y^{1\prime} = -y^2y^3, \qquad y^1(0) = 0$$
$$y^{2\prime} = -y^1y^3, \qquad y^2(0) = 1$$
$$y^{3\prime} = -.51y^1y^2, \qquad y^3(0) = 1$$

integrated to $t = 60$. The root mean square of the errors in the three y's is plotted. The Bessel equation defines $J_{16}(t) = y^1(t)$ by

$$y^{1\prime} = y^2$$
$$y^{2\prime} = \left(\frac{256}{t^2} - 1\right)y^1 - \frac{y^2}{t}$$
$$y^1(6) = 1.20195 \cdot 10^{-6}, \qquad y^2(6) = 2.98648 \cdot 10^{-6}$$

The errors were taken relative to the largest value of y^1 at about $t = 18.1$ where $y^1 \simeq 0.2612$. The average of the absolute errors at $t = 6132$, 6134, 6136, and 6138 is plotted.

We can see a slight advantage to the rational function extrapolation process for two of these examples.

The effect of reducing the error control parameter EPS is shown in Table 6.8 for rational function extrapolation. The requested error versus the actual errors times 10^9 is shown for the system

$$y^{1\prime} = -y^1, \qquad y^1(0) = 1$$
$$y^{2\prime} = y^3, \qquad y^2(0) = 0$$
$$y^{3\prime} = -y^2, \qquad y^3(0) = 1$$
$$y^{4\prime} = 2t, \qquad y^4(0) = 0$$
$$y^{5\prime} = 10t^9, \qquad y^5(0) = 0$$

Table 6.8 ERRORS e_i IN y_i VERSUS EPS PARAMETER

EPS	e_1	e_2	e_3	e_4	e_5
0.001	3235353	5135388	2981195	0	1450333913
0.0001	7724	2891	2832	0	32023507
0.00001	17	15	4	0	18740
0.000001	17	15	4	0	18740
0.0000001	0	0	0	0	6
0.00000001	0	0	0	0	6
0.000000001	0	0	0	0	6
0.0000000001	0	0	0	0	1

In summary, it can be said that the extrapolation method looks promising. If Figure 6.1 is compared with Figure 5.2, the advantage over Runge-Kutta methods can be seen. However, there are many open questions concerning stability, error estimates, and error bounds.

PROBLEMS

1. Integrate $y' = y$, $y(0) = 1$ over the interval [0, 1] using
   (a) midpoint rule (Eq. 2.5).
   (b) Heun rule (Eq. 2.8).
   (c) trapezoidal rule (Eq. 2.7).
   Perform polynomial extrapolation on these using step sizes $h = 1, \frac{1}{2}, \frac{1}{3}, \frac{1}{4}, \frac{1}{6}, \frac{1}{8}, \frac{1}{12}$ based on the assumptions: that the answers are power series in h with the linear terms absent; that the answers are power series in h^2. What are your conclusions?

2. If polynomial extrapolation of the sort given by Eq. (6.4) is used, draw up a table similar to Table 6.1 showing the worst case of round-off errors if the ratio h_i/h_{i+1} is 2 and it is assumed that the round-off error in $y(t, h_i)$ is bounded by $2^i e$.

3. Use a polynomial extrapolation on the equation $y' = -2^{10}y$, $y(0) = 1$ over the range [0, 1] and experimentally determine the approximate values of H for which the method is absolutely stable if the final answer is taken to be R^i_m for $i + m \leq 5$, where R^i_m is given by (6.4), Euler's method is used for the basic integration and the step sizes used are $2^{-i}H$.

4. The Euler method is used over the interval H with step sizes H, $H/2$, and $H/4$ to determine R^0_0, R^1_0, and R^2_0 as approximations to the solution of the differential equation $y' = \lambda y$. R^0_2 is now calculated by Eq. (6.4). Derive the equation that defines the region of absolute stability in the $H\lambda$-plane.

7 MULTIVALUE OR MULTISTEP METHODS—INTRODUCTION

The methods discussed so far have required a knowledge of the differential equations and initial values only. Consequently, given an approximation to the value of $y(t)$ at $t = t_{n-1}$, say y_{n-1}, they have provided a technique for computing $y_n \cong y(t_n)$. They could therefore be called *one-value methods* in that they only required one value of the dependent value. They are usually called *one-step methods* because they only require the value at one mesh point to compute the value at the next. Once the values at a number of points of the numerical approximation have been computed, they could be used to aid in the computation of later points (e.g., by polynomial extrapolation) much as intermediate values are used in a Runge-Kutta method. Information from previous points, say for $t \leq t_{n-1}$, can be saved and represent knowledge about the solution at $t = t_{n-1}$. In the next four chapters we are going to discuss methods which require several pieces of information about the dependent variable at time, $t = t_{n-1}$ in order to compute the equivalent pieces of information at $t = t_n$. Consequently, we call these methods *multivalue methods* since they use more than one value of the dependent variable. Often these methods use the values of the dependent variable and its derivative at k different mesh points $t_{n-1}, t_{n-2}, \ldots, t_{n-k}$. They are therefore commonly called *multistep methods*; in this case, k-step methods.

In this chapter we will present a notation for a class of multivalue methods. Two special cases of the multivalue methods will then be examined: the explicit and implicit multistep methods. In particular, we will examine the Adams-Bashforth and Adams-Moulton methods which are examples of these cases. We will discuss only the single equation $y' = f(y)$. The extension to a system is obvious in all cases.

7.1 MULTIVALUE METHODS

After approximations at a number of points, say $t_{n-k}, t_{n-k+1}, \ldots,$ and t_{n-1}, have been calculated, we have values of $y_{n-k}, y_{n-k+1}, \ldots, y_{n-1}, hy'_{n-k}, hy'_{n-k+1}$, and hy'_{n-1}. We could use this information, or some subset of it, to help in the determination of y_n and $hy'_n = hf(y_n)$. Let us write $\mathbf{y}_{n-1}$ for the column vector

$$[y_{n-1}, y_{n-2}, \ldots, y_{n-k}, hy'_{n-1}, hy'_{n-2}, \ldots, hy'_{n-k}]^T$$

(The components that are not used in calculating y_n could be dropped.) The objective of a multivalue method is to find a numerical approximation for $\mathbf{y}_n$ from $\mathbf{y}_{n-1}$ and the differential equation, where $\mathbf{y}_n$ is the column vector $[y_n, y_{n-1}, \ldots, y_{n-k+1}, hy'_n, hy'_{n-1}, \ldots, hy'_{n-k-1}]^T$. This process can be applied repetitively to compute $\mathbf{y}_1, \mathbf{y}_2, \mathbf{y}_3, \ldots, \mathbf{y}_N$ once $\mathbf{y}_0$ is given.

The problem of computing $\mathbf{y}_0$ is one which has often led people to avoid multivalue methods in favor of one-value methods, however, for large problems in which good accuracy is required, the increased speed of multivalue methods over one-value methods is significant. Since we are only given y_0 as an initial value, the other components of $\mathbf{y}_0$, which we will call the *starting values*, must be computed. A common technique is to use a one-value method such as the Runge-Kutta method to compute $y_1, y_2, \ldots, y_{k-1}$, then to compute $hy'_i = hf(y_i)$, $0 \le i < k$, and hence to form $\mathbf{y}_{k-1}$ as a starting value in order to calculate $\mathbf{y}_k, \mathbf{y}_{k+1}, \ldots$. We will assume that some such technique has been used, although in Chapter 9 we will present a program which avoids the "starting problem" of multivalue methods.

A multivalue method consists of two processes which we call *prediction* and *correction*. In the predictor process, an approximation to $\mathbf{y}_n$ is computed from $\mathbf{y}_{n-1}$ by linear extrapolation. We call this approximation $\mathbf{y}_{n,(0)}$ and it is given by

$$\mathbf{y}_{n,(0)} = B\mathbf{y}_{n-1} \tag{7.1}$$

B is any suitable matrix of constants. For example, it could be

$$\begin{bmatrix} y_{n,(0)} \\ y_{n-1} \\ \cdot \\ \cdot \\ \cdot \\ y_{n-k+1} \\ hy'_{n,(0)} \\ \cdot \\ \cdot \\ \cdot \\ hy'_{n-k+1} \end{bmatrix} = \left[\begin{array}{cccc|cccc} \alpha_1 & \alpha_2 & \cdots & \alpha_k & \beta_1 & \beta_2 & \cdots & \beta_k \\ 1 & & & & & & & \\ & \ddots & & 0 & & 0 & & \\ 0 & & & & & & & \\ & \cdots & 1 & 0 & & & & \\ \hline \gamma_1 & \gamma_2 & & \gamma_k & \delta_1 & \delta_2 & \cdots & \delta_k \\ & & & & 1 & & & \\ & & & & & \ddots & 0 & \\ & 0 & & & 0 & & & \\ & & & & & & 1 & 0 \end{array}\right] \begin{bmatrix} y_{n-1} \\ y_{n-2} \\ \cdot \\ \cdot \\ \cdot \\ y_{n-k} \\ hy'_{n-1} \\ \cdot \\ \cdot \\ \cdot \\ hy'_{n-k} \end{bmatrix}$$

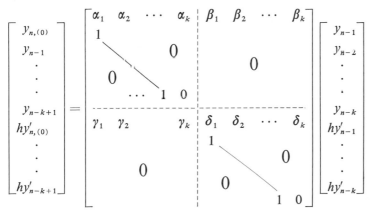

where the α_i, β_i, γ_i, and δ_i are constants such that

$$\sum_{i=1}^{k} (\alpha_i y_{n-i} + \beta_i h y'_{n-i})$$

is an approximation to y_n, and

$$\sum_{i=1}^{k} (\gamma_i y_{n-i} + \delta_i h y'_{n-i})$$

is an approximation to hy'_n.

The prediction process does not make use of the differential equation in any way, so the correction process corrects the approximate values if they do not satisfy the differential equation at $t = t_n$. The differential equation is written as

$$0 = G(\mathbf{y}_n) = -(\mathbf{y}_n)_k + hf((\mathbf{y}_n)_0) = -hy'_n + hf(y_n)$$

where by $(\mathbf{y})_i$ we mean the ith component of the vector $\mathbf{y}$. (Vector numbering starts from 0.) The amount by which $\mathbf{y}_{n,(0)}$ does not satisfy the differential equation is $G(\mathbf{y}_{n,(0)})$. A vector multiple of this scalar is added to $\mathbf{y}_{n,(0)}$ to correct it by the process

$$\mathbf{y}_{n,(1)} = \mathbf{y}_{n,(0)} + \mathbf{c}G(\mathbf{y}_{n,(0)}) \tag{7.2}$$

This process can be repeated by

$$\mathbf{y}_{n,(m+1)} = \mathbf{y}_{n,(m)} + \mathbf{c}G(\mathbf{y}_{n,(m)}) \qquad m = 1, 2, \ldots \tag{7.3}$$

for a fixed number of iterations or until there is no further change in $\mathbf{y}_{n,(m)}$. The value used for $\mathbf{y}_n$ is then $\mathbf{y}_{n,(M)}$, where M is either fixed, or large enough to get *convergence*, that is, such that $G(\mathbf{y}_{n,(M)})$ is zero to the accuracy desired. We say that M is the number of *corrector iterations*.

7.2 EXPLICIT MULTISTEP METHODS— THE ADAMS-BASHFORTH METHOD

Suppose that we take B to have the form given above with $\gamma_i = \delta_i = 0$, $1 \leq i \leq k$, and the vector $\mathbf{c}$ to be zero in all positions except for the kth, where we place a one. In that case (7.1) is

$$y_{n,(0)} = \sum_{i-1}^{k} (\alpha_i y_{n-i} + \beta_i h y'_{n-i})$$
$$hy'_{n,(0)} = 0$$

while (7.2) gives

$$y_{n,(1)} = y_{n,(0)}$$
$$hy'_{n,(1)} = hf(y_{n,(0)})$$

Additional iterations of (7.3) have no further effect so we can take $M = 1$ and get

$$\begin{aligned} y_n &= \sum_{i=1}^{k} (\alpha_i y_{n-i} + \beta_i h y'_{n-i}) \\ h y'_n &= h f(y_n) \end{aligned} \tag{7.4}$$

This is called an *explicit multistep method* because it provides an explicit way of computing y_n and hy'_n from the values of y and its derivative at preceding points. It requires only a vector inner product and one evaluation of f for each step.

The Adams-Bashforth method [Bashforth and Adams (1883)] is a particular example of this method.

We will derive it in three different ways to form a background for a later discussion of more general methods. The simplest derivation is by integration as follows. Integrate

$$y' = f(y, t) = f(t)$$

to get

$$\int_{t_{n-1}}^{t_n} y'\, dt = \int_{t_{n-1}}^{t_n} f(t)\, dt$$

or

$$y(t_n) = y(t_{n-1}) + \int_{t_{n-1}}^{t_n} f(t)\, dt \tag{7.5}$$

We can approximate $f(t)$ by an interpolating polynomial through a number of known values at $t = t_{n-1}, t_{n-2}, \ldots$, say $f_{n-1}, f_{n-2}, \ldots$. We will use the Newton backward difference formula. [See Hildebrand (1956) Section 4.3.] If $f(t)$ has a continuous kth derivative, $t_m = t_0 + mh$, $f_m = f(t_m)$, and backward differences are given by

$$\nabla^{q+1} f_m = \nabla^q f_m - \nabla^q f_{m-1}$$

where $\nabla^0 f_m = f_m$, then

$$\begin{aligned} f(t) = f_m &+ \frac{(t - t_m)\nabla f_m}{h} + (t - t_m)(t - t_{m-1})\frac{\nabla^2 f_m}{2!\, h^2} \\ &+ \cdots + (t - t_m)\ldots(t - t_{m-k+2})\frac{\nabla^{k-1} f_m}{(k-1)!\, h^{k-1}} \\ &+ (t - t_m)\ldots(t - t_{m-k+1})\frac{f^{(k)}(\xi)}{k!} \end{aligned} \tag{7.6}$$

where $f^{(k)}(\xi)$ is the kth derivative of f evaluated at some point in an interval containing t, t_{m-k+1}, and t_m. If we set $s = (t - t_{n-1})/h$ and $m = n - 1$, (7.6) becomes

$$\begin{aligned} f(t) = \binom{-s}{0} f_{n-1} - \binom{-s}{1}\nabla f_{n-1} + \cdots + (-1)^{k-1}\binom{-s}{k-1}\nabla^{k-1} f_{n-1} \\ + (-1)^k h^k \binom{-s}{k} f^{(k)}(\xi) \end{aligned}$$

where

$$\binom{s}{q} = \frac{s(s-1)\dots(s-q+1)}{q!} \quad \text{and} \quad \binom{s}{0} = 1$$

Substituting this in (7.5), we get

$$y(t_n) = y(t_{n-1}) + \int_{t_{n-1}}^{t_n} \left[\sum_{j=0}^{k-1} (-1)^j \binom{-s}{j} \nabla^j f_{n-1} + (-1)^k h^k \binom{-s}{k} y^{(k+1)}(\xi)\right] dt$$

or

$$y(t_n) = y(t_{n-1}) + h \sum_{j=0}^{k-1} \gamma_j \nabla^j f_{n-1} + (-1)^k h^{k+1} \int_0^1 \binom{-s}{k} y^{(k+1)}(\xi)\, ds \tag{7.7}$$

where

$$\gamma_j = (-1)^j \int_0^1 \binom{-s}{j} ds \tag{7.8}$$

If the last term in (7.7) is ignored, we are left with the k-step Adams-Bashforth formula

$$y_n = y_{n-1} + h \sum_{j=0}^{k-1} \gamma_j \nabla^j f_{n-1} \tag{7.9}$$

which expresses an approximation to the value of $y(t)$ at t_n in terms of the value at t_{n-1} and the backward differences of the derivatives at t_{n-1}. The backward differences can be expressed in terms of the values at preceding points by

$$\nabla^q f_{n-1} = \sum_{i=0}^{q} (-1)^i \binom{q}{i} f_{n-1-i}$$

so that (7.9) can be restated as

$$y_n = y_{n-1} + h \sum_{i=1}^{k} \beta_{ki} f_{n-i} \tag{7.10}$$

where

$$\beta_{ki} = (-1)^{i-1} \sum_{j=i-1}^{k-1} \gamma_j \binom{j}{i-1} \tag{7.11}$$

Equation (7.10) expresses y_n in terms of information at $t_{n-1}, t_{n-2}, \dots, t_{n-k}$, so it is seen to be a k-step method, although we could also call it a $(k+1)$-value method since $(k+1)$ items of information about the nature of the solution are used in the computation. It corresponds to (7.4) when we set $\alpha_1 = 1, \alpha_2 = \alpha_3 = \cdots = \alpha_k = 0$, and $\beta_i = \beta_{ki}$. Formula (7.9) could have been used in place of (7.10) by constructing a table of differences of f. The methods are equivalent, the only differences arising in the number of arithmetic operations and the round-off error. These topics will be discussed in Chapter 9.

Later we will formally define the local truncation error to be the difference between the true solution and the solution provided by one step of the method starting from exact values. As in one-step methods, the order r will be the integer such that the local truncation error is $0(h^{r+1})$. We will see that the order of a k-step Adams-Bashforth method is k since the term neglected in going from (7.7) to (7.9) is $0(h^{k+1})$. Notice, however, that only one evaluation of f is required per time step regardless of k. This is in contrast to the higher order one-step methods such as the R-K methods.

EXAMPLE

From (7.8) we see that

$$\gamma_0 = 1$$

$$\gamma_1 = \tfrac{1}{2}$$

so that a second order formula is

$$y_n = y_{n-1} + h(f_{n-1} + \tfrac{1}{2}\nabla f_{n-1})$$

or

$$y_n = y_{n-1} + h(\tfrac{3}{2}f_{n-1} - \tfrac{1}{2}f_{n-2}) \tag{7.12}$$

The equation $y' = -y$ was integrated from $y(0) = 1$ to $t = 1$ with $h = 2^{-k}$, $k = 1, 2, \ldots, 7$ by the FORTRAN program below. The results are shown in Table 7.1.

```
      WRITE (6, 4)
      DO 2 K = 1, 7
      H = 2.0**(-K)
      N = 2**K - 1
      Y = EXP(-H)
      FOLD = -H
      DO 1 I = 1, N
      F = - H*Y
      Y = Y + 1.5*F - 0.5*FOLD
1     FOLD = F
      ERROR = EXP(- 1.0) - Y
      ERRBYH = ERROR/H**2
2     WRITE (6, 3) H, Y, ERROR, ERRBYH
      RETURN
3     FORMAT (4E20.5)
4     FORMAT ('1', 13X, 'H', 19X, 'Y', 17X, 'ERROR', 13X, 'ERROR/H**2'/)
      END
```

Table 7.1 INTEGRATION OF $y' = -y$ BY SECOND ORDER ADAMS-BASHFORTH METHOD

H	*Y*	*ERROR*	*ERROR/H**2*
0.50000E 00	0.40163E 00	−0.33753E–01	−0.13501E 00
0.25000E 00	0.37628E 00	−0.83983E–02	−0.13437E 00
0.12500E 00	0.37014E 00	−0.22579E–02	−0.14451E 00
0.62500E–01	0.36846E 00	−0.58228E–03	−0.14906E 00
0.31250E–01	0.36803E 00	−0.14663E–03	−0.15015E 00
0.15625E–01	0.36791E 00	−0.34988E–04	−0.14331E 00
0.78125E–02	0.36788E 00	−0.47684E–05	−0.78125E–01

It can be seen from the last column that the error in the answer is $0(h^2)$. (The sudden reduction for the final h is due to fortunate cancellation of round-off errors which are about 10^{-6} as this calculation was done in single precision on an IBM 360.)

7.2.1 Generating Functions for the Coefficients

Equations (7.8) and (7.11) can be used to determine the coefficients γ_j and β_{kj} but they do not give the most convenient form for this. The method of *generating functions* is frequently the most convenient. Define the function $G(t)$ by

$$G(t) = \sum_{j=0}^{\infty} \gamma_j t^j$$

The summation is absolutely convergent for $|t| < 1$ since γ_j is bounded by 1 from (7.8). Consequently,

$$\begin{aligned} G(t) &= \sum_{j=0}^{\infty} (-1)^j \int_0^1 \binom{-s}{j} ds\, (t)^j \\ &= \int_0^1 \sum_{j=0}^{\infty} (-t)^j \binom{-s}{j} ds \\ &= \int_0^1 (1-t)^{-s}\, ds \\ &= -\frac{1}{\log(1-t)} [(1-t)^{-s}]_0^1 \\ &= -\frac{t}{(1-t)\log(1-t)} \end{aligned}$$

Therefore,

$$-\frac{\log(1-t)}{t} G(t) = \frac{1}{1-t}$$

or

$$(1 + \tfrac{1}{2}t + \tfrac{1}{3}t^2 + \cdots)(\gamma_0 + \gamma_1 t + \gamma_2 t^2 + \cdots) = 1 + t + t^2 + \cdots$$

Equating the coefficients of t^m, we get

$$\gamma_m + \frac{1}{2}\gamma_{m-1} + \frac{1}{3}\gamma_{m-2} + \cdots + \frac{1}{m+1}\gamma_0 = 1$$

which provides a recursive formula for the coefficients. Using this, we find the following values of γ_m:

Table 7.2 COEFFICIENTS FOR ADAMS-BASHFORTH METHODS

m	0	1	2	3	4	5
γ_m	1	$\frac{1}{2}$	$\frac{5}{12}$	$\frac{3}{8}$	$\frac{251}{720}$	$\frac{95}{288}$

Applying (7.11) to these, we get the following values for β_{kj}:

Table 7.3 COEFFICIENTS FOR ADAMS-BASHFORTH METHODS

i	1	2	3	4	5	6
β_{1i}	1					
$2\beta_{2i}$	3	-1				
$12\beta_{3i}$	23	-16	5			
$24\beta_{4i}$	55	-59	37	-9		
$720\beta_{5i}$	1901	-2774	2616	-1274	251	
$1440\beta_{6i}$	4277	-7923	9982	-7298	2877	-475

It should be noted that most of the β_{ki} exceed 1, which will have the effect of amplifying any round-off errors.

7.2.2 Two Other Techniques for Deriving the Adams-Bashforth Methods

One alternate technique for deriving methods is that of undetermined coefficients. A form for the method is assumed, say

$$y_n = \alpha_1 y_{n-1} + h\beta_1 f_{n-1} + h\beta_2 f_{n-2} \tag{7.13}$$

and we choose the α and β so that the value of y_n computed from (7.13) using correct values for y_{n-1}, f_{n-1}, and f_{n-2} differ as little as possible from the true solution. In this case we would like the error to be $0(h^3)$, so we expand $y(t_{n-q})$ and $y'(t_{n-q})$ by Taylor's series with an $0(h^3)$ remainder term to get

$$\begin{aligned} y(t_n) = {} & \alpha_1\left[y(t_n) - hy'(t_n) + \frac{h^2}{2}y''(t_n) + \frac{h^3}{2}\int_0^{-1}(1+\tau)^2 y^{(3)}(t_n+\tau h)\,d\tau\right] \\ & + \beta_1 h\left[y'(t_n) - hy''(t_n) - h^2\int_0^{-1}(1+\tau)y^{(3)}(t_n+\tau h)\,d\tau\right] \\ & + \beta_2 h\left[y'(t_n) - 2hy''(t_n) - h^2\int_0^{-2}(2+\tau)y^{(3)}(t_n+\tau h)\,d\tau\right] \end{aligned} \tag{7.14}$$

Equating terms in h^0, h^1, and h^2, we get

$$\begin{bmatrix} 1 \\ 0 \\ 0 \end{bmatrix} = \begin{bmatrix} 1 & 0 & 0 \\ -1 & 1 & 1 \\ \frac{1}{2} & -1 & -2 \end{bmatrix} \begin{bmatrix} \alpha_1 \\ \beta_1 \\ \beta_2 \end{bmatrix}$$

which has the solution

$$\alpha_1 = 1, \qquad \beta_1 = \tfrac{3}{2}, \qquad \beta_2 = -\tfrac{1}{2}$$

As expected, this leads to the Adams-Bashforth method of order two. A still simpler way of deriving these equations is to require that the method be exact for all polynomials of degree two (in this case) or less. Thus, we can substitute the functions $y = 1, s$, and s^2, where $s = (t - t_n)/h$. This is evidently equivalent to requiring that the first three terms of the Taylor's series vanish, and also leads to the same equations.

7.2.3 Truncation Error in the Adams-Bashforth Methods

Why, then, do we consider different ways of deriving these formulas? The last way is the simplest method for deriving the equations. The method may be used to find the coefficients for the general method of the form

$$y_n = \sum_{i=1}^{k} \alpha_i y_{n-i} + h \sum_{i=1}^{k} \beta_i f_{n-i} \tag{7.15}$$

and is probably the simplest way of generating the linear equations that α and β must satisfy. (Later we will see that other considerations may restrict our freedom in the choice of the α and β coefficients.)

The second method makes it possible to put a bound on the local truncation error of the method. It can be applied to a general method like (7.15). If the integral form of the remainder theorem is used in the Taylor's series method, an exact form of the remainder can be obtained. From (7.14), for example, we have

$$\begin{aligned} y(t_n) - \alpha_1 y(t_{n-1}) - h[\beta_1 y'(t_{n-1}) + \beta_2 y'(t_{n-2})] \\ = \frac{h^3}{2}\left[\int_0^{-1} (\alpha_1(1+\tau)^2 - 2\beta_1(1+\tau))y^{(3)}(t_n + \tau h)\, d\tau \right. \\ \left. - \int_0^{-2} 2\beta_2(2+\tau)y^{(3)}(t_n + \tau h)\, d\tau \right] \end{aligned}$$

Writing the left-hand side as $L_n(y(t_n))$, we have

$$|L_h(y(t_n))| = \frac{h^3}{2}\left|\int_0^{-2} G(\tau) y^{(3)}(t_n + \tau h)\, d\tau\right|$$

where

$$G(\tau) = \begin{cases} \alpha_1(1+\tau)^2 - 2\beta_1(1+\tau) - 2\beta_2(2+\tau) & 0 \geq \tau \geq -1 \\ -2\beta_2(2+\tau) & -1 > \tau \geq -2 \end{cases}$$

Therefore, the local truncation error is bounded by

$$\begin{aligned} |L_h(y(t_n))| &\leq -\frac{h^3}{2}\int_0^{-2} |G(\tau)|\,|y^{(3)}(t_n + h\tau)|\, d\tau \\ &\leq -\frac{h^3}{2}\left[\int_0^{-2} |G(\tau)|\, d\tau\right] \max_{t_n \geq t \geq t_{n-2}} |y^{(3)}(t)| \end{aligned}$$

or

$$|L_h(y(t_n))| \leq Mh^3\, |y^{(3)}(\xi)| \tag{7.16}$$

where

$$M = -\tfrac{1}{2}\int_0^{-2} |G(\tau)|\, d\tau$$

For the Adams-Bashforth two-step method,

$$G(\tau) = \tau^2 \qquad 0 \geq \tau \geq -1$$

$$G(\tau) = 2 + \tau \qquad -1 > \tau \geq -2$$

Therefore,

$$M = -\tfrac{1}{2}\int_0^{-2} |G(\tau)|\, d\tau = \tfrac{5}{6} \tag{7.17}$$

The first method for the derivation of the Adams-Bashforth method will give us this bound directly. Referring to the remainder form of the Newton backward difference formula used in (7.7), we get

$$L_n(y(t_n)) = (-1)^k h^{k+1} \int_0^1 \binom{-s}{k} y^{(k+1)}(\xi)\, ds$$

where ξ is a continuous function of s. Since $\binom{-s}{k}$ does not change sign in $[0, 1]$ we see that $L_n(y(t_n))$ takes the form

$$\begin{aligned}(-1)^k h^{k+1} \int_0^1 \binom{-s}{k} y^{(k+1)}(\xi(s))\, ds &\\ &= (-1)^k h^{k+1} y^{(k+1)}(\xi) \int_0^1 \binom{-s}{k} ds \\ &= \gamma_k h^{k+1} y^{(k+1)}(\xi) \qquad \text{where } \xi \in (t_{n-1}, t_n)\end{aligned} \tag{7.18}$$

by the second mean value theorem. This is an exact expression for the error which can be converted into the bound (7.16). We can always get an error bound similar to (7.16) for the general equation of the form (7.15). Unfortunately, an exact expression for the error involving a derivative as in (7.18) can only be given for certain methods. However, we can always get an error term of the form

$$C_{k+1} h^{k+1} y_n^{(k+1)} + 0(h^{k+2})$$

by carrying one additional term in the Taylor's series and a higher order remainder term. C_{k+1} can easily be found in the general case from

$$C_{k+1} = \frac{L_h(t^{k+1})}{h^{k+1}(k+1)!} \tag{7.19}$$

7.3 IMPLICIT MULTISTEP METHODS— THE ADAMS-MOULTON METHOD

Let us take the vector **c** in Eqs. (7.2) and (7.3) to be $[\beta_0^*, 0, \ldots, 0, 1, 0, \ldots, 0]^T$ where the 1 appears in the kth position. From (7.2)

$$\begin{aligned} y_{n,(1)} &= y_{n,(0)} + \beta_0^*(hf(y_{n,(0)}) - hy'_{n,(0)}) \\ &= \sum_{i=1}^{k} [(\alpha_i - \beta_0^*\gamma_i)y_{n-i} + (\beta_i - \beta_0^*\delta_i)hy'_{n-i}] + \beta_0^* hf(y_{n,(0)}) \end{aligned}$$

or

$$y_{n,(1)} = \sum_{i=1}^{k} (\alpha_i^* y_{n-i} + \beta_i^* h y'_{n-i}) + \beta_0^* h f(y_{n,(0)}) \tag{7.20}$$

where

$$\alpha_i^* = \alpha_i - \beta_0^* \gamma_i \quad \text{and} \quad \beta_i^* = \beta_i - \beta_0^* \delta_i$$

Then (7.3) and (7.20) imply that

$$\begin{aligned} hy'_{n,(m+1)} &= hf(y_{n,(m)}) \\ y_{n,(m+1)} &= y_{n,(m)} + \beta_0^*(hf(y_{n,(m)}) - hy'_{n,(m)}) \\ &= y_{n,(1)} + \beta_0^*(hf(y_{n,(m)}) - hy'_{n,(1)}) \\ &= \sum_{i=1}^{k} (\alpha_i^* y_{n-i} + \beta_i^* h y'_{n-i}) + \beta_0^* h f(y_{n,(m)}) \end{aligned} \tag{7.21}$$

If (7.21) is iterated to convergence, we get

$$\begin{aligned} y_n &= \sum_{i=1}^{k} (\alpha_i^* y_{n-i} + \beta_i^* h y'_{n-i}) + \beta_0^* h f(y_n) \\ hy'_n &= hf(y_n) \end{aligned} \tag{7.22}$$

Equations (7.22) define an *implicit multistep method.* It is implicit because, in general, Eq. (7.22) is a nonlinear equation as it involves the function f and must be solved for y_n. The predictor-corrector processes (7.1) and (7.3) are one way of solving it. The Adams-Moulton method [Moulton (1926)] is a particular example of an implicit method.

It can also be derived in a number of ways. The simplest way is to set $m = n$ in (7.6) and substitute into (7.5) to get

$$y(t_n) = y(t_{n-1}) + \int_{t_{n-1}}^{t_n} \left[\sum_{j=0}^{k-1} (-1)^j \binom{-s+1}{j} \nabla^j f_n + (-1)^k h^k \binom{-s+1}{k} y^{(k+1)}(\xi) \right] ds$$

from which we get the method

$$y_n = y_{n-1} + h \sum_{j=0}^{k-1} \gamma_j^* \nabla^j f_n \tag{7.23}$$

where

$$\gamma_j^* = (-1)^j \int_0^1 \binom{-s+1}{j} ds$$

The error term is

$$h^{k+1}(-1)^k \int_0^1 \binom{-s+1}{k} y^{(k+1)}(\xi)\, ds = \gamma_k^* h^{k+1} y^{(k+1)}(\xi)$$

by the second mean value theorem. By substituting for $\nabla^j f_n$ in terms of $f_n, f_{n-1}, f_{n-2}, \ldots,$ we get the form

$$y_n = y_{n-1} + h \sum_{i=0}^{k-1} \beta_{ki}^* f_{n-i}$$

where

$$\beta_{ki}^* = (-1)^j \sum_{j=i}^{k-1} \binom{j}{i} \gamma_j^*$$

Henrici (1962), Section 5.1.2, gives the following values for γ_j^* (which can be found by use of generating functions) and for β_{ki}^*:

Table 7.4 COEFFICIENTS FOR THE ADAMS-MOULTON METHOD

m	0	1	2	3	4	5
γ_m^*	1	$-\frac{1}{2}$	$-\frac{1}{12}$	$-\frac{1}{24}$	$-\frac{19}{720}$	$-\frac{3}{160}$

Table 7.5 COEFFICIENTS FOR THE ADAMS-MOULTON METHOD

i	0	1	2	3	4	5
β_{1i}^*	1					
$2\beta_{2i}^*$	1	1				
$12\beta_{3i}^*$	5	8	-1			
$24\beta_{4i}^*$	9	19	-5	1		
$720\beta_{5i}^*$	251	646	-264	106	-19	
$1440\beta_{6i}^*$	475	1427	-798	482	-173	27

There are two important differences to note between the Adams-Bashforth and the Adams-Moulton methods. The first is that the coefficients of the latter are smaller. This not only leads to smaller round-off errors, but more importantly, to smaller truncation errors at the same order, since γ_i^* is smaller than γ_i and the truncation errors are $\gamma_k h^{k+1} y^{(k+1)}$ and $\gamma_k^* h^{k+1} y^{(k+1)}$, respectively. The second difference is that for the same order the Adams-Moulton method uses information from fewer points. In other words, the k-step Adams-Bashforth method is of kth order while the k-step Adams-Moulton method is of $(k + 1)$th order. The reason for this is apparent when we consider the method of undetermined coefficients for deriving these formulas. We try the formula

$$y_n = \alpha_1^* y_{n-1} + h \sum_{i=0}^{k} \beta_i^* f_{n-i} \tag{7.24}$$

We have $(k + 2)$ unknowns, so we can hope to make the Taylor's series agree to h^{k+1} getting an order $k + 1$ Adams-Moulton method. For the Adams-Bashforth method we take $\beta_0^* = 0$, reducing the number of unknowns by one so we can only hope to achieve order k.

7.4 PREDICTOR-CORRECTOR MULTISTEP METHODS

The most common method for solving the implicit equation (7.22) is the predictor-corrector process (7.1), (7.2), and (7.3). These are

$$y_{n,(0)} = \sum_{i=1}^{k} (\alpha_i y_{n-i} + \beta_i h y'_{n-i})$$

and

$$y_{n,(m+1)} = \sum_{i=1}^{k} (\alpha_i^* y_{n-i} + \beta_i^* h y'_{n-i}) + \beta_0^* h f(y_{n,(m)}) \tag{7.25}$$

when multistep methods are used.

If we subtract (7.22) from (7.25), we get

$$\begin{aligned} y_{n,(m+1)} - y_n &= \beta_0^* h(f(y_{n,(m)}) - f(y_n)) \\ &= \beta_0^* h \frac{\partial f(\xi_n)}{\partial y}(y_{n,(m)} - y_n) \end{aligned}$$

provided that f has a continuous partial derivative with respect to y.

Hence, if h is chosen sufficiently small that $|h\beta_0(\partial f/\partial y)| < 1$ within the region of interest, $|y_{n,(m+1)} - y_n| < |y_{n,(m)} - y_n|$ and iteration (7.25) will converge to the solution of (7.22). In practice, the predictor $y_{n,(0)}$ is a good approximation, so very few (two or three) corrections are required.

The predictor formula is an explicit multistep method of the type discussed in Section 7.2. By its order we mean the order of approximation of $y_{n,(0)}$ to y_n when $\mathbf{y}_{n-1}$ contains exact values. The order of the corrector formula is the order of the method when the corrector is iterated to convergence, that is, it is the order of method (7.22).

The predictor and corrector formulas need not be of the same order. Each additional corrector application will increase the order of solution by one until the corrector is reached. Thus, if the predictor has order q and the corrector order r, we have

$$\begin{aligned} y(t_n) &= \sum_{i=1}^{k} [\alpha_i y(t_{n-i}) + h\beta_i y'(t_{n-i})] + 0(h^{q+1}) \\ &= y_{n,(0)} + 0(h^{q+1}) \qquad \text{(predictor formula)} \end{aligned}$$

$$\begin{aligned} y(t_n) &= \sum_{i=1}^{k} [\alpha_i^* y(t_{n-i}) + \beta_i^* h y'(t_{n-i})] + \beta_0^* h f(y(t_n)) + 0(h^{r+1}) \\ &= y_{n,(1)} + h\beta_0^*[f(y(t_n)) - f(y_{n,(0)})] + 0(h^{r+1}) \\ &= y_{n,(1)} + 0(h^{q+2}) + 0(h^{r+1}) \qquad \text{(first application of corrector formula)} \end{aligned}$$

and for the mth application of the corrector,

$$y(t_n) = y_{n,(m)} + 0(h^{q+m+1}) + 0(h^{r+1})$$

However, in practice we do not wish to correct many times, because each additional corrector step requires an additional function evaluation for the derivative. Usually q is taken to be $r - 1$ or r and about two corrector steps are used. [See Hull and Creemer (1963).]

There are three separate processes occuring: the prediction step which we will call P, the evaluation of the derivative based on the latest value of y called E, and the correction step called C. There is a choice of whether to finish with a C or an E step. There is some evidence that a final evaluation is superior. This will be discussed in Chapters 8 and 9. An m iteration predictor-corrector method is referred to as a $P(EC)^m$ method if it ends with a correction, or a $P(EC)^mE$ method if it ends with an evaluation. The whole class of methods is called the class of PC methods.

PROBLEMS

1. Justify the statement preceding Eq. (7.19).
2. Derive the generating function for the coefficients of the Adams-Moulton method.
3. (a) Derive the general form of the method

$$y_n = y_{n-1} + h \sum_{i=0}^{k} \bar{\gamma}_i \nabla^i f_{n-2}$$

by finding a generating function for the $\bar{\gamma}_i$.
(b) What is the truncation error?

4. (a) Derive the general form and the truncation error of the method

$$y_n = y_{n-1} + h \sum_{i=0}^{k} \tilde{\gamma}_i \nabla^i f_{n+1}$$

(b) Can you suggest any ways to use this method?

5. Integrate by computer the differential equation

$$y' = t^3 - y + 3t^2$$

from $y(0) = 0$ to $t = 1$ using $h = 2^{-k}$, $k = 1, 2, \ldots, 5$. Use each of the following methods:
(a) One-, two-, and three-step Adams-Bashforth method.
(b) One- and two-step $P(EC)^1$ Adams-Bashforth-Moulton method (corrector order one greater than predictor).
(c) One- and two-step Adams-Moulton method. (Solve for y_n explicitly since f is linear in y.)
Print the errors at $t = 1$ and comment on them.

8 GENERAL MULTISTEP METHODS, ORDER, AND STABILITY

In the previous chapter we examined two particular multistep methods that are very effective for integrating many differential equations. In this chapter we are going to examine general k-step methods of the form

$$\sum_{i=0}^{k} (\alpha_i y_{n-i} + h\beta_i f_{n-i}) = 0 \tag{8.1}$$

and discuss the important subject of stability.

One way in which we were able to derive the coefficients of Adams method was by requiring that they are exact for polynomials of order $\leq r$. There are $2k + 2$ unknowns in (8.1). There is an arbitrary normalizing factor so we can set $\alpha_0 = -1$, leaving $2k + 1$ unknowns. Consequently, we expect to be able to choose the α and β so that this method is exact for polynomials of order up to $2k$. This is possible.† However, we will see later that such methods

†Suppose that we know $y_{n-k}, \ldots, y_n, y'_{n-k}, \ldots, y'_n$. There exists a unique polynomial of degree $2k + 1$ passing through this function, and agreeing with its derivatives at $t = t_{n-k}, \ldots, t_n$. It is called the Hermite interpolation formula, and is given in the following way. Let

$$\phi(\xi) = (\xi - t_{n-k}) \cdots (\xi - t_n)$$

and

$$\phi_i(\xi) = \frac{\phi(\xi)}{\xi - t_{n-i}}$$

Then

$$P(\xi) = \sum_{i=0}^{k} \frac{\phi_i(\xi)}{\phi_i^2(t_{n-i})}\left[\phi_i(\xi) y_{n-i} - \phi(\xi)\left(y'_{n-i} - 2\sum_{j,\, j\neq i} \frac{y_{n-i}}{t_{n-i} - t_{n-j}}\right)\right]$$

is the required polynomial, as can be seen by substitution. If the coefficient of ξ^{2k+1} vanishes, then we have a polynomial of degree $2k$ passing through the points y_{n-i} and their are never useful for $k > 2$, and only marginally useful when $k = 2$. If we were only concerned with local truncation error and the problem had well-behaved derivatives, we would be tempted to use k-step methods of maximal order $2k$. However, we will see later that for $k > 2$ such methods cause the small truncation errors committed in one step to be unacceptably amplified in later steps due to instability. However there are stable k-step methods of order $k + 1$ (the Adams-Moulton method, for example) and order $k + 2$ if k is even.

We have been talking about making Eq. (8.1) exact for polynomials of various orders, and, in the last chapter, calling this the order of the method. In the next section we discuss the implications of the order of a multistep method and then turn to an analysis of the maximum order two-step method to understand the problems of stability. The chapter finishes with a look at the class of three-step methods to see the trade-offs possible between accuracy and stability.

8.1 THE ORDER OF A MULTISTEP METHOD

Define

$$L_h(y(t)) = \sum_{i=0}^{k} (\alpha_i y(t - hi) + h\beta_i y'(t - hi))$$

We saw in Chapter 7 that we could pick the α and β to make some terms in the Taylor's series for $L_h(y)$ vanish.

---

derivatives y'_{n-i}. The condition for this coefficient to vanish is a condition relating the values of y_n and y'_n to the other y_{n-i} and y'_{n-i}. The coefficient of ξ^{2k+1} vanishes if

$$-\sum_{i=0}^{k} \frac{1}{\phi_i^2(t_{n-i})}\left(y'_{n-i} - 2\sum_{j,j\neq i} \frac{y_{n-i}}{t_{n-i} - t_{n-j}}\right) = 0$$

This is equivalent to the left-hand side of (8.1) if we set

$$\alpha_i = \frac{2}{\phi_i^2(t_{n-i})} \sum_{j,j\neq i} \frac{1}{t_{n-i} - t_{n-j}}$$

and

$$\beta_i = -\frac{1}{\phi_i^2(t_{n-i})h}$$

For these values of α and β any polynomial of degree $2k$ or less satisfies (8.1). It is necessary to show that $\alpha_0 \neq 0$. This follows immediately because $\phi_0(t_n) \neq 0$ and $1/(t_n - t_{n-j}) > 0$ for $j \geq 1$. By multiplying both α and β by $h^{2k+1}(k!)^2$ and setting $t_{n-i} - t_{n-j} = h(j - i)$, we get

$$\alpha_i = \begin{cases} 2\binom{k}{i}^2 \sum_{j=i+1}^{k-i} \frac{1}{j} & i \leq \frac{k}{2} \\ -2\binom{k}{i}^2 \sum_{j=k-i+1}^{i} \frac{1}{j} & i \geq \frac{k}{2} \end{cases} \tag{8.2}$$

$$\beta_i = -\binom{k}{i}^2$$

DEFINITION 8.1

The order of the operator L_h is the largest r such that if $y(t)$ has a continuous $(r+1)$th derivative, then

$$L_h(y(t)) = 0(h^{r+1}) \tag{8.3}$$

If we assume a continuous $(r+2)$th derivative for y, then we can substitute the Taylor's series for y and y' with $0(h^{r+2})$ remainders. If the terms in $h^0, h^2, \ldots, h^{r+1}$ are gathered together, we will get

$$L_h(y(t)) = \sum_{q=0}^{r+1} C_q h^q y^{(q)}(t) + 0(h^{r+2})$$

where

$$C_q = \begin{cases} \sum_{i=0}^{k} \alpha_i & q = 0 \\ \sum_{i=0}^{k} \left[\frac{(-i)^q}{q!}\alpha_i + \frac{(-i)^{q-1}}{(q-1)!}\beta_i \right] & q > 0 \end{cases} \tag{8.4}$$

The linear equations $C_q = 0$, $q \le r$, are the equations which determine an rth order method. We note that C_{r+1} depends only on the coefficients of the method, not on y or the point of expansion t. C_{r+1} is the truncation error coefficient of L_h. Equation (8.1), and hence L_h, can be scaled by an arbitrary constant. It will be shown in Chapter 10 that the amount of truncation error introduced at each step is

$$\frac{C_{r+1}}{\sum_{i=0}^{k} \beta_i} h^{r+1} y^{(r+1)} + 0(h^{r+2})$$

Therefore the natural normalization is to make

$$\sum_{i=0}^{k} \beta_i = 1 \tag{8.5}$$

We will assume that this has been done when we discuss the error coefficient C_{r+1}.

If a method of order r is used to integrate an equation whose solution is a polynomial of degree no greater than r, the solution will be exact except for round-off errors, since the truncation error is $\sum_{r+1}^{\infty} C_q h^q y^{(q)}(t) = 0$. However, a glance forward at Table 8.6 will show that round-off errors may wreck the solution in some cases. The order simply tells us how the local truncation error behaves as a function of h. The coefficient C_{r+1} gives us a way of comparing the error in two different methods of the same order.

The order r and the error coefficient C_{r+1} can be expressed in a more convenient way as follows. Define the polynomials

$$\begin{aligned} \rho(\xi) &= \sum_{i=0}^{k} \alpha_i \xi^{k-i} \\ \sigma(\xi) &= \sum_{i=0}^{k} \beta_i \xi^{k-i} \end{aligned} \tag{8.6}$$

The maximum degree of ρ and σ is the step number of the method. Normally degree $(\sigma) \leq$ degree (ρ). If strict inequality holds, the method is explicit. Consider the function $y(t) = e^{\lambda t}$.

$$\begin{aligned} L_h(y(t)) &= \sum_{i=0}^{k} (\alpha_i + \lambda h \beta_i) e^{\lambda(t-hi)} \\ &= \sum_{i=0}^{k} e^{\lambda(t-hk)}(\alpha_i + h\lambda\beta_i)(e^{h\lambda})^{k-i} \\ &= e^{\lambda(t-hk)}[\rho(e^{h\lambda}) + h\lambda\sigma(e^{h\lambda})] \end{aligned} \tag{8.7}$$

Since the method is of order r and since $e^{hk} = 1 + 0(h)$,

$$\begin{aligned} L_h(y(t)) &= C_{r+1} h^{r+1} y^{(r+1)} + 0(h^{r+2}) \\ &= C_{r+1}(h\lambda)^{r+1} e^{\lambda t} + 0(h^{r+2}) \\ &= C_{r+1}(h\lambda)^{r+1} e^{\lambda(t-hk)} + 0(h^{r+2}) \end{aligned} \tag{8.8}$$

Therefore, from (8.7) and (8.8),

$$\rho(e^{h\lambda}) + h\lambda\sigma(e^{h\lambda}) = C_{r+1}(h\lambda)^{r+1} + 0(h^{r+2}) \tag{8.9}$$

Write $h\lambda = \log(1 + z)$ and note that $h\lambda = z + 0(z^2)$. Then

$$\rho(1 + z) + \log(1 + z)\sigma(1 + z) = C_{r+1} z^{r+1} + 0(z^{r+2}) \tag{8.10}$$

Consequently, Eq. (8.10) is a necessary condition for a method to have order r. It can also be seen to be a sufficient condition by noting that it implies that $L_h(e^{\lambda t}) = \sum_{q=0}^{\infty} C_q (h\lambda)^q e^{\lambda t} = 0(h^{r+1})$, which in turn implies that $C_0 = C_1 = \cdots = C_r = 0$.

If the equation is normalized so that (8.5) is true, $\sigma(1) = 1$. Expanding (8.10) as a power series in z, we get

$$\begin{aligned} &\rho(1) + z\rho'(1) + 0(z^2) + [z + 0(z^2)][\sigma(1) + 0(z)] \\ &\qquad = \rho(1) + z[\rho'(1) + \sigma(1)] + 0(z^2) = C_{r+1} z^{r+1} + 0(z^{r+2}) \end{aligned}$$

from which we see that order $\geq 0 \Rightarrow \rho(1) = 0$ and order $\geq 1 \Rightarrow \rho'(1) + \sigma(1) = 0$.

8.1.1 Determination of α if β is Given and Vice Versa

If the polynomial σ is given, Eq. (8.10) shows how a unique polynomial $\rho(\xi)$ of degree k can be found such that the method has order $\geq k$. This is obtained by setting $r(z)$ to be the terms in z^j $(0 \leq j \leq k)$ in the expansion of $-\log(1 + z)\sigma(1 + z)$. $\rho(\xi)$ is then given by $\rho(\xi) = r(\xi - 1)$.

EXAMPLE

If $\sigma(\xi) = \frac{3}{2}\xi - \frac{1}{2}$ and $k = 2$, we get

$$\begin{aligned} \rho(1 + z) &= -\log(1 + z)\left(\frac{3}{2}(1 + z) - \frac{1}{2}\right) + 0(z^3) \\ &= -\left(z - \frac{z^2}{2}\right)\left(\frac{3}{2}z + 1\right) + 0(z^3) \\ &= -z - z^2 = -(1 + z)^2 + (z + 1) \end{aligned}$$

Hence $\rho(\xi) = -\xi^2 + \xi$, which gives the second order Adams-Bashforth method

$$-y_n + y_{n-1} + \frac{3h}{2}y'_{n-1} - \frac{h}{2}y'_{n-2} = 0$$

If, on the other hand, $\rho(\xi)$ is given, there exists a $\sigma(\xi)$ of degree k such that the method is of order $\geq k + 1$. We find this by dividing (8.10) with $r = k + 1$ by $\log(1 + z)$ to get

$$\sigma(1 + z) = -\frac{z}{\log(1 + z)}\frac{\rho(1 + z)}{z} + C_{k+2}z^{k+1} + 0(z^{k+2})$$

Since $z/\log(1 + z)$ is analytic at $z = 0$, $\rho(1 + z)/z$ must be analytic near $z = 0$ if $k \geq 0$, that is,

$$\rho(1) = \sum_{i=0}^{k} \alpha_i = 0$$

so we can find σ by setting $s(z)$ equal to terms in z^j $(0 \leq j \leq k - 1)$ in the expansion of

$$-\frac{z}{\log(1 + z)}\frac{\rho(1 + z)}{z}$$

and writing $\sigma(\xi) = s(\xi - 1)$.

EXAMPLE

If $\rho(\xi) = -\xi^2 + \xi$ and $k = 2$, we get

$$\begin{aligned}
\sigma(1 + z) &= \frac{(1 + z)^2 - (1 + z)}{\log(1 + z)} + 0(z^3) \\
&= \frac{1 + z}{1 - \frac{z}{2} + \frac{z^2}{3}} + 0(z^3) \\
&= (1 + z)\left(1 + \frac{z}{2} - \frac{z^2}{3} + \frac{z^2}{4}\right) + 0(z^3) \\
&= 1 + \frac{3z}{2} + \frac{5z^2}{12} \\
&= \frac{5}{12}(1 + z)^2 + \frac{2}{3}(1 + z) - \frac{1}{12}
\end{aligned}$$

Hence, $\sigma(\xi) = \frac{5}{12}\xi^2 + \frac{2}{3}\xi - \frac{1}{12}$, which gives the third order Adams-Moulton method:

$$y_n = y_{n-1} + \tfrac{5}{12}hy'_n + \tfrac{2}{3}hy'_{n-1} - \tfrac{1}{12}hy'_{n-2}$$

8.1.2 The Principle Root of a Method

If we apply a multistep method to $y' = \lambda y$, we get the recurrence relation

$$\sum_{i=0}^{k} (\alpha_i + h\lambda\beta_i)y_{n-i} = 0$$

The general solution of equations of this form will be discussed in detail in Chapter 10. Here we just need to note that we can find a solution of the form $y_n = A\xi^n$ if ξ is a root of the equation

$$\sum (\alpha_i + h\lambda\beta_i)\xi^{k-i} = \rho(\xi) + h\lambda\sigma(\xi) = 0$$

Since the solution of $y' = \lambda y$ is $y = Ae^{\lambda t} = A(e^{h\lambda})^n$, we expect one root to approximate $e^{h\lambda}$ so that y_n can approximate $y(t_n)$. We call that root the *principal root* ξ_1. Other roots ξ_i, which also give rise to solutions $A\xi_i^n$, are called *extraneous roots*. The size of the extraneous roots affects the stability of the method, which will be discussed by an example in the next section.

We can show that if $\rho'(1) \neq 0$, the principal root is

$$\xi_1 = e^{h\lambda} - \frac{C_{r+1}}{\rho'(1)}(h\lambda)^{r+1} + 0(h^{r+2}) \tag{8.11}$$

Let us assume a root of the form $e^{h\lambda} + \gamma$, where γ is to be determined. We have

$$\begin{aligned} 0 &= \rho(e^{h\lambda} + \gamma) + h\lambda\sigma(e^{h\lambda} + \gamma) \\ &= \rho(e^{h\lambda}) + \gamma\rho'(e^{h\lambda}) + h\lambda\sigma(e^{h\lambda}) + 0(\gamma^2 + \gamma h) \end{aligned}$$

In view of (8.9) we can write this as

$$0 = C_{r+1}(h\lambda)^{r+1} + \gamma\rho'(e^{h\lambda}) + 0(h^{r+2}) + 0(\gamma^2 + \gamma h)$$

Since $\rho'(e^{h\lambda}) = \rho'(1 + 0(h)) = \rho'(1) + 0(h)$, it follows that one root has the form (8.11) if $\rho'(1) \neq 0$. If $r \geq 1$, then we showed at the end of Section 8.1 that $\rho'(1) + \sigma(1) = 0$, and we have already required that $\sigma(1) = 1$ for normalization; thus, we may finally write

$$\xi_1 = e^{h\lambda} + C_{r+1}(h\lambda)^{r+1} + 0(h^{r+2}) \tag{8.12}$$

8.2 MILNE'S METHOD

To get the two-step method of maximum possible order we require $C_0 = C_1 = C_2 = C_3 = C_4 = 0$ or, from (8.4),

$$\begin{aligned} \alpha_0 + \alpha_1 + \alpha_2 &= 0 \\ -\alpha_1 - 2\alpha_2 + \beta_0 + \beta_1 + \beta_2 &= 0 \\ \frac{\alpha_1}{2!} + \frac{4\alpha_2}{2!} - \beta_1 - 2\beta_2 &= 0 \\ -\frac{\alpha_1}{3!} - \frac{8\alpha_2}{3!} + \frac{\beta_1}{2!} + \frac{4\beta_2}{2!} &= 0 \\ \frac{\alpha_1}{4!} + \frac{16\alpha_2}{4!} - \frac{\beta_1}{3!} - \frac{8\beta_2}{3!} &= 0 \end{aligned}$$

These have the solution $\alpha_0 = -\alpha_2 = -1$, $\alpha_1 = 0$, $\beta_0 = \beta_2 = \frac{1}{3}$, $\beta_1 = \frac{4}{3}$, which leads to

$$y_n = y_{n-2} + \tfrac{1}{3}h(f_n + 4f_{n-1} + f_{n-2}) \tag{8.13}$$

Table 8.1 SOLUTION OF $y' = y$ BY MILNE'S METHOD

Time	y	*Error*
0.0	0.1000000000E 01	0.0
0.1	0.1105171204E 01	0.0
0.2	0.1221402168E 01	−0.95367E-06
0.3	0.1349857330E 01	−0.95367E-06
0.4	0.1491822243E 01	−0.19073E-05
0.5	0.1648717880E 01	−0.28610E-05
0.6	0.1822114944E 01	−0.38147E-05
0.7	0.2013747215E 01	−0.47684E-05
0.8	0.2225535393E 01	−0.47684E-05
0.9	0.2459595680E 01	−0.66757E-05
1.0	0.2718274117E 01	−0.66757E-05
2.0	0.7389023781E 01	0.95367E-05
3.0	0.2008541870E 02	0.10681E-03
4.0	0.5459768677E 02	0.47302E-03
5.0	0.1484117584E 03	0.19989E-02
6.0	0.4034238281E 03	0.65918E-02
7.0	0.1096617187E 04	0.21729E-01
8.0	0.2980913330E 04	0.74707E-01
9.0	0.8102945312E 04	0.23047E 00
10.0	0.2202604687E 05	0.71484E 00

This is known as *Milne's method* and can be seen to be similar to Simpson's rule for quadrature. In the example below, $y' = y$ is integrated from $y(0) = 1$ to $t = 10$ by Milne's method using step size $h = 0.1$. The answers and errors have been printed in Table 8.1 for $t = 0(0.1)1$ and $t = 2(1)10$. The accuracy is seen to be good for a relatively small amount of work. [The value of $y(0.1)$ was calculated from the exponential subroutine. Computations were performed in single precision on an IBM 360, giving about seven decimal digits.] In contrast, Table 8.2 shows similar results for $y' = -y$ from $y(0) = 1$ to $t = 10$ with the same step size. The final answer has only one decimal digit of accuracy remaining, in contrast to better than four decimal digits for $y' = y$. It is evident that this particular method is not so suitable for the second problem. To see if the choice of step size is particularly bad, the same problem has been integrated using $h = 10^{-i}$, $i = 1, 2, \ldots, 5$. The results are shown in Table 8.3.

As the step is reduced, the error certainly does not decrease like h^4, as might be expected. Evidently the rounding errors are significant, even for relatively few steps.

8.2.1 Stability of Milne's Method for $y' = \lambda y$

Evidently the errors in Milne's method grow when $\lambda = -1$ so we naturally suspect a stability problem. Let us examine the effect of a perturbation

Table 8.2 SOLUTION OF $y' = -y$ BY MILNE'S METHOD

Time	y	*Error*
0.0	0.1000000000E 01	0.0
0.1	0.9048374295E 00	0.0
0.2	0.8187311888E 00	0.35763E-06
0.3	0.7408185601E 00	0.23842E-06
0.4	0.6703208089E 00	0.65565E-06
0.5	0.6065312028E 00	0.41723E-06
0.6	0.5488125086E 00	0.77486E-06
0.7	0.4965859056E 00	0.47684E-06
0.8	0.4493299127E 00	0.83447E-06
0.9	0.4065702558E 00	0.47684E-06
1.0	0.3678804040E 00	0.83447E-06
2.0	0.1353359818E 00	−0.59605E-07
3.0	0.4978756607E-01	−0.70781E-07
4.0	0.1831618324E-01	0.23097E-06
5.0	0.6738595665E-02	0.49546E-06
6.0	0.2479620045E-02	0.79698E-06
7.0	0.9130770341E-03	0.11637E-05
8.0	0.3371220082E-03	0.16459E-05
9.0	0.1257221593E-03	0.23067E-05
10.0	0.4862506466E-04	0.32228E-05

Table 8.3 VALUE OF $y(10)$ FOR $y' = -y$, $y(0) = 1$ BY MILNE'S METHOD

h	y	*Error*
0.10000E 00	0.48625E-04	0.32251E-05
0.10000E-01	0.48211E-04	0.28114E-05
0.10000E-02	0.38639E-04	−0.67607E-05
0.10000E-03	0.32934E-04	0.37534E-04
0.10000E-04	0.57846E-04	0.12446E-04

in the numerical solution at one step on future steps. Setting $f = \lambda y$ in (8.13) and considering the difference e_n between two different solutions of (8.13), we get

$$e_n = e_{n-2} + \tfrac{1}{3}\lambda h(e_n + 4e_{n-1} + e_{n-2}) \tag{8.14}$$

This is a homogeneous second order linear difference equation for e_n. Equations of this sort can always be solved by looking for solutions of the form $e_n = \xi^n$. Substituting into (8.14), we get

$$(1 - \tfrac{1}{3}h\lambda)\xi^2 - \tfrac{4}{3}h\lambda\xi - (1 + \tfrac{1}{3}h\lambda) = 0 \tag{8.15}$$

Note that this is precisely $\rho(\xi) + h\lambda\sigma(\xi) = 0$. If this equation has two distinct roots ξ_1 and ξ_2, any e_n of the form $a\xi_1^n + b\xi_2^n$ can be seen to be a solution of (8.14). If $1 \pm \frac{1}{3}h\lambda \neq 0$, fixing any two consecutive e_i and e_{i+1} is sufficient to uniquely determine all other e_i. Thus, all solutions of (8.14) can be represented by $a\xi_1^n + b\xi_2^n$. (a and b can always be chosen so that $a\xi_1^i + b\xi_2^i = e_i$, $a\xi_1^{i+1} + b\xi_2^{i+1} = e_{i+1}$.) If $\xi_1 = \xi_2$, then the general solution of (8.14) can be written as $(a + bn)\xi_1^n$.

We see immediately that if either $|\xi_i| > 1$, the perturbation is growing exponentially, if $\xi_1 = \xi_2$ and $|\xi_1| = 1$, it is growing linearly, while if both $|\xi_i| < 1$, it is decreasing. Equation (8.15) can be solved for ξ as a power series in $h\lambda$ by the method of undetermined coefficients† to get

$$\begin{aligned}\xi_1 &= 1 + h\lambda + \frac{(h\lambda)^2}{2} + \frac{(h\lambda)^3}{6} + \frac{(h\lambda)^4}{24} + \frac{(h\lambda)^5}{72} + 0(h^6)\\ &= e^{h\lambda} + \frac{1}{180}(h\lambda)^5 + 0(h^6)\\ \xi_2 &= -\left[1 - \frac{h\lambda}{3} + \frac{(h\lambda)^2}{18} + \frac{5(h\lambda)^3}{54} + 0(h^4)\right]\\ &= -e^{-h\lambda/3} + 0(h^3)\end{aligned}$$

The agreement of the root ξ_1 to $e^{h\lambda}$ to $0(h^5)$ is expected because the computed solution will follow the true solution $e^{h\lambda}$ to $0(h^5)$ in a fourth order method. Any errors introduced may have the effect of putting the result onto another member of the family of solutions as shown in Figure 1.1; consequently, we expect some part of an introduced error to behave like $e^{\lambda t} = (e^{h\lambda})^n \cong \xi_1^n$.

The presence of a second root ξ_2 is a phenomenon that does not happen in one-step methods and is due to the fact that additional information is being saved. For $\lambda > 0$ the second component of a solution $\xi_2^n \cong (-1)^n e^{-\lambda t/3}$ is decaying and presents no problems. If, however, $\lambda < 0$, the second component dominates both the first and the true solution. From the definition of absolute stability in Chapter 1 we see that Milne's method is absolutely unstable unless Re $(\lambda) = 0$. However, when $\lambda > 0$ this instability is due to the instability of the problem itself and is not serious, whereas if $\lambda < 0$ the instability is introduced by the second component due to the method.

8.3 STABILITY OF GENERAL MULTISTEP METHODS

In Chapter 1 stability was defined as the boundedness of the effects of a perturbation in the starting values for all $h \leq h_0$. If we consider perturbations to the numerical solution of the linear problem $y' = \lambda y + f(t)$, Eq. (8.1)

†Set $\xi_1 = a_0 + a_1 h\lambda + a_2(h\lambda)^2 + \cdots$, $\xi_2 = b_0 + b_1 h\lambda + b_2(h\lambda)^2 + \cdots$ and equate the coefficients of $(h\lambda)^n$ in

$$\left(1 - \frac{h\lambda}{3}\right)(\xi - \xi_1)(\xi - \xi_2) = \left(1 - \frac{h\lambda}{3}\right)\xi^2 - \frac{4h\lambda}{3}\xi + \left(1 + \frac{h\lambda}{3}\right)$$

leads to an error equation of the form

$$\sum_{i=0}^{k} (\alpha_i + h\lambda\beta_i)e_{n-i} = 0 \tag{8.16}$$

Once again we look for solutions of the form $e_n = \xi^n$ and find that

$$\sum_{i=0}^{k} (\alpha_i + h\lambda\beta_i)\xi^{k-i} = 0$$

The ξ are then the roots of

$$\rho(\xi) + h\lambda\sigma(\xi) = 0 \tag{8.17}$$

If all of the roots ξ_j, $j = 1, \ldots, k$ are distinct, the general solution of (8.16) can be written as

$$e_n = \sum_{j=1}^{k} \gamma_j \xi_j^n \tag{8.18}$$

If some of the roots are equal, then this must be modified. For example, if ξ_i is an m-fold root, then the term $(\gamma_i + \gamma_{i+1}n + \gamma_{i+2}n^2 + \cdots + \gamma_{i+m-1}n^{m-1})\xi_i^m$ will occur.

It is evident that if any $|\xi_j| > 1$ the perturbation is growing, and we say that the method is absolutely unstable for that $h\lambda$. If we are interested in an unbounded range of t, then the problem will be unstable.

The solutions of (8.17) are functions of $h\lambda$, that is, $\xi_i = \xi_i(h\lambda)$. The roots of a polynomial are continuous functions of its coefficients. Therefore, if $|\xi_i(0)| < 1$, there exists an h_0 for any fixed λ such that $|\xi_i(h\lambda)| \leq 1$ when $h \leq h_0$. If there are some simple roots ξ_i such that $|\xi_i(0)| = 1$ [and indeed the principal root ξ_1 must be of that form if the order r is at least zero since $\xi_1(h\lambda) = e^{h\lambda} + 0(h^{r+1})$], then $\xi_i(h\lambda) = \xi_i(0) + 0(h)$ for h sufficiently small. [The roots of (8.18) are differentiable functions of $h\lambda$ in any region of $h\lambda$ where they are distinct.] Therefore, we have

$$|\xi_i^n(h\lambda)| \leq |\xi_i(0) + Kh|^n \leq (1 + Kh)^n \leq e^{Khn} \leq e^{Kt} \leq e^{Kb}$$

for $t \leq b$. Hence, we suspect that the roots of $\rho(\xi) = 0$ must obey some such condition for stability. In Chapter 10 it will be proved that *a necessary and sufficient condition for the stability of a multistep method of order* ≥ 1 *is that the roots of* $\rho(\xi) = 0$ *be inside the unit circle or simple on the unit circle.*

This will be called the *root condition* and we will refer to such $\rho(\xi)$ as *stable polynomials.*

We saw in Milne's method that more than one root on the unit circle can lead to bad behavior in some problems. Hence, we define the following terms:

DEFINITION 8.2

A method is strongly stable if all roots of $\rho(\xi) = 0$ are inside the unit circle except for the root $\xi = 1$.

DEFINITION 8.3

A method is weakly stable if it is stable but has more than one root on the unit circle.

The disastrous behavior of an unstable formula can be seen in the following example of a third order formula:

$$y_n = -4y_{n-1} + 5y_{n-2} + 4hf_{n-1} + 2hf_{n-2}$$

which is the most accurate two-step explicit formula. If the equation $y' = 0$ is solved with initial values $y_0 = 0$, $y_1 = \epsilon$ (a small rounding error!), we get the following results:

Table 8.4 EFFECT OF INSTABILITY

n	y
0	0
1	ϵ
2	-4ϵ
3	21ϵ
4	-104ϵ
5	521ϵ
6	-2604ϵ

This behavior is independent of h so it is evident that there is no hope for the computed solution to converge to the true solution as $h \to 0$.

In Chapter 10 we will formally define *convergence* (it will mean that the computed solution can be made arbitrarily close to the true solution by picking h small enough) and show that stability and an order ≥ 1 are necessary and sufficient conditions for convergence. Thus we see that stability is an important concept in that it guarantees that there is a way of getting any desired accuracy. In practice, however, we are working with a finite h and are interested in the question of how small h must be in order to achieve a given accuracy. For these problems we will often be concerned with the concept of absolute stability, which is discussed in the next section.

8.3.1 Absolute Stability

Let us integrate $y' = \lambda(y - t^3) + 3t^2$, $y(0) = 0$ with the third order Adams-Bashforth method using $h = \frac{1}{8}$; in one case with $\lambda = -1$ and in a second case with $\lambda = -100$. In both cases the solution is t^3. (For this problem a third order method would be exact were it not for rounding errors.)

We will forget the starting problem since we know that

$$f_{-2} = y'(-\tfrac{1}{4}) = 0.1875$$

$$f_{-1} = y'(-\tfrac{1}{8}) = 0.046875$$

$$f_0 = y'(0) = 0$$

$$y_0 = y(0) = 0$$

The third order Adams-Bashforth method for $h = \frac{1}{8}$ is

$$y_n = y_{n-1} + \tfrac{1}{96}(23f_{n-1} - 16f_{n-2} + 5f_{n-3}) \tag{8.19}$$

Calculations were done to eight decimal digits correctly rounded. For $\lambda = -1$ we get

Table 8.5 SOLUTION OF $y' = t^3 - y + 3t^2$ BY THE ADAMS-BASHFORTH METHOD

$n = 8t$	y	$Error = y - t^3$	y'	$\frac{y'}{96}$
−2			0.1875 0000	0.0019 5313
−1			0.0468 7500	0.0004 8828
0	0.0000 0000		0.0000 0000	0.0000 0000
1	0.0019 5317	+0.0000 0004	0.0468 7496	0.0004 8828
2	0.0156 2501	+0.0000 0001	0.1874 9999	0.0019 5312
3	0.0527 3429	−0.0000 0009	0.4218 7509	0.0043 9453
4	0.1249 9996	−0.0000 0004	0.7500 0004	0.0078 1250
5	0.2441 4058	−0.0000 0005	1.1718 7505	0.0122 0703
6	0.4218 7492	−0.0000 0008	1.6875 0008	0.0175 7813
7	0.6699 2193	+0.0000 0005	2.2968 7495	0.0239 2578
8	0.9999 9994	+0.0000 0006		

The errors introduced are round-off errors. Notice what happens to the −0.0000 0004 round-off error introduced in step 1. It is multiplied by $\lambda = -1$ in calculating y', and divided by 96 to get $(h/12)y'$. Consequently, it is lost in the round-off error. The next value (y_2) will therefore contain the same error from y_1 plus a new rounding error, in this case −0.0000 0003 for a total of +0.0000 0001. This process continues with rounding errors contributing an average of less than 0.0000 0001 at each step. If we now perform the integration for $\lambda = -100$, we get

Table 8.6 SOLUTION OF $y' = 100t^3 - 100y + 3t^2$ BY THE ADAMS-BASHFORTH METHOD

$n = 8t$	y_n	$Error = y_n - t^3$	y'	$\frac{y'}{96}$
−2			0.1875 0000	0.0019 5313
−1			0.0468 7500	0.0004 8828
0	0.0000 0000	0.0000 0000	0.0000 0000	0.0000 0000
1	0.0019 5317	+0.0000 0004	0.0468 7100	0.0004 8824
2	0.0156 2409	−0.0000 0091	0.1875 9100	0.0019 5407
3	0.0527 5568	+0.0000 2130	0.4197 4500	0.0043 7234
4	0.1244 9563	−0.0005 0437	0.8004 3700	0.0083 3789
5	0.2540 8001	+0.0099 3938	0.1779 3700	0.0018 5351
6	0.1851 6620	−0.2367 0880	25.3583 8000	0.2641 4979
7	6.2726 4466	+5.6027 2278		

Notice what happens to the same round-off error in y_1 this time. It is multiplied by $-\frac{100}{96}$ in calculating $hy'/12$. This is then multiplied by 23 to form y_2, which consequently contains an additional error of $(+0.0000\ 0004) \times 100 \times \frac{23}{96}$ or about 0.0000 0096 extra. The other errors reduce this by 5 in the last place to get $-0.0000\ 0091$. This phenomenon occurs at each stage so that the error is multiplied by about -24 at each step.

The polynomials ρ and σ for the third order Adams-Bashforth method are

$$\rho(\xi) = \xi^3 - \xi^2 = \xi^2(\xi - 1)$$

$$\sigma(\xi) = -\tfrac{1}{12}(23\xi^2 - 16\xi + 5)$$

If $h\lambda = -\frac{1}{8}$, Eq. (8.17) is

$$\xi^3 - \tfrac{73}{96}\xi^2 - \tfrac{16}{96}\xi + \tfrac{5}{96} = 0$$

whose roots are all less than 1.† Hence, round-off errors will not cause large errors in later steps. On the other hand, if $h\lambda = -\frac{100}{8}$ Eq. (8.17) becomes

$$\xi^3 + \tfrac{2204}{96}\xi^2 - \tfrac{1600}{96}\xi + \tfrac{500}{96} = 0$$

One root of this will exceed $\frac{1}{3}(\frac{2204}{96}) \gg 1$ in absolute value, so small round-off errors will be amplified rapidly.

The numerical results for this problem will not be of any great value unless h is chosen so that the perturbations due to the solutions of (8.17) do not grow faster than the solution which is $ce^{\lambda t} + t^3$ for general starting conditions. If $\mathrm{Re}\,(\lambda) \leq 0$, no root of (8.17) should exceed 1 in absolute value, whereas if $\mathrm{Re}\,(\lambda) > 0$, the principal root will have to be a reasonable approximation to $e^{h\lambda}$ and none of the extraneous roots should be larger than the principal root. This leads us to definitions of absolute and relative stability for the test equation $y' = \lambda y$.

DEFINITION 8.4

> A multistep method is absolutely stable for those values of $h\lambda$ where roots of (8.17) are ≤ 1 in absolute value.

DEFINITION 8.5

> A method is relatively stable where the extraneous roots of (8.17) are $\leq$ the principal root in absolute value.

(We have not concerned ourselves with the size of the principal root in relation to $e^{h\lambda}$ since that is an accuracy criterion.)

†This can be seen by arguing that

$$\begin{aligned} |\xi^3| &= \tfrac{1}{96}\,|73\xi^2 + 16\xi - 5| \\ &\leq \tfrac{1}{96}(73 + 16 + 5)\max\{|\xi^2|, 1\} \\ &= \tfrac{94}{96}\max\{|\xi^2|, 1\} \\ &\Rightarrow |\xi| < 1 \end{aligned}$$

Had we started this problem from a different initial value, say $y(0) = 1$, the solution would have been

$$y(t) = e^{\lambda t} + t^3$$

The first component $e^{\lambda t}$ changes by $e^{h\lambda}$ in each step. For $\lambda = -1$ and $h = \frac{1}{8}$ this is $e^{-1/8} \cong 0.88279$, while the largest root of (8.17) is $\cong 0.88274$. Consequently, we would get good accuracy for this starting value with the third order Adams-Bashforth method. (**Exercise:** Perform this computation.) For $\lambda = -100$ and $h = \frac{1}{8}$, the single step change is $e^{-100/8} \cong 0.0000044$, whereas we have seen that one root of (8.17) is much larger than 1. In order to accurately approximate the e^{-100t} term, it is necessary to use small steps so that $100h$ is in the region where one root of $\rho(\xi) - 100h\sigma(\xi) = 0$ is a sufficiently accurate approximation to e^{-100h}. Thus we could argue that we have not a stability problem but an accuracy problem caused by an excessively large h. However, by the time we have integrated to $t = \frac{1}{4}$ with a small step size for accuracy, the e^{-100t} term will be insignificant to ten digits so it is no longer necessary to accurately represent it. In that case, we are interested in methods which are absolutely stable for large values of $-h\lambda$.

A Desirable Stability Criterion

If we consider the problem

$$y' = \lambda(y - F(t)) + F'(t) \tag{8.20}$$

which has the solution

$$y = ce^{\lambda t} + F(t)$$

we see that even for a single equation we cannot say exactly what form of stability we need. In order to retain a certain number of digits of accuracy in the solution, perturbations should not grow faster than the solution. If $F(t)$ grows faster than $e^{\lambda t}$ and the initial condition is such that c is not large compared to $F(0)$, we require that h be such that:

1. The truncation errors in integrating $y' = F'(t)$ are not large.
2. The roots of (8.17) are such that $|\xi_i|^n$ does not grow faster than $F(nh)$.

The second requirement is neither an absolute nor a relative stability criterion Most multistep methods cause the error to oscillate in sign as they approach instability; thus, in practice we must try and control the step so that such errors are insignificant. Fortunately, the presence of such errors tends to make most step control procedures think that there are large derivatives present and hence reduce the step so that lack of a universal criterion for stability is not so much a hindrance to the solution of problems but rather to the classification of the properties of methods.

In order to get an idea of the stability properties of methods, we frequently look at their regions of absolute stability. The larger the region, the larger

h can be made within the truncation error restriction. Absolute stability regions are shown in Figure 8.1 for the Adams-Bashforth methods of orders one through six and in Figure 8.2 for Adams-Moulton methods of orders three through six. The first order Adams-Moulton method is the backward Euler method

$$y_{n+1} = y_n + hy'_{n+1}$$

which is stable except in the circle $|1 - h\lambda| < 1$. The second order Adams-Moulton method is the trapezoidal rule, which is stable in the negative half plane.

It can be seen that the region of stability for the implicit Adams-Moulton methods is larger by a factor of ten or more than that of the explicit Adams-Bashforth method. The truncation errors are also smaller for the implicit methods, so the implicit methods can be used with a step size that is several times larger than that of the explicit methods. This increase in step size usually more than offsets the additional effort in solving the corrector, which may require two or three function evaluations.

Systems of Linear Equations

If we consider the system of equations

$$\mathbf{y}' = A\mathbf{y} \tag{8.21}$$

where A is a constant matrix which we will assume is diagonalizable by S, we can transform them into an equivalent set

$$\mathbf{z}' = \Lambda\mathbf{z}$$

where $z = Sy$ and $\Lambda = SAS^{-1}$ is a diagonal matrix with elements λ_i. Thus, the solution of (8.21) is

$$\mathbf{y} = S^{-1}e^{\Lambda t}S\mathbf{y}_0$$

where $e^{\Lambda t}$ is the diagonal matrix with elements $e^{\lambda_i t}$. If we apply a multistep method to (8.21), we get the vector equation for the perturbations

$$\sum_{i=0}^{k} (\alpha_i + hA\beta_i)\mathbf{e}_{n-i} = 0$$

By multiplying this by S and writing $\mathbf{q}_n = S\mathbf{e}_n$, we get

$$\sum_{i=0}^{k} (\alpha_i + h\Lambda\beta_i)\mathbf{q}_{n-i} = 0$$

which is a set of independent equations for the components of $\mathbf{q}_n$, each of the form (8.16). Thus we will get components of the form ξ_{ij}^n in each component of $\mathbf{e}_n$ where ξ_{ij}, $j = 1, \ldots, k$ are the k roots of (8.17) with $\lambda = \lambda_i$. Consequently, we will be concerned with the size of all of the $|\xi_{ij}|$ relative to the principal root corresponding to the λ_i with the largest real part, say λ_1. This is a relative stability criterion for that $h\lambda_1$, but not for the other $h\lambda_i$. *For most*

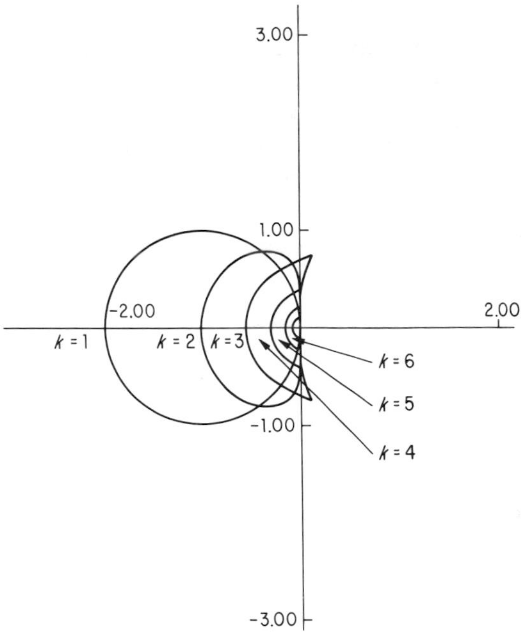

Fig. 8.1 Stability regions for Adams-Bashforth methods. Method of order k is stable inside region indicated left of origin.

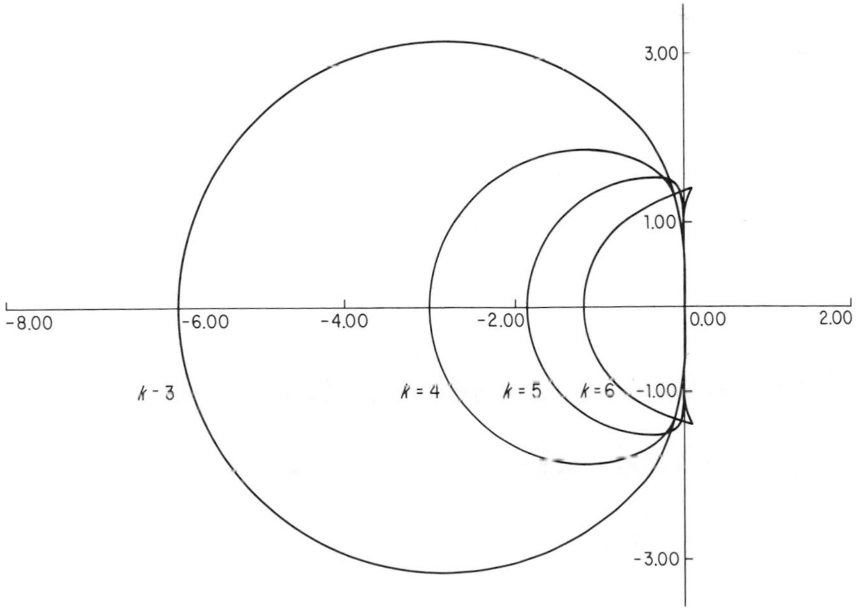

Fig. 8.2 Stability regions for Adams-Moulton method. Method of order k is stable inside region indicated.

methods and problems, the accuracy requirement is such that relative stabiity is not the limiting factor. The need to approximate $e^{\lambda h}$ accurately is more severe than keeping the extraneous roots less than the principal one. If one or more of the extraneous roots is near the unit circle for $h\lambda = 0$, then this statement is less likely to be true. In other words, weakly stable or nearly weakly stable methods are likely to give relative stability problems, strongly stable methods with roots well inside the unit circle do not.

8.4 THE CLASS OF THREE-STEP METHODS OF ORDER FOUR

The most general three-step method has the form

$$\alpha_0 y_n + \alpha_1 y_{n-1} + \alpha_2 y_{n-2} + \alpha_3 y_{n-3} + h\beta_0 f_n + h\beta_1 f_{n-1} + h\beta_2 f_{n-2} + h\beta_3 f_{n-3} = 0$$

If we require it to have fourth order, the α and β must satisfy the five equations

$$\begin{aligned}
\alpha_0 + \alpha_1 + \alpha_2 + \alpha_3 &= 0\\
3\alpha_0 + 2\alpha_1 + \alpha_2 + \beta_0 + \beta_1 + \beta_2 + \beta_3 &= 0\\
9\alpha_0 + 4\alpha_1 + \alpha_2 + 6\beta_0 + 4\beta_1 + 2\beta_2 &= 0\\
27\alpha_0 + 8\alpha_1 + \alpha_2 + 27\beta_0 + 12\beta_1 + 3\beta_2 &= 0\\
81\alpha_0 + 16\alpha_1 + \alpha_2 + 108\beta_0 + 32\beta_1 + 4\beta_2 &= 0
\end{aligned}$$

If we also normalize so that $\sigma(1) = 1$, we get the additional equation

$$\beta_0 + \beta_1 + \beta_2 + \beta_3 = 1$$

These six equations in eight unknowns can be solved in terms of two free parameters, which we will take to be β_0 and β_3. We then get

$$\begin{aligned}
\alpha_0 &= \tfrac{1}{12}(-1 - 30\beta_0 + 6\beta_3)\\
\alpha_1 &= \tfrac{1}{12}(-9 + 54\beta_0 + 18\beta_3)\\
\alpha_2 &= \tfrac{1}{12}(9 - 18\beta_0 - 54\beta_3)\\
\alpha_3 &= \tfrac{1}{12}(1 - 6\beta_0 + 30\beta_3)\\
\beta_1 &= \tfrac{1}{12}(6 + 12\beta_0 - 24\beta_3)\\
\beta_2 &= \tfrac{1}{12}(6 - 24\beta_0 + 12\beta_3)
\end{aligned}$$

while we find that the error coefficient is given by

$$C_5 = \tfrac{1}{120}(-1 + 10\beta_0 + 10\beta_3)$$

The polynomial $\rho(\xi)$ is

$$\begin{aligned}
&-\tfrac{1}{12}[\xi^3(1 + 30\beta_0 - 6\beta_3) + \xi^2(9 - 54\beta_0 - 18\beta_3)\\
&\qquad - \xi(9 - 18\beta_0 - 54\beta_3) - (1 - 6\beta_0 - 30\beta_3)]\\
&= -\tfrac{1}{12}(\xi - 1)[\xi^2(1 + 30\beta_0 - 6\beta_3)\\
&\qquad + \xi(10 - 24\beta_0 - 24\beta_3) + (1 - 6\beta_0 + 30\beta_3)]
\end{aligned}$$

We are interested in those values of β_0 and β_3 for which the quadratic factor has roots inside the unit circle. The easiest way to find this is to find the boundary where $|\xi| = 1$ is a root. We find that $\xi = -1$ is a root if $\beta_0 + \beta_3 = \frac{1}{6}$. $\xi = 1$ is not a root for any finite value of β_0 and β_3 since the sum of the coefficients is 12, independent of β_0 and β_3. The only other possibility is that a complex conjugate pair on the unit circle occurs. For this the coefficients of ξ^2 and 1 in the quadratic factor must be equal and the discriminant must be negative. This occurs when

$$\beta_0 = \beta_3 \qquad \text{and} \qquad 27(\beta_0 - \beta_3)^2 \leq 12(\beta_0 + \beta_3) - 2$$

This is the section of $\beta_0 = \beta_3$ where $\beta_0 \geq \frac{1}{12}$. The line along which one extraneous root is zero is given by

$$1 - 6\beta_0 + 30\beta_3 = 0$$

while at $\beta_0 = \frac{3}{8}$, $\beta_3 = \frac{1}{24}$, both extraneous roots are zero. This is the Adams-Moulton method. Figure 8.3 shows the region of stability in the (β_0, β_3)-plane and shows the lines of constant truncation error.

The line where the error coefficient is zero corresponds to a higher order method. It is outside of the stable region so we conclude that a stable three-step method cannot exceed order four. If the method is explicit, $\beta_0 = 0$. However, this line is also outside of the stable region so there does not exist

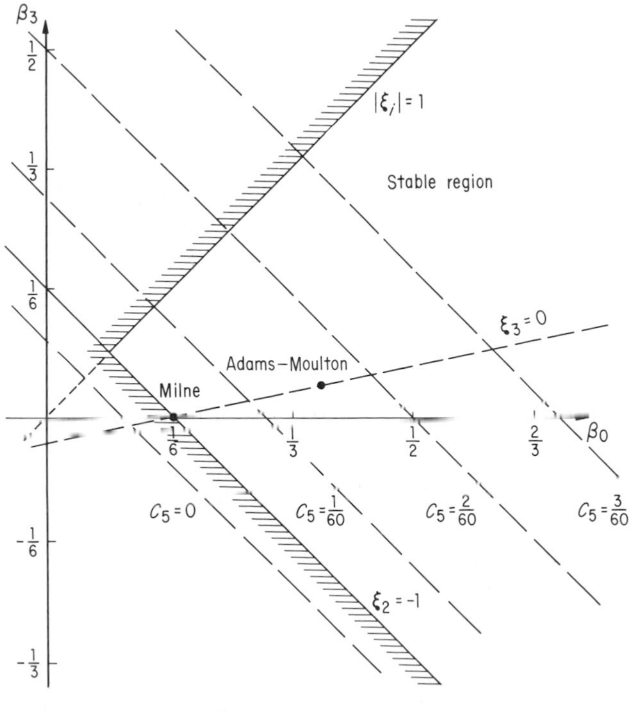

Fig. 8.3 Values of β_0, β_3 for which three-step method is stable.

a stable explicit three-step fourth order method. (The Adams-Bashforth three-step method has third order.)

The truncation error coefficient is reduced by moving toward the line $\beta_0 + \beta_3 = \frac{1}{6}$, where $\xi_2 = -1$. However, this takes us toward a weakly stable method. We saw for Milne's method that the second root behaved like $e^{-h\lambda/3}$. If we can be sure that there are no λ's present such that $|e^{-h\lambda/3}|$ is significant compared to the growth of the solution, Milne's method is a good choice as it has the smallest truncation error, and since $\alpha_3 = \beta_3 = 0$, it is only a two-step method. However, it is not recommended for general problems because of the critical stability problem. (The Adams-Moulton method is probably the best general choice for ordinary problems. This will be discussed further in Chapter 9.)

It is not desirable to move out too far to larger β_0 and β_3 because large values will adversely affect the round-off errors.

Similar "pictures" exist for k-step methods but there are more free parameters. There is a $(k - 1)$-dimensional region within which k-step methods of order $k + 1$ are stable. A fundamental result of Dahlquist (1956), which will be proved in Chapter 10, states that if k is odd, no points within this region have higher than $(k + 1)$th order, but if k is even, there is a $[(k - 2)/2]$-dimensional subregion within which the order is $k + 2$ (but no higher). However, in this region of order $k + 2$ all roots of ρ are on the unit circle so the methods are only weakly stable. Any point within the $(k - 1)$-dimensional region can be chosen for a k-step method, the choice should be made so that the truncation error is as small as possible consistent with stability not being a problem. Regrettably, it is not usually possible to make such an intelligent choice, since the required information about the problem and its solution is not known until the problem has been integrated.

PROBLEMS

1. If $\rho(\xi) = \xi^3 - \xi^2 + \xi/4 - 1/4$, find a $\sigma(\xi)$ such that:
   (a) $\sigma(\xi)$ is of second degree and the method has third order;
   (b) $\sigma(\xi)$ is of third degree and the method has fourth order. What are the error coefficients for these two methods?

2. If $\rho(\xi) = \xi^4 - 1$, find a $\sigma(\xi)$ of degree four such that the method has maximum order. What is that order and what is the error coefficient?

3. If $\sigma(\xi) = \xi^2$, find $\rho(\xi)$ such that:
   (a) $\rho(\xi)$ is of second degree and the order is two;
   (b) $\rho(\xi)$ is of third degree and the order is three. Are these methods stable?

4. What is the region of absolute stability of the method you found in Problem 3(a)?

5. Determine the coefficients of the maximum order three-step method. Is it stable?

6. Find the class of two-step third order methods in terms of the parameter β_0. For what range of β_0 are these methods stable? Express the error coefficient as a function of β_0. Draw a diagram showing the range of β_0, the error, the Adams-Moulton method, and the Milne method.

7. Consider the class of general three-step methods for which $\xi_2 = -1$ as shown in Figure 8.3. Show that for $\beta_0 > \frac{1}{12}$ one root of $\rho(\xi) + h\lambda\sigma(\xi) = 0$ is $\xi = -1 + Ah\lambda + 0(h^2)$, where A is independent of β_0. What is A?

8. Derive coefficients for the following method to give it as high an order as possible:

$$y_{n-(1/2)} = \alpha_1 y_{n-1} + \alpha_2 y_{n-2} + \beta_1 hy'_{n-1} + \beta_2 hy'_{n-2}$$

[$y_{n-(1/2)}$ is an approximation to $y(t_n - h/2)$].

$$hy'_{n-(1/2)} = hf(y_{n-(1/2)})$$

$$y_n = \alpha_1^* y_{n-1} + \alpha_2^* y_{n-2} + \gamma hy'_{n-(1/2)} + \beta_1^* hy'_{n-1} + \beta_2^* hy'_{n-2}$$

$$hy'_n = hf(y_n)$$

What is the order of the truncation error in y_n? Is the method stable?

9 MULTIVALUE METHODS

In the previous two chapters we examined multistep methods and discussed their stability. However, we did the latter only for "predictor only" or "corrector only" methods; that is, methods in which an explicit equation is used or an implicit equation is solved exactly (within round-off). In practice we use a finite (sometimes fixed) number of corrector iterations following an explicit prediction. In this chapter we will study the behavior of these $P(EC)^M$ and $P(EC)^ME$ methods as a subclass of the general multivalue methods. The properties of the general multivalue methods will be discussed and some will be seen to have important additional benefits. First, they give simple generalizations of multistep methods which will no longer have their order restricted by stability requirements, hence k-step methods can now be of order $2k$. Secondly, they show in what way different organizations of the computation are equivalent so that, for example, comparisons can be made between Adams' method using backward differences and Adams' method using previous derivative values. One particular organization will be seen to be applicable directly to higher order equations. Thirdly, they provide a mechanism for studying the operations of step changing and error estimation.

We recall that a multivalue predictor-corrector method is given by

$$\mathbf{y}_{n,(0)} = B\mathbf{y}_{n-1} \tag{9.1}$$

$$\mathbf{y}_{n,(m+1)} = \mathbf{y}_{n,(m)} + \mathbf{c}G(\mathbf{y}_{n,(m)}) \qquad m \geq 0 \tag{9.2}$$

EXAMPLE

The third order Adams-Bashforth predictor is

$$y_{n,(0)} = y_{n-1} + \frac{h}{12}(23f_{n-1} - 16f_{n-2} + 5f_{n-3})$$

and the fourth order Adams-Moulton corrector is

$$y_{n,(m+1)} = y_{n-1} + \frac{h}{24}(9f(y_{n,(m)}) + 19f_{n-1} - 5f_{n-2} + f_{n-3})$$

Hence,

$$y_{n,(1)} = y_{n,(0)} + \tfrac{3}{8}h[f(y_{n,(0)}) - (3hf_{n-1} - 3hf_{n-2} + hf_{n-3})]$$

Since y_{n-1} and y_{n-2} do not need to be saved, we can take $\mathbf{y}_n = [y_n, hy'_n, hy'_{n-1}, hy'_{n-2}]^T$, whence B and $\mathbf{c}$ are

$$B = \begin{bmatrix} 1 & \frac{23}{12} & -\frac{16}{12} & \frac{5}{12} \\ 0 & 3 & -3 & 1 \\ 0 & 1 & 0 & 0 \\ 0 & 0 & 1 & 0 \end{bmatrix}, \qquad \mathbf{c} = \begin{bmatrix} \frac{3}{8} \\ 1 \\ 0 \\ 0 \end{bmatrix}$$

9.1 BEHAVIOR OF THE ERROR

We want to examine the error of the method.

DEFINITION 9.1

If $\mathbf{y}(t_n)$ is the correct value of $\mathbf{y}_n$, and if we calculate

$$\begin{aligned} \tilde{\mathbf{y}}_{n,(0)} &= B\mathbf{y}(t_{n-1}) \\ \tilde{\mathbf{y}}_{n,(m+1)} &= \tilde{\mathbf{y}}_{n,(m)} + \mathbf{c}G(\tilde{\mathbf{y}}_{n,(m)}) \\ \tilde{\mathbf{y}}_n &= \tilde{\mathbf{y}}_{n,(M)} \end{aligned} \tag{9.3}$$

then the local truncation error is $\mathbf{d}_n$, where

$$\mathbf{d}_n = \tilde{\mathbf{y}}_n - \mathbf{y}(t_n)$$

Note that this defines the truncation error for a predictor-corrector scheme for a solution $\mathbf{y}(t)$ of the differential equation $G(\mathbf{y}) = 0$.

We define the global error at the nth step by

$$\mathbf{e}_n = \mathbf{y}_n - \mathbf{y}(t_n)$$

but we define $\mathbf{e}_{n,(m)}$ as $\mathbf{y}_{n,(m)} - \tilde{\mathbf{y}}_{n,(m)}$. Subtract (9.3) from (9.2) and use the mean value theorem to get

$$\mathbf{e}_{n,(m+1)} = \mathbf{e}_{n,(m)} + \mathbf{c}\frac{\partial G}{\partial \mathbf{y}}(\xi_m)\mathbf{e}_{n,(m)}$$

where ξ_m is a point between $\mathbf{y}_{n,(m)}$ and $\tilde{\mathbf{y}}_{n,(m)}$. (Note that $\partial G/\partial \mathbf{y}$ is a row vector.) Thus we have

$$\mathbf{e}_{n,(M)} = \prod_{i=0}^{M-1}\left(I + \mathbf{c}\frac{\partial G}{\partial \mathbf{y}}(\xi_i)\right)\mathbf{e}_{n,(0)} \tag{9.4}$$

(Note that the matrix multiplication should place the terms with larger subscripts on the left.)

From the definitions we get

$$\mathbf{e}_{n,(0)} = \mathbf{y}_{n,(0)} - \tilde{\mathbf{y}}_{n,(0)} = B(\mathbf{y}_{n-1} - \mathbf{y}(t_{n-1})) = B\mathbf{e}_{n-1}$$

and

$$\mathbf{e}_n = \mathbf{y}_n - \mathbf{y}(t_n) = \mathbf{y}_n - \tilde{\mathbf{y}}_n + \tilde{\mathbf{y}}_n - \mathbf{y}(t_n) = \mathbf{e}_{n,(M)} + \mathbf{d}_n$$

Substituting the last two equations into (9.4), we get

$$\mathbf{e}_n = S_n \mathbf{e}_{n-1} + \mathbf{d}_n$$

for $n \geq 1$, where

$$S_n = \prod_{i=0}^{M-1} \left(I + \mathbf{c}\frac{\partial G}{\partial \mathbf{y}}(\boldsymbol{\xi}_i)\right) B \tag{9.5}$$

From this we see that

$$\mathbf{e}_N = \sum_{i=0}^{N} \prod_{j=i+1}^{N} S_j \mathbf{d}_i \tag{9.6}$$

where $\mathbf{d}_0$ is the error in the initial values.

We naturally expect stability to be related to the "size" of the matrix S_j. For example, if $\| S_j \| \leq 1$ for all j, then $\| e_n \| \leq \sum_{i=0}^{N} \| \mathbf{d}_i \|$.

9.1.1 Stability of Predictor-Corrector Methods

In Chapter 8 we stated that stability is equivalent to requiring the root condition to hold for $\rho(\xi)$, and defined absolute and relative stability in terms of the roots of (8.17). For multivalue schemes we require similar conditions on the eigenvalues of the matrices S_j in Eq. (9.6). We will examine these conditions and show that they are equivalent to the previous conditions when the corrector equation is solved exactly (as it is if the iteration is continued to convergence). First we note that in the case that $f(y) = \lambda y$, $\partial G/\partial \mathbf{y}$ takes the form

$$\left[h\frac{\partial f}{\partial y}, 0, \ldots, -1, 0, \ldots, 0\right] = h\lambda\, \boldsymbol{\delta}_0^T - \boldsymbol{\delta}_k^T$$

where $\boldsymbol{\delta}_i$ is the column vector with a one in the ith position (numbering from 0) and zeros elsewhere. $\boldsymbol{\delta}_i^T$ is its transpose. From (9.5) we see that S_j depends only on $h\lambda$, so we get

$$S_j = S(h\lambda) = \left[\begin{array}{cccc|cccc} 1 + h\lambda\beta_0^* & & 0 & & -\beta_0^* & & & \\ & 1 & & & & & 0 & \\ 0 & & \ddots & & & & & \\ & & & 1 & & & & \\ \hline h\lambda & & & & 0 & & & \\ & & & & 1 & & 0 & \\ & & 0 & & & \ddots & & \\ & & & & 0 & & & 1 \end{array}\right]^M B$$

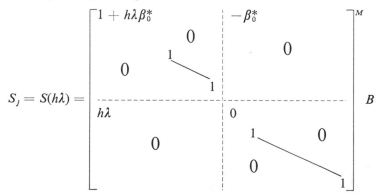

where all other elements are zero. If we set

$$L_m = 1 + h\lambda\beta_0^* + \cdots + (h\lambda\beta_0^*)^m$$

we have

$$S(h\lambda) = \left[\begin{array}{ccc|ccc} L_M & & 0 & -\beta_0^* L_{M-1} & & \\ & 1 & & & & \\ 0 & & \ddots & & 0 & \\ & & 1 & & & \\ \hline h\lambda L_{M-1} & & & -\beta_0^* h\lambda L_{M-2} & & \\ & 0 & & & 1 & 0 \\ & & & 0 & & \ddots \\ & & & & & 1 \end{array}\right] B$$

$$= \left[\begin{array}{ccccc|ccc} a_1 & a_2 & \cdots & a_k & & b_1 & \cdots & b_k \\ 1 & & & & & & & \\ & \ddots & 0 & & & & 0 & \\ & 0 & \ddots & & & & & \\ & & & 1 & 0 & & & \\ \hline c_1 & c_2 & \cdots & c_k & & d_1 & \cdots & d_k \\ & & & & & 1 & & \\ & & 0 & & & & \ddots & 0 \\ & & & & & 0 & & \ddots \\ & & & & & & 1 & 0 \end{array}\right]$$

where

$$a_i = L_M\alpha_i - \beta_0^* L_{M-1}\gamma_i$$
$$b_i = L_M\beta_i - \beta_0^* L_{M-1}\delta_i$$
$$c_i = h\lambda(L_{M-1}\alpha_i - \beta_0^* L_{M-2}\gamma_i)$$

and

$$d_i = h\lambda(L_{M-1}\beta_i - \beta_0^* L_{M-2}\delta_i)$$

The stability of the method is determined by the eigenvalues of this matrix. In the region in which they are inside the unit circle or simple† on the unit circle, perturbations are not increasing so that method is absolutely stable. Consequently, we define stability regions for multivalue methods as:

†We will use the word "simple" to mean that the eigenvalue corresponds to a linear elementary divisor. Repeated roots on the unit circle with independent eigenvectors do not cause growth of the perturbation.

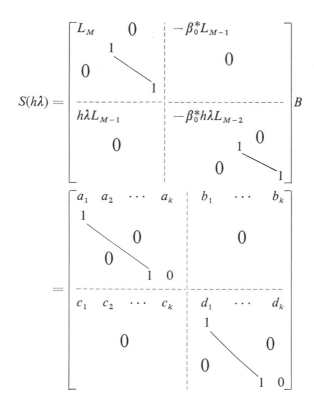

DEFINITION 9.2

> The region of absolute stability consists of those values of $h\lambda$ for which the eigenvalues of $S(h\lambda)$ are inside the unit circle or simple on the unit circle.

DEFINITION 9.3

> The region of relative stability consists of those values of $h\lambda$ for which the extraneous eigenvalues of $S(h\lambda)$ are less in magnitude than the principal eigenvalue. The principal eigenvalue is that one which approximates $e^{h\lambda}$ most closely.

We will show that if the corrector is iterated to convergence, these eigenvalues are precisely the roots of $\rho^*(\xi) + h\lambda\sigma^*(\xi) = 0$.

If the corrector is iterated to convergence,

$$L_{M-2} = L_{M-1} = \cdots = L = \frac{1}{1 - h\lambda\beta_0^*}$$

(This implies that $|h\lambda\beta_0^*| < 1$.) In this case the kth row of $S(h\lambda)$ is $h\lambda$ times the zero-th row of $S(h\lambda)$. If we perform the similarity transformation QSQ^{-1}, where

$$Q = \left[\begin{array}{ccc|ccc} 1 & & 0 & & & \\ & \ddots & & & 0 & \\ 0 & & 1 & & & \\ \hline -h\lambda & & 0 & 1 & & 0 \\ & \ddots & & & \ddots & \\ 0 & & -h\lambda & 0 & & 1 \end{array}\right]$$

we get

$$\tilde{S} = QS(h\lambda)Q^{-1} = \left[\begin{array}{cccc|cccc} & \cdots\epsilon_i\cdots & & & & \cdots b_i\cdots & & \\ 1 & & 0 & & & & & \\ & \ddots & & & & 0 & & \\ 0 & & 1 & 0 & & & & \\ \hline & & & & 0 & & 0 & \\ & & & & 1 & & & \\ & 0 & & & & \ddots & & \\ & & & & 0 & & 1 & 0 \end{array}\right]$$

where

$$\begin{aligned}\epsilon_i &= L\alpha_i - \beta_0^* L\gamma_i + h\lambda(L\beta_i - \beta_0^* L\delta_i)\\ &= L[\alpha_i - (\alpha_i - \alpha_i^*) + h\lambda(\beta_i - (\beta_i - \beta_i^*)\\ &= L(\alpha_i^* + h\lambda\beta_i^*)\end{aligned}$$

The eigenvalues of $\tilde{S}$ are the eigenvalues of the upper left-hand partition and k zeros. The upper left-hand matrix is the companion matrix† of the polynomial

$$\begin{aligned}-\xi^k + \sum_{i=1}^{k} \xi^{k-i}\epsilon_i &= L\left[\sum_{i=1}^{k} \xi^{k-i}(\alpha_i^* + h\lambda\beta_i^*) - \frac{1}{L}\xi^k\right]\\ &= L[\rho^*(\xi) + h\lambda\sigma^*(\xi)]\end{aligned}$$

since $1/L = 1 - h\lambda\beta_0^*$ and $\alpha_0^* = -1$; so the eigenvalues of S are the same as the zeros of the corrector polynomial $\rho^*(\xi) + h\lambda\sigma^*(\xi)$. If a finite number of corrections are made, the corrector stability is influenced by the predictor. However, if $h\lambda = 0$, the first corrector is the same as all future correctors, so the stability is determined by the eigenvalues of $S = S(0) = (I - \mathbf{c}\boldsymbol{\delta}_k^T)B$, which are the roots of $\xi^k\rho^*(\xi) = 0$.

When $h\lambda \neq 0$ and the corrector is not iterated to convergence, we have twice as many eigenvalues as we had roots of the polynomial previously. The additional k roots arise in a $P(EC)^M$ method because there are $2k$ different values, $y_{n-1}, \ldots, y_{n-k}$, and $hy'_{n-1}, \ldots, hy'_{n-k}$ being saved from step to step. The error in hy'_i is not directly related to that in y_i because the y_i is corrected after the final evaluation of hy'_i. One root arises for each "independent" number being computed because each can contain an "independent" error. If either $h\lambda = 0$ or the corrector is iterated to convergence, we know that $hy'_i = hf(y_i)$, so that their errors are not independent (if round-off is ignored). Consequently, we only expect to get k roots. The remaining k are zero, which means that even if errors are introduced by initial values or by round-off so that for $i < k$, $hy'_i \neq hf(y_i)$, the effect disappears after a few steps.

A P(EC)^{M}E method corresponds to performing an additional corrector iteration using the vector $\mathbf{c} = [0, \ldots, 0, 1, 0, \ldots]^T$. The value of hy'_n is corrected without changing y_n. In this case S_n contains an additional factor. For the equation $y' = \lambda y$, S_n takes the form

†A companion matrix has the form

$$A = \begin{bmatrix} a_1 & a_2 & \cdots & a_k \\ 1 & & 0 & \\ & \ddots & & \\ 0 & & 1 & 0 \end{bmatrix}$$

Its characteristic polynomial det $(A - \lambda I)$ is

$$(-1)^k(\lambda^k - a_1\lambda^{k-1} - a_2\lambda^{k-2} - \cdots - a_k)$$

$$
\left[\begin{array}{cccc:cccc}
a_1 & \cdots & \cdots & a_k & b_1 & \cdots & \cdots & b_k \\
1 & & & 0 & & & & \\
 & \ddots & & \vdots & & 0 & & \\
0 & & 1 & 0 & & & & \\
\hdashline
c_1 & \cdots & \cdots & c_k & d_1 & \cdots & \cdots & d_k \\
 & & & & 1 & & & 0 \\
 & 0 & & & & \ddots & & \vdots \\
 & & & & 0 & & 1 & 0
\end{array}\right]
$$

where a_i and b_i are as before,

$$c_i = h\lambda a_i \qquad \text{and} \qquad d_i = h\lambda b_i$$

Since the kth row is $h\lambda$ times the first, k roots are zero as before. Therefore, we suspect that there may be larger regions of relative or absolute stability from $\mathrm{P(EC)^ME}$ methods than from $\mathrm{P(EC)^M}$ methods because there are k fewer eigenvalues to be bounded.

It is difficult to make any general statement about the way in which the predictor affects the stability of the corrector when only a finite number of corrector iterations is used. There have been some efforts reported in the literature to use predictors to increase the stability range of the corrector. See, for example, Stetter (1968) and Crane and Klopfenstein (1965).

9.2 EQUIVALENT METHODS

In the previous section we characterized multistep methods by a matrix B and a vector $\mathbf{c}$. The information from the current and previous mesh points was retained in a vector $\mathbf{y}_n$. The basis for the predictor step was an extrapolation process (usually of a polynomial form) from the retained information. It is equally reasonable to save other linear combinations of the components of $\mathbf{y}_n$ as long as we can recover the original information. For example, we saw that in the Adams' methods we could either retain derivative values at previous mesh points or could retain backward differences. The one set is linear combinations of the other.

Let T be any nonsingular matrix with the dimension of $\mathbf{y}_n$. Define

$$
\begin{aligned}
\mathbf{a}_n &= T\mathbf{y}_n \\
\mathbf{a}_{n,(m)} &= T\mathbf{y}_{n,(m)}
\end{aligned}
\tag{9.7}
$$

If these are substituted in (9.2) and (9.1), we get

$$\mathbf{a}_{n,(0)} = A\mathbf{a}_{n-1} \tag{9.8}$$

$$\mathbf{a}_{n,(m+1)} = \mathbf{a}_{n,(m)} + \mathbf{l}F(\mathbf{a}_{n,(m)}) \tag{9.9}$$

where $A = TBT^{-1}$, $\mathbf{l} = T\mathbf{c}$, and $F(\mathbf{x}) = G(T^{-1}\mathbf{x})$.

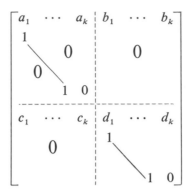

When we obtain the error propagation equation, we get

$$\mathbf{e}_n = \tilde{S}_n \mathbf{e}_{n-1} + \bar{\mathbf{d}}_n$$

where $\bar{\mathbf{d}}_n$ is the truncation error for the transformed method ($= T\mathbf{d}_n$), and

$$\begin{aligned}
\tilde{S}_n &= \prod_{i=0}^{M-1} \left(I + \mathbf{l}\frac{\partial F(\xi_i)}{\partial \mathbf{a}}\right) A \\
&= \prod_{i=0}^{M-1} \left[T\left(I + \mathbf{c}\frac{\partial G(T^{-1}\xi_i)}{\partial \mathbf{y}}\right)T^{-1}\right] TBT^{-1} \\
&= T \prod_{i=0}^{M-1} \left(I + \mathbf{c}\frac{\partial G}{\partial \mathbf{y}}(T^{-1}\xi_i)\right) BT^{-1} \\
&= TS_nT^{-1}
\end{aligned}$$

so, as expected, the stability of the "new" method is the same as the stability of the old method. It is evident that if calculations are performed exactly (without round-off errors), the two methods will give identical results. Consequently, it is useful to state:

DEFINITION 9.4

Two multivalue methods are equivalent if they are related by the transformations (9.7) to (9.9).

Note: While making a transformation to an equivalent method does not affect its stability properties or truncation error, it can affect both the round-off properties and the amount of computation.

9.2.1 Factors Affecting the Choice of Representation

In the step by step computation of a numerical solution there are a number of different operations to be executed in a practical program. The obvious operations are the multiplication by a matrix A in the prediction step (9.8) and the multiplication of the scalar F by the vector $\mathbf{l}$ followed by a vector addition at each correction step (9.9). In fact, only those components of $\mathbf{a}_n$ that are needed to form y_n and hy'_n need be updated at each correction step. The remainder can be handled by noting that

$$\mathbf{a}_{n,(M)} = \mathbf{a}_{n,(0)} + \mathbf{l}(F(\mathbf{a}_{n,(0)}) + F(\mathbf{a}_{n,(1)}) + \cdots + F(\mathbf{a}_{n,(M-1)}))$$

If we assume that y_n and hy'_n are both explicit components of $\mathbf{a}_n$, say in the zeroth and first positions, only two components of $\mathbf{a}_n$ need be updated at each correction step. The remainder can be updated once.

In addition a practical method will periodically have to change step size, change order, and provide estimates of the error terms so that the appropriate order and step size can be selected.

Change of Step

Changing step size is an interpolation process in which the saved values for the new step size must be found from the saved values for the old step size. Suppose a three-step method is being used, and y and its derivatives are known at the points t_n, $t_n - h$ and $t_n - 2h$. If the step size is to be halved, the values at t_n, $t_n - (h/2)$ and $t_n - h$ can be computed by interpolation. In this case, the computation of values at t_n and $t_n - h$ is simple, but a linear combination of the old values is needed to approximate $y(t_n - (h/2))$ and $hy'(t_n - h/2)$. In general, changing the step by the ratio α corresponds to premultiplying $\mathbf{a}_n$ by a matrix $C(\alpha)$, so we have the step changing operation represented by

$$\tilde{\mathbf{a}}_n = C(\alpha)\mathbf{a}_n \tag{9.10}$$

where $\tilde{\mathbf{a}}_n$ is the value of $\mathbf{a}_n$ adjusted for the new step size αh.

If a large number of simultaneous equations are being integrated, Eq. (9.10) can be a time consuming operation. An alternative scheme that is more economical under these circumstances is to use different integration formula for variable step sizes. We could, for example, compute the coefficients β_i that would make the integration formula

$$y_n = y_{n-1} + \beta_0 h_n f_n + \beta_1 h_{n-1} f_{n-1} + \cdots + \beta_k h_{n-k} f_{n-k}$$

of order $k + 1$, where $y_{n-i} \cong y(t_{n-i})$ and the mesh points are no longer equally spaced, so $h_j = t_j - t_{j-1}$. If the step size is changed about every k steps, the coefficients β_i would have to be recomputed for every step. However, this is independent of the number of equations, so if there is a large number of equations it will be more economic than using (9.10), which has to be applied to every dependent variable.

It should be noted that the two processes are not equivalent. In the first case, all of the old values of $\mathbf{a}_n$ may take part in forming the new values $\tilde{\mathbf{a}}_n$ by interpolation. Consequently, the value at the most distant mesh point may affect many of the saved values $\tilde{\mathbf{a}}_n$. In the second case, the most distant value is discarded after the next step and no longer has an effect.

We can illustrate this by the following example. The three-step Adams-Bashforth formula is given by

$$B = \begin{bmatrix} 1 & \frac{23}{12} & -\frac{16}{12} & \frac{5}{12} \\ 0 & 0 & 0 & 0 \\ 0 & 1 & 0 & 0 \\ 0 & 0 & 1 & 0 \end{bmatrix}, \qquad \mathbf{c} = \begin{bmatrix} 0 \\ 1 \\ 0 \\ 0 \end{bmatrix}$$

If the step is halved, the interpolation can be performed by premultiplying $\mathbf{y}_n$ by

$$C(\tfrac{1}{2}) = \begin{bmatrix} 1 & 0 & 0 & 0 \\ 0 & \frac{1}{2} & 0 & 0 \\ 0 & \frac{3}{16} & \frac{3}{8} & -\frac{1}{16} \\ 0 & 0 & \frac{1}{2} & 0 \end{bmatrix}$$

The third row uses the relationship

$$\tilde{h}y'(t-\tilde{h}) = \tfrac{3}{16}hy'(t) + \tfrac{3}{8}hy'(t-h) - \tfrac{1}{16}hy'(t-2h) + 0(h^4)$$

where $\tilde{h} = h/2$. Suppose we have a value for $\mathbf{y}_n$ based on the step size h, and we integrate the equation $y' = 0$ by first halving the step size and then performing two integration steps. The result is to multiply $\mathbf{y}_n$ by the matrix $BBC(\frac{1}{2})$, which is

$$BBC(\tfrac{1}{2}) = \begin{bmatrix} 1 & \frac{23}{192} & -\frac{26}{192} & \frac{11}{192} \\ 0 & 0 & 0 & 0 \\ 0 & 0 & 0 & 0 \\ 0 & \frac{1}{2} & 0 & 0 \end{bmatrix}$$

The first row tells us in what way the errors in the initial values affect the answer.

Alternatively, we could use the approximations

$$y\left(t+\frac{h}{2}\right) = y(t) + \frac{h}{24}(17y'(t) - 7y'(t-h) + 2y'(t-2h)) + 0(h^4)$$

and

$$y\left(t+\frac{h}{2}\right) = y(t) + \frac{h}{72}\left(64y'(t) - 33y'\left(t-\frac{h}{2}\right) + 5y'\left(t-\frac{3h}{2}\right)\right) + 0(h^4)$$

which use the points labeled $n-3$, $n-2$, and $n-1$ to get to n and $n-2$, $n-1$, and n to get to $n+1$ in Figure 9.1.

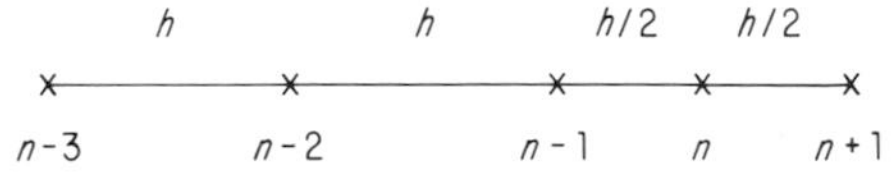

Fig. 9.1 Step changing by variable formulas.

The first step corresponds to the use of the matrix

$$B_1 = \begin{bmatrix} 1 & \frac{17}{24} & -\frac{7}{24} & \frac{2}{24} \\ 0 & 0 & 0 & 0 \\ 0 & 1 & 0 & 0 \\ 0 & 0 & 1 & 0 \end{bmatrix}$$

and the second to the use of

$$B_2 = \begin{bmatrix} 1 & \frac{64}{72} & -\frac{33}{72} & \frac{5}{72} \\ 0 & 0 & 0 & 0 \\ 0 & \frac{1}{2} & 0 & 0 \\ 0 & 0 & \frac{1}{2} & 0 \end{bmatrix}$$

(In the second step the saved values of hy' are halved to get $\tilde{h}y'$.)

The result of these two steps is equivalent to applying the matrix B_2B_1 to the initial values. This time the effect is

$$B_2B_1 = \begin{bmatrix} 1 & \frac{9}{36} & -\frac{8}{36} & \frac{3}{36} \\ 0 & 0 & 0 & 0 \\ 0 & 0 & 0 & 0 \\ 0 & \frac{1}{2} & 0 & 0 \end{bmatrix}$$

We see that $B_2B_1 \neq BBC(\frac{1}{2})$, so that the two techniques for step halving are not identical.

The result of this difference could be a difference in overall stability properties, but up to this time there has been no study of the stability of multistep methods in the presence of step changing.

Change of Order

The order of the method is determined by the coefficients of the method. Generally, higher order methods require a greater number of prior points. A sufficiently large set of prior information could be saved always. Thus the 5×5 matrix and 5 vector

$$B = \begin{bmatrix} 1 & \frac{23}{12} & -\frac{16}{12} & \frac{5}{12} & 0 \\ 0 & 0 & 0 & 0 & 0 \\ 0 & 1 & 0 & 0 & 0 \\ 0 & 0 & 1 & 0 & 0 \\ 0 & 0 & 0 & 1 & 0 \end{bmatrix}, \qquad \mathbf{c} = \begin{bmatrix} 0 \\ 1 \\ 0 \\ 0 \\ 0 \end{bmatrix}$$

is equally well the representation for the three-step Adams-Bashforth method, although it also saves the value of hy'_{n-3} in $\mathbf{y}_n$. In practice, it would cost nothing but storage space to save already computed information if the representation uses it in the form in which it is originally computed. Some representations require that the saved information be processed at each step. In that case, it would be unwise to retain unnecessary data from step to step except just prior to increasing the order. Thus, if Adams' method with backward differences is used, only those differences needed should be carried.

Estimating the Error

We have seen that the local truncation error is proportional to a derivative of the solution $y(t)$. The derivative can be estimated by using a numerical differentiation formula [see Hildebrand (1956), Section 3.3]. This is a linear combination of the saved information $\mathbf{a}_n$. It may be desirable to estimate the error in several different order methods so several different derivatives may have to be estimated. This is represented by the operation $D\mathbf{a}_n$, where D is a $q \times$ length $(\mathbf{a}_n)$ matrix if q different derivatives are to be estimated. In some representations D may be particularly simple. For example, if backward differences of hy' are retained in $\mathbf{a}_n$, the difference $\nabla^j hy'$ is an estimate of $h^{j+1}y^{(j+1)}$, so D need contain only one nonzero entry in each row.

9.2.2 Adams' Methods in the Backward Difference Representation

The **B** and **c** matrices for the three-step Adams-Bashforth-Moulton method of order four were seen to be

$$B = \begin{bmatrix} 1 & \frac{23}{12} & -\frac{16}{12} & \frac{5}{12} \\ 0 & 3 & -3 & 1 \\ 0 & 1 & 0 & 0 \\ 0 & 0 & 1 & 0 \end{bmatrix}, \qquad \mathbf{c} = \begin{bmatrix} \frac{3}{8} \\ 1 \\ 0 \\ 0 \end{bmatrix}$$

If the third order Adams-Moulton corrector is used, the reader can verify that B is unchanged and $\mathbf{c}$ becomes $[\frac{5}{12}, 1, 0, 0]^T$. If we transform to backward differences, we have

$$\begin{bmatrix} y_n \\ hy'_n \\ \nabla hy'_n \\ \nabla^2 hy'_n \end{bmatrix} = \begin{bmatrix} 1 & 0 & 0 & 0 \\ 0 & 1 & 0 & 0 \\ 0 & 1 & -1 & 0 \\ 0 & 1 & -2 & 1 \end{bmatrix} \begin{bmatrix} y_n \\ hy'_n \\ hy'_{n-1} \\ hy'_{n-2} \end{bmatrix} = T\mathbf{y}_n$$

and the new method is given by

$$A = TBT^{-1} = \begin{bmatrix} 1 & 1 & \frac{1}{2} & \frac{5}{12} \\ 0 & 1 & 1 & 1 \\ 0 & 0 & 1 & 1 \\ 0 & 0 & 0 & 1 \end{bmatrix}, \text{ and at least } \mathbf{l} = T\mathbf{c} = \begin{bmatrix} \beta_0 \\ 1 \\ 1 \\ 1 \end{bmatrix} \tag{9.11}$$

where $\beta_0 = \frac{3}{8}$ for a fourth order corrector and $\frac{5}{12}$ for a third order corrector. The number of operations is not as large as it might appear. Multiplication of $\mathbf{a}_n$ by A involves only two multiplications, five additions, and four stores, since the last three rows of A can be handled by

$$a_{n-1,3} + a_{n-1,2} \longrightarrow a_{n,2}$$
$$a_{n,2} + a_{n-1,1} \longrightarrow a_{n,1}$$

where $a_{n-1,i}$ is the ith component of $\mathbf{a}_{n-1}$.

This representation requires fewer operations per step than the representation using derivative values. It is also one of the most convenient for estimating derivatives since the backward differences provide these estimates directly. Unfortunately, it is one of the less convenient for changing step size. The matrix $C(\alpha)$ which must be used to interpolate to get the backward differences based on a new step size is nontrivial. Apart from the first two rows and columns, which are diagonal, the remainder of $C(\alpha)$ is full upper triangular.

For the 4 × 4 case above,

$$C(\alpha) = \begin{bmatrix} 1 & 0 & 0 & 0 \\ 0 & \alpha & 0 & 0 \\ 0 & 0 & \alpha^2 & \dfrac{\alpha^2(1-\alpha)}{2} \\ 0 & 0 & 0 & \alpha^3 \end{bmatrix}$$

The third row is based on

$$\tilde{h}\tilde{\nabla}y' = \alpha^2 h\nabla y' + \frac{\alpha^2(1-\alpha)h\nabla^2 y'}{2} + 0(h^4)$$

where $\tilde{h} = \alpha h$, ∇ is based on a step of size h, and $\tilde{\nabla}$ is based on a step of size $\tilde{h}$.

General Adams' Methods in Backward Difference Representation.

The general form of Adams' methods in this representation is given by

$$A = \begin{bmatrix} 1 & \gamma_0 & \gamma_1 \ldots \gamma_{k-2} \\ & 1 & 1 \ldots 1 \\ & & 1 \\ & & & \ddots \\ & & & & 1 \end{bmatrix}, \qquad \mathbf{l} = \begin{bmatrix} \gamma_q \\ 1 \\ 1 \\ \vdots \\ 1 \end{bmatrix}$$

where $q = k - 2$ or $k - 1$, depending on whether the corrector is of the same order $(k - 1)$ as the predictor or one higher (k), respectively.

An advantage to this representation is that the components of $\mathbf{a}_n$ tend to get smaller since the jth difference of hf_n is $h^{j+1}f_n^{(j+1)} + 0(h^{j+2})$. Consequently, the rounding errors, which in floating point computation are proportional to the size of the entries, are only significant in the first few components. If multiple precision has to be used, it may not be necessary to carry more than single precision in the last few components of $\mathbf{a}_n$.

The use of divided differences [see Hildebrand (1956), Chapter 2] has been proposed by Krogh (1969) as a way to simplify the step changing process. This introduces k multiplications into each step for each dependent variable, but may be preferable for large systems of equations where the overhead of computing the predictor-corrector coefficients is small.

9.2.3 The Nordsieck Form of Adams' Method

Nordsieck (1962) proposed saving approximations to $y_n, y'_n, hy''_n/2, \ldots, h^{k-2}y_n^{(k-1)}/(k-1)!$ instead of the dependent variables and their derivatives. His motive was to make step changing simple. The correspondence between Adams' method and Nordsieck's method can be shown by a transformation as follows.

The set of saved values $y_n, hy'_n, \ldots, hy'_{n-k+2}$ used in the $(k-1)$-step Adams' method uniquely determines a $(k-1)$th degree polynomial which agrees with these y and y'. This polynomial can be equally well represented by means of its value and first $(k-1)$ derivatives at t_n. Thus, we have a transformation for $(k-1)$th degree polynomials

$$\mathbf{a}_n = T\mathbf{y}_n$$

where $\mathbf{y}_n = [y_n, hy'_n, \ldots, hy'_{n-k+2}]^T$ and $\mathbf{a}_n = [y_n, hy'_n, \ldots, h^{k-1}y_n^{(k-1)}/(k-1)!]^T$ For the third order predictor, fourth order corrector method given by B and $\mathbf{c}$ preceding (9.11) we have

$$T = \begin{bmatrix} 1 & 0 & 0 & 0 \\ 0 & 1 & 0 & 0 \\ 0 & \frac{3}{4} & -1 & \frac{1}{4} \\ 0 & \frac{1}{6} & -\frac{1}{3} & \frac{1}{6} \end{bmatrix} \quad \text{and} \quad T^{-1} = \begin{bmatrix} 1 & 0 & 0 & 0 \\ 0 & 1 & 0 & 0 \\ 0 & 1 & -2 & 3 \\ 0 & 1 & -4 & 12 \end{bmatrix}$$

from which we obtain

$$A = TBT^{-1} = \begin{bmatrix} 1 & 1 & 1 & 1 \\ & 1 & 2 & 3 \\ & & 1 & 3 \\ 0 & & & 1 \end{bmatrix}, \qquad \mathbf{l} = T\mathbf{c} = \begin{bmatrix} \frac{3}{8} \\ 1 \\ \frac{3}{4} \\ \frac{1}{6} \end{bmatrix} \tag{9.12}$$

It appears that the work in premultiplying by A has been increased in order to save on the step changing process. In fact, it is not as bad as it appears at first sight since the premultiplication by A only involves additions. A is the *Pascal triangle* matrix whose (i, j) element is $\binom{j}{i}$, $k > j \geq i \geq 0$. It expresses the fact that the $(k-1)$th order predictor formula for $h^i y_n^{(i)}/i!$ is given by the Taylor's series as

$$\frac{h^i y_n^{(i)}}{i!} = \sum_{j=i}^{k-1} \frac{\binom{j}{i} h^j y_{n-1}^{(j)}}{j!} + O(h^k)$$

In view of the relation

$$\binom{j}{i} + \binom{j}{i+1} = \binom{j+1}{i+1}$$

we can calculate $A\mathbf{a}$ by the steps

$$a_{k-1} + a_{k-2} \longrightarrow a_{k-2}$$
$$\cdots$$
$$a_1 + a_0 \longrightarrow a_0$$

$$a_{k-1} + a_{k-2} \longrightarrow a_{k-2}$$
$$\cdots$$
$$a_2 + a_1 \longrightarrow a_1$$

$$a_{k-1} + a_{k-2} \longrightarrow a_{k-2}$$
$$\cdots$$
$$a_3 + a_2 \longrightarrow a_2$$

$$\cdots$$

$$a_{k-1} + a_{k-2} \longrightarrow a_{k-2}$$

where a_j is the jth element of $\mathbf{a}$. This requires $k(k-1)/2$ additions and stores. This is more than the backward difference method, which requires $2k-3$ additions and $k-1$ stores for the same step.

The matrix $C(\alpha)$ for changing the step size by α is

$$C(\alpha) = \begin{bmatrix} 1 & & & & 0 \\ & \alpha & & & \\ & & \alpha^2 & & \\ & 0 & & \ddots & \\ & & & & \alpha^{k-1} \end{bmatrix}$$

This method has the other advantages of the backward difference method in that round-off errors are small and direct estimates of the derivatives are available. It will also be seen to be useful for higher order equations to be discussed in Section 9.2.5.

9.2.4 Modified Multistep Methods

Reference has been made previously to a theorem of Dahlquist's which states that a strongly stable k-step method cannot have order greater than $k+1$. The notation we have developed for k-step methods includes methods which are not strictly multistep methods. The $\mathbf{c}$ in Eq. (9.1) contained one constant determined by the method (β_0^*), a one, and $2k-2$ elements equal to zero. What happens if we allow any elements of $\mathbf{c}$ to be nonzero?

We can best see this by an example. Let us start with the three-step Adams-Bashforth-Moulton scheme of order four given by (9.11). Instead of representing polynomials of degree three by $[y_n, hy'_n, hy'_{n-1}, hy'_{n-2}]^T$, let us use $\mathbf{a}_n = [y_n, hy'_n, y_{n-1}, hy'_{n-1}]^T$. The transformation is

$$T = \begin{bmatrix} 1 & 0 & 0 & 0 \\ 0 & 1 & 0 & 0 \\ 1 & -\frac{5}{12} & -\frac{8}{12} & \frac{1}{12} \\ 0 & 0 & 1 & 0 \end{bmatrix}$$

The method is given by

$$A = TBT^{-1} = \begin{bmatrix} -4 & 4 & 5 & 2 \\ -12 & 8 & 12 & 5 \\ 1 & 0 & 0 & 0 \\ 0 & 1 & 0 & 0 \end{bmatrix}, \qquad \mathbf{l} = T\mathbf{c} = \begin{bmatrix} \frac{3}{8} \\ 1 \\ -\frac{1}{24} \\ 0 \end{bmatrix}$$

This corresponds to the *modified multistep method* given by

$$
\begin{aligned}
y_{n,(0)} &= -4y_{n-1} + 5\bar{y}_{n-2} + 4hf_{n-1} + 2hf_{n-2}\\
D_{(0)} &= hf(y_{n,(0)}, t_n) - (-12y_{n-1} + 12\bar{y}_{n-2} + 8hf_{n-1} + 5hf_{n-2})\\
y_{n,(1)} &= y_{n,(0)} + \frac{3D_{(0)}}{8}\\
&\cdots\\
D_{(m)} &= hf(y_{n,(m)}, t_n) - hf(y_{n,(m-1)}, t_n)\\
y_{n,(m+1)} &= y_{n,(m)} + \frac{3D_{(m)}}{8}\\
&\cdots\\
y_n &= y_{n,(M)}\\
f_n &= f(y_{n,(M-1)})\\
\bar{y}_{n-1} &= y_{n-1} - \frac{D_{(0)} + D_{(1)} + \cdots + D_{(M-1)}}{24}
\end{aligned}
$$

This is identical to an ordinary multistep method until the last step, which is a correction to one of the earlier function values. It is this step which serves to stabilize the method. Since the method is equivalent to the three-step Adams-Bashforth-Moulton predictor-corrector scheme, we know that it is stable and of fourth order although it is only a two-step method.

We could have started with a $(2k - 1)$-step Adams-Bashforth-Moulton method of order $2k$ and transformed it to a k-step modified multistep method of the same order. Thus, we see that stable k-step methods of order $2k$ are possible provided that we relax the definition of multistep methods to allow for additional nonzero components in $\mathbf{c}$. It is for this reason that we prefer the name *multivalue methods*.† The important characterization of a method is the number of values that are saved from step to step and not the number of steps. A k-value method for first order equations can be of kth order.

9.2.5 Higher Order Equations

As with one-step methods, multivalue methods can be applied to higher order equations by reducing them to first order systems. They can also be applied directly to the higher order equations.

Let us consider the formulation we have for multivalue methods. The information about the prior behavior of the solution is carried in a vector $\mathbf{a}_{n-1}$ which can be considered to represent a polynomial of degree $k - 1$ which approximates the solution in the neighborhood of t_{n-1} (assuming k components in $\mathbf{a}_{n-1}$). The prediction step $\mathbf{a}_{n,(0)} = A\mathbf{a}_{n-1}$ is an extrapolation using that polynomial (or some other approximation to it) to the value of $\mathbf{a}$ at t_n. If the highest possible order predictor formula $(k - 1)$ is used, $\mathbf{a}_n$ will represent the same polynomial. If the solution to the differential equation

†A term suggested by Prof. B. Parlett of Berkeley.

is a polynomial of degree $k-1$ or less and $\mathbf{a}_{n-1}$ represented that solution correctly, then $\mathbf{a}_n$ will also represent that solution apart from round-off errors. If the solution of the differential equation is not a polynomial of degree $\leq k-1$, there will also be truncation errors of size $0(h^k)$. It is evident that the predictor by itself cannot be stable because it pays no heed to the differential equation at all. Consequently, we correct for the errors, both those propagated from earlier steps and those introduced through round-off and truncation errors, by measuring whether the approximating polynomial given by $\mathbf{a}_{n,(0)}$ satisfies the equation at t_n. This is done by the function $F(\mathbf{a}_{n,(0)})$, which is zero if the differential equation is satisfied. Multiples of $F(\mathbf{a}_{n,(0)})$ are added to $\mathbf{a}_{n,(0)}$ when F is not zero by the corrector equation $\mathbf{a}_{n,(m+1)} = \mathbf{a}_{n,(m)} + \mathbf{l}F(\mathbf{a}_{n,(m)})$. If this process converges, it converges to an $\mathbf{a}_n$ such that $F(\mathbf{a}_n) = 0$. The stability of the process is governed by the matrix

$$S_n = \prod_{i=0}^{M-1} \left(I + \mathbf{l}\frac{\partial F}{\partial \mathbf{a}}(\xi_i)\right) A$$

and we have seen that there exist $\mathbf{l}$ for first order equations such that the process is stable since we derived them from stable multistep methods.

We naturally ask what happens if exactly the same technique is applied to higher order equations. Suppose that we have the differential equation

$$y^{(p)} = f(y, y', y'', \ldots, y^{(p-1)}, t)$$

Let us use a representation in which the first $p+1$ components of $\mathbf{a}$ are $y, hy', h^2y''/2, \ldots, h^py^{(p)}/p!$. Define

$$F(\mathbf{a}) = \frac{h^p}{p!} f\left(a_0, \frac{a_1}{h}, \frac{2a_2}{h^2}, \ldots, \frac{(p-1)!a_{p-1}}{h^{p-1}}, t\right) - a_p$$

where a_i is the ith component of $\mathbf{a}$, numbering from 0. We can use the same prediction process, namely

$$\mathbf{a}_{n,(0)} = A\mathbf{a}_{n-1}$$

and a similar correction process given by

$$\mathbf{a}_{n,(m+1)} = \mathbf{a}_{n,(m)} + \mathbf{l}F(\mathbf{a}_{n,(m)})$$

where $F(\mathbf{a})$ is the amount by which the differential equation is not satisfied locally by $\mathbf{a}$, and $\mathbf{l}$ is chosen to achieve stability and accuracy. If we examine the error propogation, we again get the equation

$$\mathbf{e}_n = S_n \mathbf{e}_{n-1} + \mathbf{d}_n$$

where $\mathbf{d}_n$ is the local truncation error and S_n is given by

$$S_n = \prod_{i=1}^{M} \left(I + \mathbf{l}\frac{\partial F}{\partial \mathbf{a}}(\xi_i)\right) A$$

Since

$$\frac{\partial F}{\partial \mathbf{a}} = -\delta_p^T + \sum_{q=0}^{p-1} \frac{\partial f}{\partial y^{(q)}} \boldsymbol{\delta}_q^T h^{p-q} \frac{q!}{p!}$$

we find that

$$S_n = S + 0(h)$$

where

$$S = (I - \mathbf{l}\boldsymbol{\delta}_p^T)^M A$$

The stability of the method is governed by the eigenvalues of S. If $\mathbf{d}_n$ is no worse than $0(h^{p+1})$ and all of the eigenvalues of S are inside the unit circle or on the unit circle and simple except for a p-fold eigenvalue at 1, the method will converge as $h \longrightarrow 0$. This will be proved in Chapter 10 as will the following results:

If A is such that the prediction process is of the maximum possible order $(k-1)$, an $\mathbf{l}$ *can be chosen such that $k - p$ eigenvalues of*

$$S = (I - \mathbf{l}\boldsymbol{\delta}_p^T)^M A$$

take on any desired values. The other p eigenvalues of S are 1. The $k - p$ eigenvalues uniquely determine the last $k - p$ components of $\mathbf{l}$. *The first p components of* $\mathbf{l}$ *can be chosen to make the truncation error* $0(h^{k+1})$. *Such a method for a* pth *order equation will converge with a global error* $0(h^{k+1-p})$ *if the $k - p$ extraneous eigenvalues of S are inside the unit circle or simple on the unit circle.*

(If some of the derivatives are absent from f, and if the starting values are chosen judiciously, the order can be increased slightly.)

Thus we see that regardless of the representation we use, we can choose the extraneous eigenvalues to have any values we wish. The most common choice is to make them zero, in which case we have the higher order analogs of Adams' method.

S has p principal eigenvalues equal to 1 for a pth order equation. If the stability matrix for the general linear pth order differential equation

$$y^{(p)} + a_1 y^{(p-1)} + \cdots + a_p y = 0$$

is derived, the p principal roots will approximate $e^{h\lambda_i}$, $i = 1, \ldots, p$, where the λ_i are roots of

$$\lambda^p + a_1 \lambda^{p-1} + \cdots + a_p \lambda^0 = 0$$

A representation of the method in which the derivatives appear explicitly, at least through order p, is desirable so as to avoid performing linear transformations at each step in order to get the derivatives needed when F is evaluated. The values of $\mathbf{l}$ for $p = 1, 2, 3$, and 4 are given in Table 9.1 below for the representation which uses $\mathbf{a} = [y, hy', \ldots, h^{k-1}y^{(k-1)}/(k-1)!]^T$. All extraneous roots are zero, so for $p = 1$ these are equivalent to Adams' methods. The local truncation error is $0(h^{k+q})$, where q is the largest integer such that the differential equation can be written as $y^{(p)} = f(y, y', \ldots, y^{(p-q)}, t)$. (The relevance of q will be explained in Chapter 10.) The global error is $0(h^{k+q-p})$.

Table 9.1 COEFFICIENTS FOR NORMAL FORM MULTIVALUE METHODS

p	k	l_0	l_1	l_2	l_3	l_4	l_5	l_6	l_7
1	3	$\frac{5}{12}$	1	$\frac{1}{2}$					
	4	$\frac{3}{8}$	1	$\frac{3}{4}$	$\frac{1}{6}$				
	5	$\frac{251}{720}$	1	$\frac{11}{12}$	$\frac{1}{3}$	$\frac{1}{24}$			
	6	$\frac{95}{288}$	1	$\frac{25}{24}$	$\frac{35}{72}$	$\frac{5}{48}$	$\frac{1}{120}$		
	7	$\frac{19087}{60480}$	1	$\frac{137}{120}$	$\frac{5}{8}$	$\frac{17}{96}$	$\frac{1}{40}$	$\frac{1}{720}$	
	8	$\frac{5257}{17280}$	1	$\frac{49}{40}$	$\frac{203}{270}$	$\frac{49}{192}$	$\frac{7}{144}$	$\frac{7}{1440}$	$\frac{1}{5040}$
2	4	$\frac{1}{6}$	$\frac{5}{6}$	1	$\frac{1}{3}$				
	5	$\frac{19}{120}$	$\frac{3}{4}$	1	$\frac{1}{2}$	$\frac{1}{12}$			
	6	$\frac{3}{20}$	$\frac{251}{360}$	1	$\frac{11}{18}$	$\frac{1}{6}$	$\frac{1}{60}$		
	7	$\frac{863}{6048}$	$\frac{665}{1008}$	1	$\frac{25}{36}$	$\frac{35}{144}$	$\frac{1}{24}$	$\frac{1}{360}$	
	8	$\frac{1925}{14112}$	$\frac{19087}{30240}$	1	$\frac{137}{180}$	$\frac{5}{16}$	$\frac{17}{240}$	$\frac{1}{120}$	$\frac{1}{2520}$
3	5	$\frac{1}{4}$	$\frac{1}{2}$	$\frac{5}{4}$	1	$\frac{1}{4}$			
	6	$\frac{3}{80}$	$\frac{19}{40}$	$\frac{9}{8}$	1	$\frac{3}{8}$	$\frac{1}{20}$		
	7	$\frac{221}{5040}$	$\frac{9}{20}$	$\frac{251}{240}$	1	$\frac{11}{24}$	$\frac{1}{10}$	$\frac{1}{120}$	
	8	$\frac{2185}{46368}$	$\frac{863}{2016}$	$\frac{95}{96}$	1	$\frac{25}{48}$	$\frac{49}{336}$	$\frac{1}{48}$	$\frac{1}{840}$
4	6	$\frac{1}{30}$	$\frac{1}{10}$	1	$\frac{5}{3}$	1	$\frac{1}{5}$		
	7	$\frac{16}{630}$	$\frac{3}{20}$	$\frac{19}{20}$	$\frac{3}{2}$	1	$\frac{3}{10}$	$\frac{1}{30}$	
	8	$\frac{11}{630}$	$\frac{221}{1260}$	$\frac{9}{10}$	$\frac{251}{180}$	1	$\frac{11}{30}$	$\frac{1}{15}$	$\frac{1}{210}$

EXAMPLE

Consider the four-value method for second order equations obtained from Table 9.1. It saves the values y_n, hy'_n, $h^2y''_n/2$, and $h^3y'''_n/6$. The vector $\mathbf{l}$ is $[\frac{1}{6}, \frac{5}{6}, 1, \frac{1}{3}]^T$. If we transform to a representation saving y_n, y_{n-1}, $h^2y''_n/2$, and $h^2y''_{n-1}/2$ by

$$\begin{bmatrix} y_n \\ y_{n-1} \\ \dfrac{h^2y''_n}{2} \\ \dfrac{h^2y''_{n-1}}{2} \end{bmatrix} = \begin{bmatrix} 1 & 0 & 0 & 0 \\ 1 & -1 & 1 & -1 \\ 0 & 0 & 1 & 0 \\ 0 & 0 & 1 & -3 \end{bmatrix} \begin{bmatrix} y_n \\ hy'_n \\ \dfrac{h^2y''_n}{2} \\ \dfrac{h^3y'''_n}{6} \end{bmatrix}$$

or $\mathbf{y}_n = Q\mathbf{a}_n$, we get

$$B = QAQ^{-1} = \begin{bmatrix} 2 & -1 & 2 & 0 \\ 1 & 0 & 0 & 0 \\ 0 & 0 & 2 & -1 \\ 0 & 0 & 1 & 0 \end{bmatrix}$$

$$\mathbf{c} = Q\mathbf{l} = [\tfrac{1}{6}, 0, 1, 0]^T$$

This is equivalent to using the predictor equation

$$y_{n,(0)} = 2y_{n-1} - y_{n-2} + h^2 y''_{n-1}$$

This is one of the Stormer (1907 and 1921) explicit formulas for the special second order equation $y'' = f(y, t)$. (In the representation using scaled derivatives, we can apply it to the general second order equation with y' present.) The corrector equation is equivalent to

$$y_{n,(m+1)} = 2y_{n-1} - y_{n-2} + \frac{h^2}{12}[f(y_{n,(m)}) + 10y''_{n-1} + y''_{n-2}]$$

This is one of the Cowell implicit methods for the special second order equation [Cowell and Crommelin (1910)]. General forms of these methods for the special second order equation are discussed in Henrici (1962), Chapter 6. We will discuss the theory of these methods using the scaled derivative representation in Chapter 10.

9.3 AUTOMATIC CONTROL OF STEP SIZE AND ORDER

A program embodying a multivalue method will have to use techniques for starting, changing step, and changing order as necessary. In this section we will discuss these techniques and illustrate them by means of a general purpose automatic program for first order equations.

The choice of which class of equivalent methods to use depends on the problem. Little is known about the problem to be integrated at the time that a general purpose program is to be written, so the Adams' methods whose extraneous eigenvalues are zero is usually the best choice. (Other methods for special problems will be discussed in Chapter 11.) However, what is to be said in this section and the program to be given later is equally applicable to other methods if the constants are changed appropriately.

If the order of the predictor plus the number of corrector iterations exceeds the order of the corrector, the local truncation error is $C_{q+1}h^{q+1}y^{(q+1)} + 0(h^{q+2})$ for a qth order corrector. For the Adams-Moulton method, $C_{q+1} = \gamma_q^*$ (see Table 7.4). In order to estimate $y^{(q+1)}$, at least $q + 2$ values are needed. If $q + 1$ are carried in $\mathbf{a}_{n-1}$, that and the one additional value generated when $\mathbf{a}_n$ is derived provide enough for estimating the local truncation error. Since the errors for several different orders need to be compared, we will use either backward differences or scaled derivatives in $\mathbf{a}$. Since step size changing is simpler with scaled derivatives, we elect to use $\mathbf{a} = [y, hy', \ldots, h^q y^{(q)}/q!]^T$ with a qth order corrector. With $q + 1$ values in $\mathbf{a}$, the predictor can also be of qth order so the error is $C_{q+1}h^{q+1}y^{(q+1)} + 0(h^{q+2})$ even for only one corrector iteration. The $\mathbf{l}$ for these methods can be obtained from Table 9.1, where $l_1, \ldots, l_q$ are obtained from the row with $k = q$, while l_0 is obtained from the row with $k = q - 1$. (Table 9.1 gives coefficients for methods in which the corrector has order one higher than the predictor.)

With this representation the change ∇a_q in the last coefficient of $\mathbf{a}$ in each step is an estimate of $h^{q+1}y^{(q+1)}/q!$.† Consequently, if we wish to control the single step truncation error to be less than ϵ, we must select h so that

$$C_{q+1}q!\,\nabla a_q \leq \epsilon$$

where ∇a_q is the backward difference of the last component of $\mathbf{a}$. When a system of equations is to be integrated, it may be desirable to control the error in each component differently, so we control

$$C_{q+1}q!\left\|\frac{\nabla a_q}{\omega}\right\|_2 \leq \epsilon \tag{9.13}$$

where for each member of the system there is a ∇a_q component and a weight component ω. $\|\cdot\|_2$ is the L_2-norm. The L_2-norm was used because it was slightly faster to calculate on the computer used. The max-norm could be used equally well if it were faster. Note that if the L_2-norm is used, it is only necessary to form $(\|\cdot\|_2)^2$ to perform the test (9.13).

The basic step control mechanism is to execute one step and to perform the test (9.13). If the test succeeds, the step is accepted; otherwise, it is rejected. The step size to use for the next step or to repeat the rejected step is estimated to be αh, where

$$C_{q+1}q!\alpha^{q+1}\left\|\frac{\nabla a_q}{\omega}\right\|_2 = \epsilon$$

If this step size were used and if the error were exactly proportional to h^{q+1} (that is, if ∇a_q is constant from step to step), the test would just be satisfied next time. However, ∇a_q is not usually constant, so a slightly smaller step is used in order that test (9.13) can reasonably be expected to be satisfied. In the program given later, α is estimated by

$$\alpha = \frac{1}{1.2}\left[\frac{\epsilon}{C_{q+1}q!}\frac{1}{\left\|\frac{\nabla a_q}{\omega}\right\|_2}\right]^{1/q+1}$$

It is also necessary to test the step sizes that could be used in other orders. Since

$$\nabla^2 a_q \cong \frac{h^{q+2}}{q!}y^{(q+2)}$$

$$a_q \cong \frac{h^q}{q!}y^{(q)}$$

the step sizes that could be used in orders $q+1$ and $q-1$ can be estimated

†To justify it, it is necessary to show that the numerical solution y_n is given by $y(t_n) + h^s\delta(t_n, h)$, where $\delta(t_n, h)$ has at least $q + 2 - s$ continuous derivatives with respect to t. Although this cannot be proved except in special cases, the technique appears to work successfully.

to be αh, where

$$\alpha = \frac{1}{1.4}\left[\frac{\epsilon}{C_{q+2}q!}\frac{1}{\left\|\frac{\nabla^2 a_q}{\omega}\right\|_2}\right]^{1/(q+2)} \qquad \text{for order } q+1$$

$$\alpha = \frac{1}{1.3}\left[\frac{\epsilon}{C_q q!}\frac{1}{\left\|\frac{a_q}{\omega}\right\|_2}\right]^{1/q} \qquad \text{for order } q-1$$

The factors 1.3 and 1.4 are to provide a similar range in which test (9.13) will succeed. They were chosen on an ad hoc basis to bias the method in favor first of not changing order since it requires additional computer time, and then in favor of reducing the order because it is a little less work per step.

The estimates of α are made:

1. if a step fails, except then no attempt is made to increase the order.
2. $q + 1$ steps after the last change in order or stepsize. [Tests quoted by Nordsieck (1962) and confirmed by this writer indicate that if a step is increased more frequently than this, large errors can accumulate, resulting in a later step reduction. See also Problems 2 and 3 at the end of this chapter.]
3. ten steps after the α were last estimated if no step increase was made at that time. (This is to reduce the overhead of testing too frequently.)

When the α are estimated, the order corresponding to the largest is chosen and the step changed appropriately. The program given below does not increase the step at the current order q if $\alpha \leq 1.1$ since it is felt that the increase is not worth the computer time to perform it.

The program below will handle N equations and allows for Adams' or stiff methods to be used. The latter will be discussed in Chapter 11.

Starting is almost automatic. The value of hy'_0 can be calculated from the initial value and the differential equations. This is sufficient to allow a first order process to be used. The order control mechanism can then increase the order to a desirable level.

The YMAX parameter is an array containing the components of the weights used in (9.13). The elements of YMAX are updated after a completed step to contain the absolute values of the components y^i if the latter are larger than the current values of YMAX (I). This provides an error test relative to the largest value in the history of the y^i unless the user overrides it by changing YMAX before each call. The estimated single step errors are controlled to be less than the parameter EPS. Thus, the overall error is proportional to the number of steps if f is independent of y. If the equations are stable ($\partial f/\partial y$ is negative), early errors are decreased so the error is smaller; if they are unstable, the error is larger. This is because the error control is based on estimates of the local truncation error only.

Note to Program

It should be noted that in the case of stiff methods (to be described in Chapter 11) a program MATINV is called to invert a matrix. In fact, the inverse of this matrix is used to premultiply a vector. This costs approximately $5N^3/6$ additional multiplications each time that the matrix is inverted, where N is the number of equations being solved in the system. The single call of the subroutine MATINV can be replaced by a call to the first stage of a Gaussian elimination routine such as DECOMP on p. 68 of Forsythe and Moler (1967), while the loop to premultiply the vector by the inverse can be replaced by the second stage (the back substitution process) of the Gaussian elimination routine, such as SOLVE on p. 69 of Forsythe and Moler (1967). This loop occurs in the five statements ending on statement number 400. The additional cost of this change is the overhead of a call of SOLVE, which occurs about ten times as frequently as calls on DECOMP or MATINV. For N of any size (over about 5, depending on the computer used) the changed program would execute faster.

```
      SUBROUTINE DIFSUB(N,T,Y,SAVE,H,HMIN,HMAX,EPS,MF,YMAX,ERROR,KFLAG,
     1                  JSTART,MAXDER,PW)
      IMPLICIT REAL*8 (A-H,O-Z)
C**********************************************************************
C*                                                                    *
C* THIS SUBROUTINE INTEGRATES A SET OF N ORDINARY DIFFERENTIAL FIRST  *
C* ORDER EQUATIONS OVER ONE STEP OF LENGTH H AT EACH CALL.  H CAN BE  *
C* SPECIFIED BY THE USER FOR EACH STEP, BUT IT MAY BE INCREASED OR    *
C* DECREASED BY DIFSUB WITHIN THE RANGE HMIN TO HMAX IN ORDER TO      *
C* ACHIEVE AS LARGE A STEP AS POSSIBLE WHILE NOT COMMITTING A SINGLE  *
C* STEP ERROR WHICH IS LARGER THAN EPS IN THE L-2 NORM, WHERE EACH    *
C* COMPONENT OF THE ERROR IS DIVIDED BY THE COMPONENTS OF YMAX.       *
C*                                                                    *
C* THE PROGRAM REQUIRES THREE SUBROUTINES NAMED                       *
C*    DIFFUN(T,Y,DY)                                                  *
C*    MATINV(PW,N,M,J)                                                *
C*    PEDERV(T,Y,PW,M)                                                *
C* THE FIRST, DIFFUN, EVALUATES THE DERIVATIVES OF THE DEPENDENT      *
C* VARIABLES STORED IN Y(1,I) FOR I = 1 TO N, AND STORES THE          *
C* DERIVATIVES IN THE ARRAY DY.  THE SECOND IS CALLED ONLY IF THE     *
C* METHOD FLAG MF IS SET TO 1 OR 2 FOR STIFF METHODS.  IT MUST INVERT *
C* THE N BY N MATRIX STORED IN THE ARRAY PW(M,M).  IF THE INVERSION IS *
C* SUCCESFUL, J SHOULD BE SET TO 1, OTHERWISE IT SHOULD BE SET TO -1.  *
C* PEDERV IS USED ONLY IF MF IS 1, AND COMPUTES THE PARTIAL           *
C* DERIVATIVES OF THE DIFFERENTIAL EQUATIONS AS DESCRIBED UNDER THE   *
C* MF PARAMETER.                                                      *
C*                                                                    *
C* THE PROGRAM USES DOUBLE PRECISION ARITHMETIC FOR ALL FLOATING      *
C* POINT VARIABLES EXCEPT THOSE STARTING WITH P.  THE FORMER ARE      *
C* SINGLE PRECISION TO SAVE TIME AND SPACE.                           *
C*                                                                    *
C* THE TEMPORARY STORAGE SPACE IS PROVIDED BY THE CALLER IN THE       *
C* SINGLE PRECISION ARRAY PW AND THE DOUBLE PRECISION ARRAY SAVE.     *
C* THE ARRAY PW IS USED ONLY TO HOLD THE MATRIX OF THE SAME NAME, BUT  *
C* SAVE IS USED TO HOLD SEVERAL ARRAYS.  THE REGIONS USED ARE         *
C*    SAVE(J,I)   1.LE.J.LE.8  AND  1.LE.I.LE.N  IS USED TO SAVE THE  *
C*                VALUES OF Y IN CASE A STEP HAS TO BE REPEATED.      *
C*    SAVE(9,I)   IS USED MAINLY TO HOLD THE CORRECTION TERMS IN THE  *
C*                CORRECTOR LOOP.                                     *
```

```
C*     SAVE(10,I)  IS USED TO SAVE THE VALUES OF THE SUMS OF ALL OF THE *
C*                 CORRECTION TERMS IN THE PREVIOUS STEP AFTER THEY     *
C*                 HAVE BEEN ACCUMULATED IN THE ARRAY ERROR IN THE      *
C*                 CURRENT STEP.  THIS ENABLES THE BACKWARDS DIFFERENCE *
C*                 OF ERROR TO BE FORMED.  IT IS USED TO ESTIMATE THE   *
C*                 STEP SIZE FOR ONE ORDER HIGHER THAN CURRENT.         *
C*     SAVE(N1+I,1)  IS USED TO STORE THE DERIVATIVES WHEN THEY ARE     *
C*                 COMPUTED BY DIFFUN.  IT IS ALSO ACCESSED AS          *
C*                 SAVE(N2,1) AS A COMPLETE ARRAY.                      *
C*     SAVE(N5+I,1)  HOLDS THE DERIVATIVES DURING JACOBIAN EVALUATIONS. *
C*                 IT IS REFERENCED AS SAVE(N6,1) AS A COMPLETE ARRAY.  *
C*                                                                      *
C* THE PARAMETERS TO THE SUBROUTINE DIFSUB HAVE                         *
C* THE FOLLOWING MEANINGS..                                             *
C*                                                                      *
C*   N        THE NUMBER OF FIRST ORDER DIFFERENTIAL EQUATIONS.  N      *
C*              MAY BE DECREASED ON LATER CALLS IF THE NUMBER OF        *
C*              ACTIVE EQUATIONS REDUCES, BUT IT MUST NOT BE            *
C*              INCREASED WITHOUT CALLING WITH JSTART = 0.              *
C*   T        THE INDEPENDENT VARIABLE.                                 *
C*   Y        AN 8 BY N ARRAY CONTAINING THE DEPENDENT VARIABLES AND    *
C*              THEIR SCALED DERIVATIVES.  Y(J+1,I) CONTAINS            *
C*              THE J-TH DERIVATIVE OF Y(I) SCALED BY                   *
C*              H**J/FACTORIAL(J) WHERE H IS THE CURRENT                *
C*              STEP SIZE. ONLY Y(1,I) NEED BE PROVIDED BY              *
C*              THE CALLING PROGRAM ON THE FIRST ENTRY.                 *
C*                IF IT IS DESIRED TO INTERPOLATE TO NON MESH POINTS    *
C*              THESE VALUES CAN BE USED.  IF THE CURRENT STEP SIZE     *
C*              IS H AND THE VALUE AT T + E IS NEEDED, FORM             *
C*              S = E/H, AND THEN COMPUTE                               *
C*                                NQ                                    *
C*                 Y(I)(T+E) =   SUM   Y(J+1,I)*S**J                    *
C*                                J=0                                   *
C*   SAVE     A BLOCK OF AT LEAST 12*N FLOATING POINT LOCATIONS         *
C*              USED BY THE SUBROUTINES.                                *
C*   H        THE STEP SIZE TO BE ATTEMPTED ON THE NEXT STEP.           *
C*              H MAY BE ADJUSTED UP OR DOWN BY THE PROGRAM             *
C*              IN ORDER TO ACHEIVE AN ECONOMICAL INTEGRATION.          *
C*              HOWEVER, IF THE H PROVIDED BY THE USER DOES             *
C*              NOT CAUSE A LARGER ERROR THAN REQUESTED, IT             *
C*              WILL BE USED.  TO SAVE COMPUTER TIME, THE USER IS       *
C*              ADVISED TO USE A FAIRLY SMALL STEP FOR THE FIRST        *
C*              CALL.  IT WILL BE AUTOMATICALLY INCREASED LATER.        *
C*   HMIN     THE MINIMUM STEP SIZE THAT WILL BE USED FOR THE           *
C*              INTEGRATION.  NOTE THAT ON STARTING THIS MUST           *
C*              MUCH SMALLER THAN THE AVERAGE H EXPECTED SINCE          *
C*              A FIRST ORDER METHOD IS USED INITIALLY.                 *
C*   HMAX     THE MAXIMUM SIZE TO WHICH THE STEP WILL BE INCREASED      *
C*   EPS      THE ERROR TEST CONSTANT.  SINGLE STEP ERROR ESTIMATES     *
C*              DIVIDED BY  YMAX(I)  MUST BE LESS THAN THIS             *
C*              IN THE EUCLIDEAN NORM. THE STEP AND/OR ORDER IS         *
C*              ADJUSTED TO ACHEIVE THIS.                               *
C*   MF       THE METHOD INDICATOR. THE FOLLOWING ARE ALLOWED..         *
C*                      0    AN ADAMS PREDICTOR CORRECTOR IS USED.      *
C*                      1    A MULTI-STEP METHOD SUITABLE FOR STIFF     *
C*                             SYSTEMS IS USED. IT WILL ALSO WORK FOR   *
C*                             NON STIFF SYSTEMS.  HOWEVER THE USER     *
C*                             MUST PROVIDE A SUBROUTINE PEDERV WHICH   *
C*                             EVALUATES THE PARTIAL DERIVATIVES OF     *
C*                             THE DIFFERENTIAL EQUATIONS WITH RESPECT  *
C*                             TO THE Y'S.   THIS IS DONE BY CALL       *
C*                             PEDERV(T,Y,PW,M). PW IS AN N BY N ARRAY  *
C*                             WHICH MUST BE SET TO THE PARTIAL OF      *
C*                             THE I-TH EQUATION WITH RESPECT           *
C*                             TO THE J DEPENDENT VARIABLE IN PW(I,J).  *
C*                             PW IS ACTUALLY STORED IN AN M BY M       *
C*                             ARRAY WHERE M IS THE VALUE OF N USED ON  *
C*                             THE FIRST CALL TO THIS PROGRAM.          *
```

```
C*                    2   THE SAME AS CASE 1, EXCEPT THAT THIS              *
C*                           SUBROUTINE COMPUTES THE PARTIAL                *
C*                           DERIVATIVES BY NUMERICAL DIFFERENCING          *
C*                           OF THE DERIVATIVES. HENCE PEDERV IS            *
C*                           NOT CALLED.                                    *
C*   YMAX    AN ARRAY OF  N LOCATIONS WHICH CONTAINS THE MAXIMUM            *
C*             OF EACH Y SEEN SO FAR.  IT SHOULD NORMALLY BE SET TO         *
C*             1 IN EACH COMPONENT BEFORE THE FIRST ENTRY. (SEE THE         *
C*             DESCRIPTION OF EPS.)                                         *
C*   ERROR   AN ARRAY OF  N ELEMENTS WHICH CONTAINS THE ESTIMATED           *
C*             ONE STEP ERROR IN EACH COMPONENT.                            *
C*   KFLAG   A COMPLETION CODE WITH THE FOLLOWING MEANINGS..                *
C*                   +1   THE STEP WAS SUCCESFUL.                           *
C*                   -1   THE STEP WAS TAKEN WITH H = HMIN, BUT THE         *
C*                           REQUESTED ERROR WAS NOT ACHIEVED.              *
C*                   -2   THE MAXIMUM ORDER SPECIFIED WAS FOUND TO          *
C*                           BE TOO LARGE.                                  *
C*                   -3   CORRECTOR CONVERGENCE COULD NOT BE                *
C*                           ACHIEVED FOR H .GT. HMIN.                      *
C*                   -4   THE REQUESTED ERROR IS SMALLER THAN CAN           *
C*                           BE HANDLED FOR THIS PROBLEM.                   *
C*   JSTART  AN INPUT INDICATOR WITH THE FOLLOWING MEANINGS..               *
C*                   -1   REPEAT THE LAST STEP WITH A NEW H                 *
C*                    0   PERFORM THE FIRST STEP.  THE FIRST STEP           *
C*                           MUST BE DONE WITH THIS VALUE OF JSTART         *
C*                           SO THAT THE SUBROUTINE CAN INITIALIZE          *
C*                           ITSELF.                                        *
C*                   +1   TAKE A NEW STEP CONTINUING FROM THE LAST.         *
C*             JSTART IS SET TO NQ, THE CURRENT ORDER OF THE METHOD         *
C*             AT EXIT.  NQ IS ALSO THE ORDER OF THE MAXIMUM                *
C*             DERIVATIVE AVAILABLE.                                        *
C*   MAXDER  THE MAXIMUM DERIVATIVE THAT SHOULD BE USED IN THE              *
C*             METHOD.  SINCE THE ORDER IS EQUAL TO THE HIGHEST             *
C*             DERIVATIVE USED, THIS RESTRICTS THE ORDER. IT MUST           *
C*             BE LESS THAN 8 FOR ADAMS AND 7 FOR STIFF METHODS.            *
C*   PW      A BLOCK OF AT LEAST N**2 FLOATING POINT LOCATIONS.             *
C************************************************************************
      DIMENSION Y(8,1),YMAX(1),SAVE(10,1),ERROR(1),PW(1),
     1          A(8),PERTST(7,2,3)
C************************************************************************
C* THE COEFFICIENTS IN PERTST ARE USED IN SELECTING THE STEP AND        *
C* ORDER, THEREFORE ONLY ABOUT ONE PERCENT ACCURACY IS NEEDED.          *
C************************************************************************
      DATA PERTST /2.0,4.5,7.333,10.42,13.7,17.15,1.0,
     1             2.0,12.0,24.0,37.89,53.33,70.08,87.97,
     2             3.0,6.0,9.167,12.5,15.98,1.0,1.0,
     3             12.0,24.0,37.89,53.33,70.08,87.97,1.0,
     4             1.,1.,0.5,0.1667,0.04133,0.008267,1.0,
     5             1.0,1.0,2.0,1.0,.3157,.07407,.0139/
      DATA A(2) / -1.0 /
      IRET = 1
      KFLAG = 1
      IF (JSTART.LE.0) GO TO 140
C************************************************************************
C* BEGIN BY SAVING INFORMATION FOR POSSIBLE RESTARTS AND CHANGING       *
C* H BY THE FACTOR R IF THE CALLER HAS CHANGED H.  ALL VARIABLES        *
C* DEPENDENT ON H MUST ALSO BE CHANGED.                                 *
C* E IS A COMPARISON FOR ERRORS OF THE CURRENT ORDER NQ. EUP IS         *
C* TO TEST FOR INCREASING THE ORDER, EDWN FOR DECREASING THE ORDER.     *
C* HNEW IS THE STEP SIZE THAT WAS USED ON THE LAST CALL.                *
C************************************************************************
 100  DO 110 I = 1,N
        DO 110 J = 1,K
 110      SAVE(J,I) = Y(J,I)
      HOLD = HNEW
      IF (H.EQ.HOLD) GO TO 130
 120  RACUM = H/HOLD
```

```
      IRET1 = 1
      GO TO 750
 130  NQOLD = NQ
      TOLD = T
      RACUM = 1.0
      IF (JSTART.GT.0) GO TO 250
      GO TO 170
 140  IF (JSTART.EQ.-1) GO TO 160
C***********************************************************************
C* ON THE FIRST CALL, THE ORDER IS SET TO 1 AND THE INITIAL            *
C* DERIVATIVES ARE CALCULATED.                                         *
C***********************************************************************
      NQ = 1
      N3 = N
      N1 = N*10
      N2 = N1 + 1
      N4 = N**2
      N5 = N1 + N
      N6 = N5 + 1
      CALL DIFFUN(T,Y,SAVE(N2,1))
      DO 150 I = 1,N
 150    Y(2,I) = SAVE(N1+I,1)*H
      HNEW = H
      K = 2
      GO TO 100
C***********************************************************************
C* REPEAT LAST STEP BY RESTORING SAVED INFORMATION.                    *
C***********************************************************************
 160  IF (NQ.EQ.NQOLD) JSTART = 1
      T = TOLD
      NQ = NQOLD
      K = NQ + 1
      GO TO 120
C***********************************************************************
C* SET THE COEFFICIENTS THAT DETERMINE THE ORDER AND THE METHOD        *
C* TYPE.  CHECK FOR EXCESSIVE ORDER.  THE LAST TWO STATEMENTS OF       *
C* THIS SECTION  SET IWEVAL .GT.0 IF PW IS TO BE RE-EVALUATED          *
C* BECAUSE OF THE ORDER CHANGE, AND THEN REPEAT THE INTEGRATION        *
C* STEP IF IT HAS NOT YET BEEN DONE (IRET = 1) OR SKIP TO A FINAL      *
C* SCALING BEFORE EXIT IF IT HAS BEEN COMPLETED (IRET = 2).            *
C***********************************************************************
 170  IF (MF.EQ.0) GO TO 180
      IF (NQ.GT.6) GO TO 190
      GO TO (221,222,223,224,225,226),NQ
 180  IF (NQ.GT.7) GO TO 190
      GO TO (211,212,213,214,215,216,217),NQ
 190  KFLAG = -2
      RETURN
C***********************************************************************
C* THE FOLLOWING COEFFICIENTS SHOULD BE DEFINED TO THE MAXIMUM         *
C* ACCURACY PERMITTED BY THE MACHINE.  THEY ARE, IN THE ORDER USED..   *
C*                                                                     *
C* -1                                                                  *
C* -1/2,-1/2                                                           *
C* -5/12,-3/4,-1/6                                                     *
C* -3/8,-11/12,-1/3,-1/24                                              *
C* -251/720,-25/24,-35/72,-5/48,-1/120                                 *
C* -95/288,-137/120,-5/8,-17/96,-1/40,-1/720                           *
C* -19087/60480,-49/40,-203/270,-49/192,-7/144,-7/1440,-1/5040         *
C*                                                                     *
C* -1                                                                  *
C* -2/3,-1/3                                                           *
C* -6/11,-6/11,-1/11                                                   *
C* -12/25,-7/10,-1/5,-1/50                                             *
C* -120/274,-225/274,-85/274,-15/274,-1/274                            *
C* -180/441,-58/63,-15/36,-25/252,-3/252,-1/1764                       *
C***********************************************************************
 211  A(1) = -1.0
```

```
           GO TO 230
    212    A(1) = -0.500000000
           A(3) = -0.500000000
           GO TO 230
    213    A(1) = -0.416666666666667
           A(3) = -0.750000000
           A(4) = -0.166666666666667
           GO TO 230
    214    A(1) = -0.375000000
           A(3) = -0.916666666666667
           A(4) = -0.333333333333333
           A(5) = -0.041666666666667
           GO TO 230
    215    A(1) = -0.348611111111111
           A(3) = -1.041666666666667
           A(4) = -0.486111111111111
           A(5) = -0.104166666666667
           A(6) = -0.008333333333333333
           GO TO 230
    216    A(1) = -0.329861111111111
           A(3) = -1.141666666666667
           A(4) = -0.625000000
           A(5) = -0.177083333333333
           A(6) = -0.025000000
           A(7) = -0.001388888888888889
           GO TO 230
    217    A(1) = -0.3155919312169312
           A(3) = -1.225000000
           A(4) = -0.7518518518518519
           A(5) = -0.2552083333333333
           A(6) = -0.04861111111111111
           A(7) = -0.004861111111111111
           A(8) = -0.0001984126984126984
           GO TO 230
    221    A(1) = -1.000000000
           GO TO 230
    222    A(1) = -0.666666666666667
           A(3) = -0.3333333333333333
           GO TO 230
    223    A(1) = - 0.5454545454545455
           A(3) = A(1)
           A(4) = -0.09090909090909091
           GO TO 230
    224    A(1) = -0.480000000
           A(3) = -0.700000000
           A(4) = -0.200000000
           A(5) = -0.020000000
           GO TO 230
    225    A(1) = -0.437956204379562
           A(3) = -0.8211678832116798
           A(4) = -0.3102189781021898
           A(5) = -0.05474452554744526
           A(6) = -0.0036496350364963504
           GO TO 230
    226    A(1) = -0.4081632653061225
           A(3) = -0.9206349206349206
           A(4) = -0.416666666666667
           A(5) = -0.0992063492063492
           A(6) = -0.0119047619047619
           A(7) = -0.000566893424036282
    230    K = NQ+1
           IDOUB = K
           MTYP = (4 - MF)/2
           ENQ2 = .5/FLOAT(NQ + 1)
           ENQ3 = .5/FLOAT(NQ + 2)
           ENQ1 = 0.5/FLOAT(NQ)
           PEPSH = EPS
           EUP = (PERTST(NQ,MTYP,2)*PEPSH)**2
```

```
      E = (PERTST(NQ,MTYP,1)*PEPSH)**2
      EDWN =(PERTST(NQ,MTYP,3)*PEPSH)**2
      IF (EDWN.EQ.0) GO TO 780
      BND = EPS*ENQ3/DFLOAT(N)
 240  IWEVAL = MF
      GO TO ( 250 , 680 ),IRET
C***********************************************************************
C* THIS SECTION COMPUTES THE PREDICTED VALUES BY EFFECTIVELY           *
C* MULTIPLYING THE SAVED INFORMATION BY THE PASCAL TRIANGLE            *
C* MATRIX.                                                             *
C***********************************************************************
 250  T = T + H
      DO 260 J = 2,K
        DO 260 J1 = J,K
          J2 = K - J1 + J - 1
          DO 260 I = 1,N
 260        Y(J2,I) = Y(J2,I) + Y(J2+1,I)
C***********************************************************************
C*   UP TO 3 CORRECTOR ITERATIONS ARE TAKEN.  CONVERGENCE IS TESTED    *
C*   BY REQUIRING CHANGES TO BE LESS THAN BND WHICH IS DEPENDENT ON    *
C*   THE ERROR TEST CONSTANT.                                          *
C*      THE SUM OF THE CORRECTIONS IS ACCUMULATED IN THE ARRAY         *
C*   ERROR(I).  IT IS EQUAL TO THE K-TH DERIVATIVE OF Y MULTIPLIED     *
C*   BY  H**K/(FACTORIAL(K-1)*A(K)), AND IS THEREFORE PROPORTIONAL     *
C*   TO THE ACTUAL ERRORS TO THE LOWEST POWER OF H PRESENT. (H**K)     *
C***********************************************************************
      DO 270 I = 1,N
 270    ERROR(I) = 0.0
      DO 430 L = 1,3
          CALL DIFFUN (T,Y,SAVE(N2,1))
C***********************************************************************
C*   IF THERE HAS BEEN A CHANGE OF ORDER OR THERE HAS BEEN TROUBLE     *
C*   WITH CONVERGENCE, PW IS RE-EVALUATED PRIOR TO STARTING THE        *
C*   CORRECTOR ITERATION IN THE CASE OF STIFF METHODS.  IWEVAL IS      *
C*   THEN SET TO -1 AS AN INDICATOR THAT IT HAS BEEN DONE.             *
C***********************************************************************
       IF (IWEVAL.LT.1) GO TO 350
         IF (MF.EQ.2) GO TO 310
         CALL PEDERV(T,Y,PW,N3)
         R = A(1)*H
         DO 280 I = 1,N4
 280       PW(I) = PW(I)*R
C***********************************************************************
C* ADD THE IDENTITY MATRIX TO THE JACOBIAN AND INVERT TO GET PW.       *
C***********************************************************************
 290     DO 300 I = 1,N
 300       PW(I*(N3+1)-N3) = 1.0 + PW(I*(N3+1)-N3)
         IWEVAL = -1
         CALL MATINV(PW,N,N3,J1)
         IF (J1.GT.0) GO TO 350
         GO TO 440
C***********************************************************************
C* EVALUATE THE JACOBIAN INTO PW BY NUMERICAL DIFFERENCING.  R IS THE  *
C* CHANGE MADE TO THE ELEMENT OF Y.  IT IS EPS RELATIVE TO Y WITH      *
C* A MINIMUM OF EPS**2.                                                *
C***********************************************************************
 310     DO 320 I = 1,N
 320       SAVE(9,I) = Y(1,I)
         DO 340 J = 1,N
           R = EPS*DMAX1(EPS,DABS(SAVE(9,J)))
           Y(1,J) = Y(1,J) + R
           D = A(1)*H/R
           CALL DIFFUN(T,Y,SAVE(N6,1))
           DO 330 I = 1,N
 330         PW(I+(J-1)*N3) = (SAVE(N5+I,1) - SAVE(N1+I,1))*D
 340       Y(1,J) = SAVE(9,J)
         GO TO 290
```

```
 350      IF (MF.NE.0) GO TO 370
          DO 360 I = 1,N
 360        SAVE(9,I) = Y(2,I) - SAVE(N1+I,1)*H
          GO TO 410
 370      DO 380 I = 1,N
 380        SAVE(N5+I,1) = Y(2,I) - SAVE(N1+I,1)*H
          DO 400 I = 1,N
            D = 0.0
            DO 390 J = 1,N
 390          D = D + PW(I+(J-1)*N3)*SAVE(N5+J,1)
 400        SAVE(9,I) = D
 410      NT = N
C**********************************************************************
C* CORRECT AND SEE IF ALL CHANGES ARE LESS THAN BND RELATIVE TO YMAX. *
C* IF SO, THE CORRECTOR IS SAID TO HAVE CONVERGED.                    *
C**********************************************************************
        DO 420 I = 1,N
          Y(1,I) = Y(1,I) + A(1)*SAVE(9,I)
          Y(2,I) = Y(2,I) - SAVE(9,I)
          ERROR(I) = ERROR(I) + SAVE(9,I)
          IF  (DABS(SAVE(9,I)).LE.(BND*YMAX(I))) NT = NT - 1
 420      CONTINUE
        IF (NT.LE.0) GO TO 490
 430    CONTINUE
C**********************************************************************
C* THE CORRECTOR ITERATION FAILED TO CONVERGE IN 3 TRIES.  VARIOUS    *
C* POSSIBILITIES ARE CHECKED FOR.  IF H IS ALREADY HMIN AND           *
C* THIS IS EITHER ADAMS METHOD OR THE STIFF METHOD IN WHICH THE       *
C* MATRIX PW HAS ALREADY BEEN RE-EVALUATED, A NO CONVERGENCE EXIT     *
C* IS TAKEN. OTHERWISE THE MATRIX PW IS RE-EVALUATED AND/OR THE       *
C* STEP IS REDUCED TO TRY AND GET CONVERGENCE.                        *
C**********************************************************************
 440  T = TOLD
      IF ((H.LE.(HMIN*1.00001)).AND.((IWEVAL - MTYP).LT.-1)) GO TO 460
      IF ((MF.EQ.0).OR.(IWEVAL.NE.0)) RACUM = RACUM*0.25D0
      IWEVAL = MF
      IRET1 = 2
      GO TO 750
 460  KFLAG = -3
 470  DO 480 I = 1,N
        DO 480 J = 1,K
 480      Y(J,I) = SAVE(J,I)
      H = HOLD
      NQ = NQOLD
      JSTART = NQ
      RETURN
C**********************************************************************
C* THE CORRECTOR CONVERGED AND CONTROL IS PASSED TO STATEMENT 520     *
C* IF THE ERROR TEST IS O.K.,  AND TO 540 OTHERWISE.                  *
C* IF THE STEP IS O.K. IT IS ACCEPTED. IF IDOUB HAS BEEN REDUCED      *
C* TO ONE,  A TEST IS MADE TO SEE IF THE STEP CAN BE INCREASED        *
C* AT THE CURRENT ORDER OR BY GOING TO ONE HIGHER OR ONE LOWER.       *
C* SUCH A CHANGE IS ONLY MADE IF THE STEP CAN BE INCREASED BY AT      *
C* LEAST 1.1.  IF NO CHANGE IS POSSIBLE IDOUB IS SET TO 10 TO         *
C* PREVENT FUTHER TESTING FOR 10 STEPS.                               *
C* IF A CHANGE IS POSSIBLE, IT IS MADE AND IDOUB IS SET TO            *
C* NQ + 1     TO PREVENT FURTHER TESING FOR THAT NUMBER OF STEPS.     *
C* IF THE ERROR WAS TOO LARGE, THE OPTIMUM STEP SIZE FOR THIS OR      *
C* LOWER ORDER IS COMPUTED, AND THE STEP RETRIED.  IF IT SHOULD       *
C* FAIL TWICE MORE IT IS AN INDICATION THAT THE DERIVATIVES THAT      *
C* HAVE ACCUMULATED IN THE Y ARRAY HAVE ERRORS OF THE WRONG ORDER     *
C* SO THE FIRST DERIVATIVES ARE RECOMPUTED AND THE ORDER IS SET       *
C* TO 1.                                                              *
C**********************************************************************
 490  D = 0.0
      DO 500 I = 1,N
 500    D = D + (ERROR(I)/YMAX(I))**2
      IWEVAL = 0
```

```
      IF (D.GT.E) GO TO 540
      IF (K.LT.3) GO TO 520
C*************************************************************************
C* COMPLETE THE CORRECTION OF THE HIGHER ORDER DERIVATIVES AFTER A       *
C* SUCCESFUL STEP.                                                       *
C*************************************************************************
      DO 510 J = 3,K
        DO 510 I = 1,N
 510      Y(J,I) = Y(J,I) + A(J)*ERROR(I)
 520  KFLAG = +1
      HNEW = H
      IF (IDOUB.LE.1) GO TO 550
      IDOUB = IDOUB - 1
      IF (IDOUB.GT.1) GO TO 700
      DO 530 I = 1,N
 530    SAVE(10,I) = ERROR(I)
      GO TO 700
C*************************************************************************
C* REDUCE THE FAILURE FLAG COUNT TO CHECK FOR MULTIPLE FAILURES.         *
C* RESTORE T TO ITS ORIGINAL VALUE AND TRY AGAIN UNLESS THERE HAVE       *
C* THREE FAILURES.  IN THAT CASE THE DERIVATIVES ARE ASSUMED TO HAVE     *
C* ACCUMULATED ERRORS SO A RESTART FROM THE CURRENT VALUES OF Y IS       *
C* TRIED.                                                                *
C*************************************************************************
 540  KFLAG = KFLAG - 2
      IF (H.LE.(HMIN*1.00001)) GO TO 740
      T = TOLD
      IF (KFLAG.LE.-5) GO TO 720
C*************************************************************************
C* PR1, PR2, AND PR3  WILL CONTAIN THE AMOUNTS BY WHICH THE STEP SIZE    *
C* CHOULD BE DIVIDED AT ORDER ONE LOWER, AT THIS ORDER, AND AT ORDER     *
C* ONE HIGHER RESPECTIVELY.                                              *
C*************************************************************************
 550  PR2 = (D/E)**ENQ2*1.2
      PR3 = 1.E+20
      IF ((NQ.GE.MAXDER).OR.(KFLAG.LE.-1)) GO TO 570
      D = 0.0
      DO 560 I = 1,N
 560    D = D + ((ERROR(I) - SAVE(10,I))/YMAX(I))**2
      PR3 = (D/EUP)**ENQ3*1.4
 570  PR1 = 1.E+20
      IF (NQ.LE.1) GO TO 590
      D = 0.0
      DO 580 I = 1,N
 580    D = D + (Y(K,I)/YMAX(I))**2
      PR1 = (D/EDWN)**ENQ1*1.3
 590  CONTINUE
      IF (PR2.LE.PR3) GO TO 650
      IF (PR3.LT.PR1) GO TO 660
 600  R = 1.0/AMAX1(PR1,1.E-4)
      NEWQ = NQ - 1
 610  IDOUB = 10
      IF ((KFLAG.EQ.1).AND.(R.LT.(1.1))) GO TO 700
      IF (NEWQ.LE.NQ) GO TO 630
C*************************************************************************
C* COMPUTE ONE ADDITIONAL SCALED DERIVATIVE IF ORDER IS INCREASED.       *
C*************************************************************************
      DO 620 I = 1,N
 620    Y(NEWQ+1,I) = ERROR(I)*A(K)/DFLOAT(K)
 630  K = NEWQ + 1
      IF (KFLAG.EQ.1) GO TO 670
      RACUM = RACUM*R
      IRET1 = 3
      GO TO 750
 640  IF (NEWQ.EQ.NQ) GO TO 250
      NQ = NEWQ
      GO TO 170
 650  IF (PR2.GT.PR1) GO TO 600
```

```
      NEWQ = NQ
      R = 1.0/AMAX1(PR2,1.E-4)
      GO TO 610
  660 R = 1.0/AMAX1(PR3,1.E-4)
      NEWQ = NQ + 1
      GO TO 610
  670 IRET = 2
      R = DMIN1(R,HMAX/DABS(H))
      H = H*R
      HNEW = H
      IF (NQ.EQ.NEWQ) GO TO 680
      NQ = NEWQ
      GO TO 170
  680 R1 = 1.0
      DO 690 J = 2,K
        R1 = R1*R
        DO 690 I = 1,N
  690     Y(J,I) = Y(J,I)*R1
      IDOUB = K
  700 DO 710 I = 1,N
  710   YMAX(I) = DMAX1(YMAX(I),DABS(Y(1,I)))
      JSTART = NQ
      RETURN
  720 IF (NQ.EQ.1) GO TO 780
      CALL DIFFUN (T,Y,SAVE(N2,1))
      R = H/HOLD
      DO 730 I = 1,N
        Y(1,I) = SAVE(1,I)
        SAVE(2,I) = HOLD*SAVE(N1+I,1)
  730   Y(2,I) = SAVE(2,I)*R
      NQ = 1
      KFLAG = 1
      GO TO 170
  740 KFLAG = -1
      HNEW = H
      JSTART = NQ
      RETURN
C***********************************************************************
C* THIS SECTION SCALES ALL VARIABLES CONNECTED WITH H AND RETURNS     *
C* TO THE ENTERING SECTION.                                            *
C***********************************************************************
  750 RACUM = DMAX1(DABS(HMIN/HOLD),RACUM)
      RACUM = DMIN1(RACUM,DABS(HMAX/HOLD))
      R1 = 1.0
      DO 760 J = 2,K
        R1 = R1*RACUM
        DO 760 I = 1,N
  760     Y(J,I) = SAVE(J,I)*R1
      H = HOLD*RACUM
      DO 770 I = 1,N
  770   Y(1,I) = SAVE(1,I)
      IDOUB = K
      GO TO ( 130 , 250 , 640 ),IRET1
  780 KFLAG = -4
      GO TO 470
      END
```

The program was used to integrate $y' = -y$ from $y(0) = 1$ to $t = 20$ with E $=$ EPS $= 10^{-i}$, $i = 2, 3, \ldots, 11$. The results are shown in Table 9.2 and Figure 9.2. YMAX(1) was set to the current value of y before each step.

Table 9.2 ERROR VERSUS FUNCTION EVALUATIONS FOR ADAMS' METHODS

EPS	*Actual error at t = 20 relative to e^{-20}*	*Number of function evaluations*	*Number of steps*
0.10000D-01	0.10950D-01	201	73
0.10000D-02	0.20696D-02	262	88
0.10000D-03	−0.79186D-03	386	131
0.10000D-04	−0.35877D-04	391	132
0.10000D-05	−0.61224D-06	529	179
0.10000D-06	0.16620D-05	688	230
0.10000D-07	−0.45687D-08	773	259
0.10000D-08	0.14422D-07	734	247
0.10000D-09	0.94759D-08	767	261
0.10000D-10	0.20069D-08	946	323

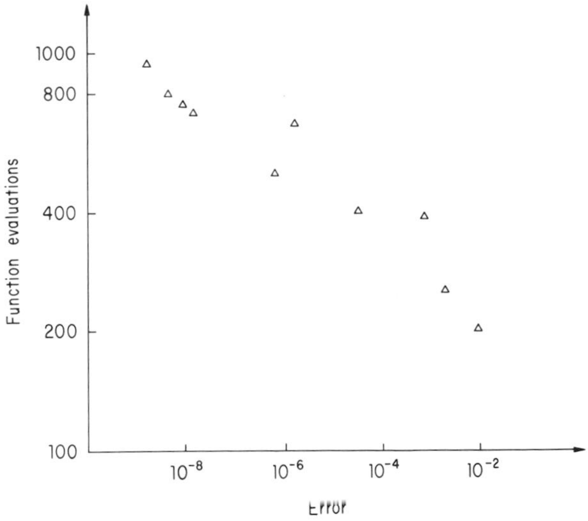

Fig. 9.2 Error for exponential equation by automatic Adams' methods.

PROBLEMS

1. If the truncation error as defined by Eqs. (9.13) is $\mathbf{d}_n$, show that the truncation error of the equations transformed by T is $T\mathbf{d}_n$.
2. The two-step Adams-Bashforth-Moulton method of order three in a representa-

tion using $\mathbf{a} = [y, hy', h^2y''/2]^T$ is given by

$$A = \begin{bmatrix} 1 & 1 & 1 \\ 0 & 1 & 2 \\ 0 & 0 & 1 \end{bmatrix}, \qquad \mathbf{l} = \begin{bmatrix} \frac{5}{12} \\ 1 \\ \frac{1}{2} \end{bmatrix}$$

What is the stability matrix S_n for the equation $y' = 0$ in this representation?

If the step size is changed by an amount α_i after the $2i$th step, $i = 1, 2, \ldots$, what is the solution in terms of the initial values? Show that the result is independent of the step sizes used. Is this true if the step size is changed after every step?

3. Repeat Problem 2 for the three-step Adams-Bashforth-Moulton method of order four. How many steps must be taken without a step change before the result is independent of the step size used?

4. Derive a predictor formula and a corrector formula of the form

$$y(t + \alpha h) = y(t) + h\beta_1 y'(t) + h\beta_2 y'(t - h) + 0(h^3)$$

and

$$y(t + \alpha h) = y(t) + h\beta_0^* y'(t + \alpha h) + h\beta_1^* y'(t) + h\beta_2^* y'(t - h) + 0(h^4)$$

for taking a step of size αh after a step of size h. Compare this with an interpolation step change method for the two-step Adams-Bashforth-Moulton integration formula, and show that the result of changing step size and performing one step is identical in the two methods.

5. Verify that the matrix B for the three-step Adams-Bashforth-Moulton method is independent of whether the corrector order is three or four. Is B the same if a second order corrector is used?

6. Equation (9.12) gives a third order predictor, fourth order corrector which we will call a P3C4 method. What are the matrices and vectors for the P3C3 Adams' method in the same representation?

10 EXISTENCE, CONVERGENCE, AND ERROR ESTIMATES FOR MULTIVALUE METHODS

The theoretical background of multistep and multivalue methods will be developed in this chapter. Generally, a single equation will be considered, although the results are equally applicable to systems of equations, even those of differing orders each integrated by a method suitable for that order.

After defining convergence and stability, we will show that stability implies the root condition, that this and an order of at least one (called *consistency*) imply convergence, that convergence implies the root condition, which, with consistency, implies stability. It will also be shown that convergence implies consistency for multistep methods, completing the chain of relations for first order equations shown in Figure 10.1.

Extensions of the multivalue methods to higher order equations were discussed in Chapter 9. In this chapter we will consider a pth order differential equation of the form

$$y^{(p)} = f(y, y', \ldots, y^{(p-q)}, t) \qquad 1 \leq q \leq p \tag{10.1}$$

that is, the highest $q - 1$ derivatives are absent (excepting, of course, the pth one). We will call this a *q-differential pth order equation.* If $q = 1$ we will refer to it as the *general pth order equation*, while if $q = p$ we call it the *special pth order equation.* The special second order equation $y'' = f(y, t)$ frequently occurs in lossless mechanical systems such as celestial motion, so is worthy of special study.

A chain of relations similar to those in Figure 10.1 will hold for pth order equations. The order of a method will be defined in such a way that the order will have to be at least p for consistency. [This differs from Henrici's (1962) definition by an amount $p - 1$. We call a method with a local error of $0(h^{r+1})$

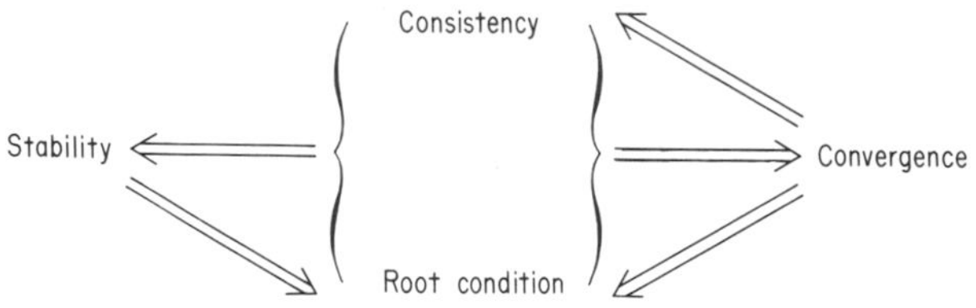

Fig. 10.1 Basic theorems.

an rth order method; he calls it an $(r + 1 - p)$th order method.] The concepts of q-stability, q-convergence, and q-root condition will be defined for the q-differential equation, and then Figure 10.1 will be seen to hold except that the result convergence $\Rightarrow$ consistency has not been shown.

Dahlquist's important theorem limiting the stable order of multistep methods for first order equations will be proved, then the existence of maximum order stable multivalue methods will be proved. The choice of the vector $\mathbf{l}$ to get stability will leave some freedom, which can be used to select the error coefficients. This will be examined in order to see what type of error bounds and estimates can be obtained.

The results will be developed for fixed order, fixed step methods. It was simple to extend the results for one-step methods to variable step size (see Section 4.6) or even to variable order methods. The same extensions are more difficult and restrictive for multivalue methods and most of the automatic methods used in practice are not yet backed by adequate theory, although there is every reason to expect that the results will be extended and that the automatic methods behave as can be expected by analogy with the fixed step and order methods.

Some of the results are most easily expressed and proved in a conventional multistep formulation, while others are more easily handled in the matrix notation developed for multivalue methods. The pth order equation (10.1) can be integrated by a multistep method written in the form

$$\sum_{i=0}^{k} \left(\alpha_i y_{n-i} + \beta_i \frac{h^p}{p!} f_{n-i}\right) = 0 \tag{10.2}$$

where $f_m = f(y_m, y'_m, \ldots, y_m^{(p-1)}, t_m)$ and the values of $y'_m, \ldots, y_m^{(p-1)}$ can be obtained from the formula

$$\frac{h^q y_m^{(q)}}{q!} = \sum_{i=0}^{k} \left(\alpha_{qi} y_{m-i} + \beta_{qi} \frac{h^p}{p!} f_{m-i}\right) \tag{10.3}$$

for $0 < q < p$. These equations are implicit if β_0 or β_{q0} are not zero.

Because of the additional work involved in evaluating (10.3), this method is not usually used except for the special pth order equation. For such equations, only (10.2), is used.

The case in which (10.3) is not required can easily be expressed in the matrix notation. If (10.2) is first applied as a predictor (with $\beta_0 = 0$ and $\alpha_0 = -1$), then used as a corrector to iterate in the form

$$y_{n,(m+1)} = \sum_{i=1}^{k} \left(\alpha_i^* y_{n-i} + \beta_i^* \frac{h^p}{p!} f_{n-i}\right) + \beta_0^* \frac{h^p}{p!} f(y_{n,(m)}, t_n) \tag{10.4}$$

we can define $\mathbf{y}_n = [y_n, y_{n-1}, \ldots, y_{n-k+1}, (h^p/p!)y_n^{(p)}, \ldots, (h^p/p!)y_{n-k+1}^{(p)}]^T$ and express the method by

$$\begin{aligned} \mathbf{y}_{n,(0)} &= B\mathbf{y}_{n-1} \\ \mathbf{y}_{n,(m+1)} &= \mathbf{y}_{n,(m)} + \mathbf{c}G(\mathbf{y}_{n,(m)}) \end{aligned} \tag{10.5}$$

where

$$B = \left[\begin{array}{cccc|cccc} \alpha_1 & \alpha_2 & & \alpha_k & \beta_1 & & & \beta_k \\ 1 & & & & & & & \\ & \ddots & & & & 0 & & \\ & & 1 & 0 & & & & \\ \hline \gamma_1 & & & \gamma_k & \delta_1 & & & \delta_k \\ & & & & 1 & & & \\ & 0 & & & & \ddots & & \\ & & & & & & 1 & 0 \end{array}\right], \qquad \mathbf{c} = \begin{bmatrix} \beta_0^* \\ 0 \\ \cdot \\ \cdot \\ \cdot \\ 0 \\ 1 \\ 0 \\ \cdot \\ \cdot \\ \cdot \\ 0 \end{bmatrix}$$

$$\gamma_i = \frac{\alpha_i - \alpha_i^*}{\beta_0^*}$$

$$\delta_i = \frac{\beta_i - \beta_i^*}{\beta_0^*}$$

$$G(\mathbf{y}_{n,(m)}) = \frac{h^p}{p!} f(y_{n,(m)}, t_n) - \frac{h^p}{p!} y_{n,(m)}^{(p)}$$

The stability analysis is identical to that for first order methods. Stability is determined by the behavior of the matrix

$$S_n = \prod_{i=0}^{M-1} \left(I + \mathbf{c}\frac{\partial G}{\partial \mathbf{y}}(\xi_i)\right) B \tag{10.6}$$

In the limit as h goes to zero

$$S = (I - \mathbf{c}\boldsymbol{\delta}_k^T)^M B \tag{10.7}$$

Whenever the vector $\mathbf{y}$ is referred to, we will be talking about this representation of a multistep method.

Equivalent methods can be derived by applying transformations as described in Section 9.2. In particular, we will use the multivalue formulation in which $\mathbf{a} = [y, hy', \ldots, h^{k-1}y^{(k-1)}/(k-1)!]^T$. In this chapter $\mathbf{a}$ will always refer to this particular form unless stated otherwise and we will refer to this as the *normal form* of a multivalue method. We will assume that there are k components in $\mathbf{a}$ unless it is otherwise stated.

The more general pth order equation (10.1) can be handled directly in this formulation since the derivatives are available, so it is not necessary to provide additional formulas such as (10.3) to evaluate them. [Not all multistep predictor-corrector methods determined by Eqs. (10.2) and (10.3) can be expressed in the matrix notation. However, if the predictor coefficients are restricted, it is possible.] We will only discuss methods for the general pth order equations which can be so expressed and hence can be transformed into the normal form. Multistep methods for the special pth order equation can always be put in this form, although if the corrector equation (10.2) is solved exactly, it is usually more convenient to deal with a multistep notation.

10.1 CONVERGENCE AND STABILITY

Convergence expresses the property that by using a sufficiently small step and accurate computation the numerical solution can be made arbitrarily close to the true solution. That is, if we want to know the answer at a fixed point t in the range $0 < t \leq b$, we can choose an N sufficiently large so that if $h = t/N$, y_N is sufficiently close to $y(t)$ for our purposes. In a multivalue or multistep method we start with an initial vector $\mathbf{a}_0$ or $\mathbf{y}_0$ which may not be entirely specified by the initial values. For example, in a two-step method for a first order equation we are given y_0 but also need to know y_1. Since any numerical technique we use will almost certainly introduce errors into y_1 and possibly also y_0, we must allow for these in the definition of convergence if it is to be practical. Hence we state:

DEFINITION 10.1

A multistep (multivalue) method for first order equations is convergent if, for any differential equation satisfying a Lipschitz condition, the computed solution $\mathbf{y}_n$ $[\mathbf{a}_n]$ converges to $\mathbf{y}(t)$ $[\mathbf{a}(t)]$ uniformly in $0 \leq t \leq b$ as $\mathbf{y}_0 \longrightarrow \mathbf{y}(0)$ $[\mathbf{a}_0 \longrightarrow \mathbf{a}(0)]$ and $n \longrightarrow \infty$ with $h = t/n$.

If we are working with a pth order equation, we have p initial values for $y_{(0)}, y'_{(0)}, \ldots, y^{(p-1)}_{(0)}$ to work with, but the method may still require additional starting values. We must require that the starting values used be such that as $h \longrightarrow 0$ they correctly represent the initial values; that is, the components of $\mathbf{a}_0$ are such that

$$q!\,\frac{(\mathbf{a}_0)_q}{h^q} \longrightarrow y^{(q)}(0) \qquad \text{for } q < p$$

$$\frac{(\mathbf{a}_0)_q}{h^{p-1}} \longrightarrow 0 \qquad \text{for } q \geq p$$

where $(\mathbf{a})_q$ is the qth component of $\mathbf{a}$.

We will handle this by defining the norm $\|\cdot\|_h^p$ as follows:

$$\|\mathbf{a}\|_h^p = \max_i \frac{|(\mathbf{a})_i|}{h^{q(i)}}$$

where

$$q(i) = \begin{cases} i & \text{if } i < p \\ p-1 & \text{if } i \geq p \end{cases}$$

We must require that the starting value $\mathbf{a}_0$ for a pth order method converge to $\mathbf{a}(0)$ in this norm, that is,

$$\|\mathbf{a}_0 - \mathbf{a}(0)\|_h^p \longrightarrow 0 \qquad \text{as } h \longrightarrow 0$$

Larger errors correspond to a change in the initial conditions, as can be seen in the following example.

EXAMPLE

Consider the method given by $\mathbf{a}_n = [y_n, hy'_n, h^2y''_n/2]^T$,

$$A = \begin{bmatrix} 1 & 1 & 1 \\ 0 & 1 & 2 \\ 0 & 0 & 1 \end{bmatrix}, \qquad \mathbf{l} = \begin{bmatrix} 0 \\ 0 \\ 1 \end{bmatrix}$$

for the second order equation $0 = f(y, y', t) - y''$, which is equivalent to

$$0 = \frac{h^2}{2} f\left(a_0, \frac{a_1}{h}, t\right) - a_2 \tag{10.8}$$

If we take the particular case $f(y, y', t) = 0$, we get

$$\mathbf{a}_N = S\mathbf{a}_{N-1} \tag{10.9}$$

where

$$S = \begin{bmatrix} 1 & 1 & 1 \\ 0 & 1 & 2 \\ 0 & 0 & 0 \end{bmatrix}$$

The solution to (10.9) is $\mathbf{a}_N = S^N\mathbf{a}_0$, where

$$S^N = \begin{bmatrix} 1 & N & 2N-1 \\ 0 & 1 & 2 \\ 0 & 0 & 0 \end{bmatrix}$$

Hence the numerical solution is

$$y_N = y_0 + Nhy'_0 + (2N-1)\frac{h^2y''_0}{2}$$

when the initial vector is $[y_0, hy'_0, h^2y''_0/2]^T$. Since $Nh = t$, we have

$$y_N = y_0 + ty'_0 + (2t - h)\frac{hy''_0}{2}$$

An error of size ϵ in component $(\mathbf{a}_0)_1$ changes the solution to

$$y_N = y_0 + t\left(y'_0 + \frac{\epsilon}{h}\right) + (2t - h)\frac{hy''_0}{2}$$

so it is evident that ϵ must behave as $o(h)$ for convergence. Initial errors in $h^2y''_0/2$ must also behave as $o(h)$ while errors in y_0 need only to go to zero as $o(1)$.

Although we require the starting values to converge to the initial values in the way given by the norm $\|\cdot\|_h^p$, we are only interested in those components of $\mathbf{a}_n$ which appear in f. Hence, we state:

Definition 10.2

A multivalue method is q-convergent for pth order equations if, for any pth order q-differential equation satisfying a Lipschitz condition, the computed solution $\mathbf{a}_n$ is such that

$$\|\mathbf{a}_n - \mathbf{a}(t)\|_h^{p-q+1} \longrightarrow 0$$

uniformly for $0 \le t \le b$ as $\|\mathbf{a}_0 - \mathbf{a}(0)\|_h^p \longrightarrow 0$ and $n \longrightarrow \infty$ with $h = t/n$.

We will call a 1-convergent method simply *convergent*.

10.1.1. Stability

In Chapter 1 stability was defined to mean that a small perturbation in the initial values could cause only a bounded change in the answer as h was reduced to zero. Since each step in a one-step method is effectively a new initial value problem, this serves to bound changes due to small perturbations at any step in the computation. A similar requirement is desirable in multistep methods, but we must realize that for pth order equations some changes may be equivalent to making large changes in the equivalent initial value problem. For first order equations we can adopt the definition unchanged, but we saw in the example leading to Eq. (10.9) that a fixed change in hy'_0 causes an arbitrarily large in y_N as h goes to zero. Thus, we use the following definitions of stability:

Definition 10.3

A multistep (multivalue) method is stable for first order equations if, for any first order equation satisfying a Lipschitz condition, there exist constants K and h_0 such that

$$\begin{aligned} \|\mathbf{y}_n - \mathbf{y}_n^*\| &\le K\|\mathbf{y}_0 - \mathbf{y}_0^*\| \\ [\|\mathbf{a}_n - \mathbf{a}_n^*\| &\le K\|\mathbf{a}_0 - \mathbf{a}_0^*\|] \end{aligned} \tag{10.10}$$

for all $0 \le t \le b$ and all $h = (t/n) \in (0, h_0)$, where $\mathbf{y}_n$ and $\mathbf{y}_n^*[\mathbf{a}_n$ and $\mathbf{a}_n^*]$ are two numerical solutions.

We will see that for pth order equations we will need a form of stability in which starting value perturbations cause bounded changes in those derivatives which occur in the differential equation. Hence, we define:

DEFINITION 10.4

A multivalue method is q-stable for pth order equations if for any q-differential pth order equation which satisfies a Lipschitz condition, there exist constants K and h_0 such that

$$\|\mathbf{a}_n - \mathbf{a}_n^*\|_h^{p-q+1} \le K\|\mathbf{a}_0 - \mathbf{a}_0^*\|_h^p \tag{10.11}$$

for all $0 \le t \le b$ and all $h = (t/n) \in (0, h_0)$.

The definition of stability given above is a desirable one rather than a convenient one. We want a method to be relatively insensitive to small changes but we want a practical technique for testing for stability. This is the root condition.

DEFINITION 10.5

A multistep method for first order equations satisfies the root condition if all the roots of $\rho(\xi) = 0$ are inside the unit circle or on the unit circle and simple. A multivalue method for first order equations satisfies the root condition if all the eigenvalues of $S = (I - \mathbf{l}\boldsymbol{\delta}_1^T)^M A$ are inside the unit circle and correspond to linear elementary divisors.†

For pth order equations we define the q-root condition. It corresponds to q-stability.

DEFINITION 10.6

A multivalue method for pth order equations satisfies the q-root condition if the eigenvalues of $S = (I - \mathbf{l}\boldsymbol{\delta}_p^T)^M A$ are inside the unit circle or on the unit circle. Those on the unit circle may not correspond to elementary divisors of rank greater than p. If there is an eigenvalue

†If the matrix S is reduced to its Jordan canonical form by a similarity transformation, repeated eigenvalues may give rise to m by m blocks on the diagonal of the form

$$\begin{bmatrix} \xi_i & 1 & & 0 \\ & \ddots & \ddots & \\ & & \ddots & 1 \\ 0 & & & \xi_i \end{bmatrix}$$

This is an *elementary divisor* of *rank m*. If m is one, it is a *linear elementary divisor*.

on the unit circle of rank p, no other eigenvalues on the unit circle may have rank greater than q.

We will see that consistency will require that the eigenvalue 1 have rank p, so this effectively limits extraneous eigenvalues to rank q. We will also call the 1-root condition the *strict root condition.*

The following two theorems indicate the necessity of the root conditions.

THEOREM 10.1

If a multistep method for first order equations is stable, the polynomial $\rho(\xi)$ satisfies the root condition.

THEOREM 10.2

If a multivalue method is q-stable, it must satisfy the q-root condition.

Two lemmas are useful for proving these results and will be needed later. The first is:

LEMMA 10.1

The solution of the difference equation

$$-y_n + \alpha_1 y_{n-1} + \cdots + \alpha_k y_{n-k} = 0 \qquad \alpha_k \neq 0 \tag{10.12}$$

can be expressed as

$$y_n = \sum_{i=1}^{s} \sum_{j=1}^{m_i} C_{ij} n^{j-1} \xi_i^n \tag{10.13}$$

where the ξ_i are the s different roots of

$$\rho(\xi) = -\xi^k + \sum_{i=1}^{k} \alpha_i \xi^{k-i} = 0$$

and ξ_i has multiplicity m_i. The C_{ij} are uniquely determined by the initial conditions for y_i, $0 \leq i < k$.

Proof: Substituting (10.13) into the left-hand side of (10.12) with $\alpha_0 = -1$, we get

$$\sum_{i=1}^{s} \sum_{j=1}^{m_i} C_{ij} \sum_{l=0}^{k} \alpha_l \xi_i^{n-l} (n-l)^{j-1} = \sum_{i=1}^{s} \sum_{j=1}^{m_i} C_{ij} \left(\left(\xi \frac{d}{d\xi} \right)^{j-1} (\xi^{n-k} \rho(\xi)) \right)_{\xi = \xi_i}$$

If ξ_i is an m_i fold root of ρ, then

$$\xi^{n-k} \rho(\xi) = (\xi - \xi_i)^{m_i} Q(\xi)$$

where $Q(\xi)$ is a polynomial in ξ. Hence,

$$\left(\left(\xi \frac{d}{d\xi} \right)^{j-1} (\xi^{n-k} \rho(\xi)) \right)_{\xi = \xi_i} = 0 \qquad \text{for } j \leq m_i$$

Therefore, y_n given by (10.13) is a solution of (10.12). Equation (10.12) uniquely determines $y_n, n \geq k$, when $y_l, 0 \leq l < k$, are given, so it remains to be shown that these y_i uniquely determine the C_{ij}. This follows from the fact that (10.13) for $n = 0, 1, \ldots, k - 1$ is a nonsingular system of linear equations for the C_{ij} in terms of the y_l. [Henrici (1962), p. 214, gives the value of the determinant of the system as

$$\prod_{1 \leq \mu < \nu \leq s} (\xi_\mu - \xi_\nu)^{m_\mu + m_\nu} \prod_{\nu=1}^{s} (m_\nu - 1)!!$$

where $0!! = 1$ and $k!! = k!((k-1)!!)$. This determinant is nonzero.]

LEMMA 10.2

If the matrix S has eigenvalues ξ_i corresponding to sets of elementary divisors of rank m_i, then

1. *if all $|\xi_i| < 1$, all elements of $S^n \longrightarrow 0$ as $n \longrightarrow \infty$;*
2. *if any $|\xi_i| > 1$, some elements of $S^n \geq 0(\xi_i^n)$† as $n \longrightarrow \infty$;*
3. *if all $|\xi_i| \leq 1$, and the largest m_i such that $|\xi_i| = 1$ is m, some elements of $S^n = 0(n^{m-1})$ as $n \longrightarrow \infty$, but none are larger.*

Proof: We express S in its Jordan form by

$$T^{-1}ST = \Delta = \begin{bmatrix} \xi_1 & 1 & & & & & & \\ & \ddots & \ddots & & & & & \\ & & \ddots & 1 & & & & \\ & & & \xi_1 & & & & \\ & & & & \xi_2 & 1 & & \\ & & & & & \ddots & 1 & \\ & & & & & & \xi_2 & \\ & & & & & & & \ddots \\ & & & & & & & & \xi_s \end{bmatrix}$$

Hence,

$$S^n = (T\Delta T^{-1})^n = T\Delta^n T^{-1}$$

An elementary divisor of order m with eigenvalue ξ will lead to the diagonal block in Δ^n of

$$\begin{bmatrix} \xi^n & n\xi^{n-1} & \cdots & \binom{n}{m-1}\xi^{n-m+1} \\ & \ddots & \ddots & \vdots \\ & & \ddots & n\xi^{n-1} \\ 0 & & & \xi^n \end{bmatrix}$$

†The notation $a(x) \geq 0(x)$ will be used to mean that there exists a constant K such that $|a(x)/Kx| \geq 1$ for all x in a neighborhood of its limiting value. In this case, it means that some elements of S^n go to infinity at least as fast as $|\xi_i^n|$.

1. If $|\xi| < 1$, all elements of this $\longrightarrow 0$, hence if all $|\xi_i| < 1$, all elements of $\Delta^n \longrightarrow 0$.
2. If $|\xi_i| > 1$, then $S^n\mathbf{x} = \xi_i^n\mathbf{x} = 0(\xi_i^n)$ in some components, where $\mathbf{x}$ is the eigenvector corresponding to ξ_i. Hence, some components of S^n are $\geq 0(\xi_i^n)$.
3. If $|\xi_i| = 1$, then the top right-hand element of the block of Δ^n corresponding to this ξ_i is $0(n^{m_i - 1})$.

Since T^{-1} is nonsingular, there exists an $\mathbf{x}$ such that $T^{-1}\mathbf{x} = [0, \ldots, 0, 1, 0, \ldots, 0]^T$, where the one appears in the position corresponding to the rightmost element of this block. Hence,

$$S^n\mathbf{x} = T\Delta^n T^{-1}\mathbf{x} = T\left[\begin{array}{c|c|c} & 0 & \\ \hline & \begin{matrix} \xi_i^n & \binom{n}{m_i - 1}\xi_i^{n-m+1} \\ & \ddots \\ & \xi_i^n \end{matrix} & \\ \hline & 0 & \end{array}\right]\begin{bmatrix} 0 \\ \vdots \\ 0 \\ \vdots \\ 1 \\ \vdots \\ 0 \end{bmatrix}$$

$$= T\begin{bmatrix} 0 \\ \vdots \\ 0 \\ \binom{n}{m_i - 1}\xi_i^{n-m_i+1} \\ \vdots \\ \xi_i^n \\ 0 \\ \vdots \\ 0 \end{bmatrix}$$

which has elements of $0(n^{m_i-1})$. No vector $\mathbf{x}$ can lead to $S^n\mathbf{x}$ larger than $0(n^{m-1})$.

Proof of Theorem 10.1: If the root condition is not satisfied because a root $\tilde{\xi}$ of $\rho(\xi)$ is such that $|\tilde{\xi}| > 1$, the solution of $y' = 0$ with $\mathbf{y}_0 = 0$ gives $\mathbf{y}_N = 0$, while if $\mathbf{y}_0^*$ is such that the starting values correspond to the solution $y_i^* = \tilde{\xi}^i$, $0 \leq i < k$, then $\mathbf{y}_N^*$ is unbounded as $N \longrightarrow \infty$. If $\tilde{\xi}$ is a root of modulus one and multiplicity $m > 1$, starting values $\mathbf{y}_0^*$ due to $y_i^* = i^{m-1}\tilde{\xi}^i$, $0 \leq i < k$, lead to the solution

$$|y_N| = |N^{m-1}\tilde{\xi}^N|$$

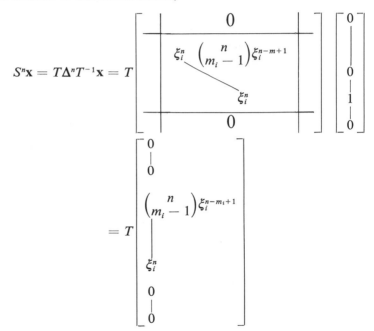

which is unbounded as $N \longrightarrow \infty$. Thus in either case $\|\mathbf{y}_0 - \mathbf{y}_0^*\|$ is bounded but $\| \mathbf{y}_N - \mathbf{y}_N^* \|$ is unbounded, so violating the root condition violates stability for multistep methods. Q.E.D.

Proof of Theorem 10.2: Consider perturbations $\delta \mathbf{a}_0 = \mathbf{a}_0^* - \mathbf{a}_0$ in the starting values such that $\| \delta \mathbf{a}_0 \|_h^p = 0(1)$ for the problem $y^{(p)} = 0$ with $\mathbf{a}_0 = [0, \dots, 0]^T$. The solution is

$$\mathbf{a}_N^* = S^N \delta \mathbf{a}_0$$

From Lemma 10.2, if an eigenvalue ξ of S is greater than 1, some components of S^N are $\geq 0(\xi^N)$; $\mathbf{a}_N^*$ is therefore unbounded in some components. If an eigenvalue ξ of modulus one has rank m, some component of S^N is $0(N^{m-1})$, hence some components of $\mathbf{a}_N^*$ are $0(N^{m-1}h^{p-1}) = 0(N^{m-p})$. These are unbounded if $m > p$.

Since $\| \mathbf{a}_n^* \|_h^{p-q+1}$ must be $0(1)$ for q-stability, the "worst" behavior that S^n may have (in the sense of largest) in powers of n is as shown in the representation below

$$
\underset{\mathbf{a}_n^*}{0\left(\begin{bmatrix} 1 \\ n^{-1} \\ \cdots \\ n^{q-p} \\ \hdashline n^{q-p} \\ \cdots \\ n^{q-p} \\ \hdashline n^{q-p} \\ \cdots \\ n^{q-p} \end{bmatrix}\right)} =
$$

$$
0\left(\overset{S^n}{\left[\begin{array}{cccc:cccc:ccc}
1 & n & \cdots & n^{p-q} & n^{p-q+1} & & \cdots & n^{p-1} & n^{p-1} & \cdots & n^{p-1} \\
 & 1 & \cdots & n^{p-q-1} & n^{p-q} & & \cdots & n^{p-2} & n^{p-2} & \cdots & n^{p-2} \\
 & 0 & \cdots & & \cdots & & & \cdots & \cdots & & \cdots \\
 & & & 1 & n & & \cdots & n^{q-1} & n^{q-1} & \cdots & n^{q-1} \\
\hdashline
 & & & 1 & n & & \cdots & n^{q-1} & n^{q-1} & \cdots & n^{q-1} \\
 & 0 & & \cdots & \cdots & & & \cdots & \cdots & & \cdots \\
 & & & 1 & n & & \cdots & n^{q-1} & n^{q-1} & \cdots & n^{q-1} \\
\hdashline
 & & & 1 & n & & \cdots & n^{q-1} & n^{q-1} & \cdots & n^{q-1} \\
 & 0 & & \cdots & \cdots & & & \cdots & \cdots & & \cdots \\
 & & & 1 & n & & \cdots & n^{q-1} & n^{q-1} & \cdots & n^{q-1}
\end{array}\right]}
\overset{\delta \mathbf{a}_0}{\begin{bmatrix} 1 \\ n^{-1} \\ \cdots \\ n^{q-p} \\ \hdashline n^{q-1-p} \\ \cdots \\ n^{1-p} \\ \hdashline n^{1-p} \\ \cdots \\ n^{1-p} \end{bmatrix}}\right)
$$

This will be violated if S has two or more eigenvalues on the unit circle, one of which has rank p and one rank greater than q.† Q.E.D.

A simple extension of the proof of Theorem 10.2 yields

THEOREM 10.3

A q-convergent multivalue method satisfies the q-root condition.

This implies that the multistep method (10.2) for the special pth order equation cannot have more than p-fold roots on the unit circle since it is equivalent to a multivalue method.

10.1.2 Order

The order of a multistep method was defined for first order equations in Definition 8.1. This definition can be extended easily to the multistep method for the special pth order equation $y^{(p)} = f(y, t)$ given by (10.2).

DEFINITION 10.7

If the operator L_h is defined by

$$L_h(y(t)) = \sum_{i=0}^{k} \left(\alpha_i y(t - hi) + \frac{h^p}{p!}\beta_i y^{(p)}(t - hi)\right)$$

then the order r is the largest integer r such that

$$L_h(y(t)) = 0\,(h^{r+1})$$

whenever $y \in C_{r+1}$.

If we assume that $y \in C_{r+2}$, we can substitute Taylor's series with remainder terms of $0\,(h^{r+2})$ for $y\,(t - hi)$ and $h^q y^{(q)}(t - hi)$ to get

$$L_h(y(t)) = \sum_{q=0}^{r+1} C_q h^q y^{(q)}(t) + 0\,(h^{r+2}) \tag{10.14}$$

where

†This can be seen by assuming that there are at least two eigenvalues on the unit circle of ranks p and $\tilde{q}$ with $p \geq \tilde{q} > q$. If S is written as $T\Delta T^{-1}$ and only components of size $n^{\tilde{q}-1}$ are considered in Δ^n, we can find the corresponding components in S. Since $\tilde{q} > q$, they may only appear in rows 0 to $p - \tilde{q}$ to maintain the above form. There are exactly $p - \tilde{q} + 2$ components in Δ^n of size $n^{\tilde{q}-1}$. They appear in different rows and columns. They select $p - \tilde{q} + 2$ columns of T and rows of T^{-1}. Since only $p - \tilde{q} + 1$ rows of the result may be nonzero, either those columns of T or rows of T^{-1} must be linearly dependent, but T is nonsingular, hence $\tilde{q} \not> q$.

$$C_q = \begin{cases} \sum_{i=0}^{k} \frac{(-i)^q}{q!}\alpha_i & q < p \\ \sum_{i=0}^{k}\left[\frac{(-i)^q}{q!}\alpha_i + \frac{(-i)^{q-p}}{p!(q-p)!}\beta_i\right] & r+1 \geq q \geq p \end{cases} \tag{10.15}$$

This shows that the order is determined by the coefficients of the method. If we define the polynomials ρ and σ as before, we see that

$$C_q = \begin{cases} \frac{1}{q!}\left[\left(\xi\frac{d}{d\xi}\right)^q(\xi^{-k}\rho(\xi))\right]_{\xi=1} & q < p \\ \frac{1}{q!}\left[\left(\xi\frac{d}{d\xi}\right)^q(\xi^{-k}\rho(\xi)) + \binom{q}{p}\left(\xi\frac{d}{d\xi}\right)^{q-p}(\xi^{-k}\sigma(\xi))\right]_{\xi=1} & q \geq p \end{cases} \tag{10.16}$$

From this it follows that if the order $r \geq p$,

$$\begin{aligned} \rho(1) &= 0 \\ \rho'(1) &= 0 \\ &\cdots \\ \rho^{(p-1)}(1) &= 0 \\ \rho^{(p)}(1) + \sigma(1) &= 0 \end{aligned} \tag{10.17}$$

Conversely, if (10.17) holds, the order is $\geq p$ since $C_q = 0$ for $q \leq p$. Note that this defines the order of a method based on a corrector only; that is, one in which the formula is explicit or the corrector is iterated to convergence.

The order of a multivalue method or a PC multistep method apparently has to be defined in terms of the solution of a differential equation. We can define it as follows.

DEFINITION 10.8

If $\mathbf{a}(t)$ is the correct value of the vector $\mathbf{a}$ for some h at time t, and we define

$$\begin{aligned} \tilde{\mathbf{a}}_{(0)} &= A\mathbf{a}(t-h) \\ \tilde{\mathbf{a}}_{(m+1)} &= \tilde{\mathbf{a}}_{(m)} + \mathbf{l}F(\tilde{\mathbf{a}}_{(m)}) \\ \tilde{\mathbf{a}}(t) &= \tilde{\mathbf{a}}_{(M)} \end{aligned} \tag{10.18}$$

then the order of the method for pth order equations is the largest r such that if F represents any differential equation of order p with a solution $y \in C_{r+1}$,

$$\tilde{\mathbf{a}}(t) - \mathbf{a}(t) = 0(h^{r+1}) \tag{10.19}$$

For first order equations we have seen that the order of the predictor can be less than the order of the corrector, and that the order of the result is increased by one for each corrector iteration up to the order of the corrector. In a similar manner, a multistep explicit predictor formula can be used for the special pth order equation, and its order will be increased by p for each corrector iteration up to the order of the corrector.

The order of a multivalue method may depend on the equation being integrated. For example, the predictor equation could be of order q and the corrector $r > p + q$. A single corrector iteration could then lead to an order $q + 1$ method. However, for the differential equation $y^{(p)} = f(t)$, the method will have order r since the predictor formula is not used. Thus, the order as given in Definition 10.8 is the minimum over all equations with smooth solutions. If the class of equations is restricted (for example, to q-differential equations), the order of the method could be higher. It is evident that the order of the method for a given class is uniquely determined by the coefficients of the method, but we have not yet expressed the order in terms of algebraic relations on those coefficients. This will be delayed until Section 10.4 because it is complex to determine the upper limit of the order, and relatively unimportant to practical schemes. In this section we will derive some necessary conditions on the coefficients that will be useful in convergence proofs.

If a predictor-corrector scheme can be expressed as a multivalue scheme [as it can be if $p = 1$, if f depends on y and t only, or if restrictions on the predictors for $y', \ldots, y^{(p-1)}$ are satisfied], the order of the predictor can be readily seen from the form of the matrix A when the method is put in the normal form. If the predictor has order r' and the corrector order is at least r', the first $r' + 1$ columns of A take the form

$$\left[\begin{array}{cccc} 1 & 1 & \cdots & 1 \\ & 1 & \cdots & r' \\ & & \ddots & r' \\ & 0 & & 1 \\ \hdashline & & 0 & \end{array}\right]$$

That is, they form the Pascal triangle matrix in the upper section and zeros in the lower section.

Lemma 10.3

If a multivalue method in normal form for pth order equations has order $\geq p - 1$, *then the first p columns of*

$$S = (I - \mathbf{1}\boldsymbol{\delta}_p^T)^M A$$

form the Pascal triangle matrix in the upper section and zero in the lower section. If the order of the predictor is not $\geq p - 1$, *then* $l_p = 1$.

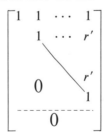

Proof: We consider the problem

$$y^{(p)} = p!$$

We set $\mathbf{a}(t)$ to contain $t^{p-i}h^i\binom{p}{i}$ in its ith component. For this problem we have

$$F(\mathbf{a}) = h^p - a_p$$

where a_p is the pth component of $\mathbf{a}$. Thus, the pth component of $\tilde{\mathbf{a}}_{(m+1)} = \tilde{\mathbf{a}}_{(m)} + \mathbf{l}F(\tilde{\mathbf{a}}_{(m)})$ is $(1 - l_p)(\tilde{\mathbf{a}}_{(m)})_p + l_p h^p$ so that

$$F(\tilde{\mathbf{a}}_{(m+1)}) = (1 - l_p)F(\tilde{\mathbf{a}}_{(m)}) = (1 - l_p)^{m+1}F(\tilde{\mathbf{a}}_{(0)})$$

Hence,

$$\tilde{\mathbf{a}}_{(M)} = \tilde{\mathbf{a}}_{(0)} + \mathbf{l}\sum_{i=0}^{M-1} (1 - l_p)^i F(\tilde{\mathbf{a}}_{(0)})$$

To reduce confusion, write $\tilde{\mathbf{l}}$ for $\mathbf{l}(1 - (1 - l_p)^M)/l_p$ if $l_p \neq 0$, or $\mathbf{l}M$ if l_p is zero. Consequently,

$$\tilde{\mathbf{a}}(t) = \tilde{\mathbf{a}}_{(M)} = \tilde{\mathbf{a}}_{(0)} + \tilde{\mathbf{l}}F(\mathbf{a}_{(0)}) = \tilde{\mathbf{l}}h^p + (I - \tilde{\mathbf{l}}\boldsymbol{\delta}_p^T)A\mathbf{a}(t - h)$$

If the order is at least $p - 1$, $\tilde{\mathbf{a}}(t)$ can differ from $\mathbf{a}(t)$ by at most $0(h^p)$. Hence, the first p columns of $(I - \tilde{\mathbf{l}}\boldsymbol{\delta}_p^T)A$ must be of the form described since the first p elements of $\mathbf{a}(t - h)$ are larger than $0(h^p)$. It can easily be seen that

$$S = (I - \mathbf{l}\boldsymbol{\delta}_p^T)^M A = (I - \tilde{\mathbf{l}}\boldsymbol{\delta}_p^T)A$$

and the first part of the result follows.

If the order of the predictor is $< p - 1$, then some row of A contains non-Pascal triangle entries to the left of the pth element. *The effect of premultiplying A by $(I - \mathbf{l}\boldsymbol{\delta}_p^T)$ is to subtract l_i times the pth row from the ith row.* If $l_p = 1$, the pth row is zero after the first such step, but if $l_p \neq 1$, it will remain nonzero for any finite number of iterations. Consequently, if there are nonzero elements to the left of the diagonal of the pth row (that is, in columns 0 to $p - 1$), $l_p = 1$ in order that they be zero in S. If there are non-Pascal triangle elements to the left of the pth column in any rows, there must be nonzero elements in the same columns of the pth row.

This result can be extended easily to:

LEMMA 10.4

If a multivalue method for pth order equations in normal form has order $r \leq k - 1$, then the first $r + 1$ columns of $(I - \mathbf{l}\boldsymbol{\delta}_p^T)^M A$ agree with the first $r + 1$ columns of $(I - \mathbf{l}\boldsymbol{\delta}_p^T)^M \tilde{A}$, where $\tilde{A}$ is the Pascal triangle matrix.

The proof is left for Problem 4.

In addition to providing an indicator of how the error behaves, a minimum order for a method is a necessary condition, as the following theorems show.

THEOREM 10.4

If the multistep method (10.2) is convergent for pth order equations, then the order of (10.2) is at least p.

Proof: We will use Eq. (10.17) as a characterization of a pth order method. We note that $\xi = 1$ is a $(j + 1)$-fold root of $\rho(\xi) = 0$ if and only if it is a j-fold root of $\rho(\xi)$ and

$$\sum_{i=0}^{k} \alpha_i p_j(m + i) = 0 \tag{10.20}$$

where p_j is any polynomial of degree j.† Consider the equation $y^{(p)} = 0$ with the initial values $y^{(q)}(0) = 0$, $0 \leq q < p$, except for $y^{(j)}(0) = j!$, $j < p$. The solution to this problem is $y = t^j$. We examine the solution of (10.2) starting with the correct initial values $y_i = z_i h^j$, where $z_i = i^j$. For this we get the solution $y_n = z_n h^j$, where

$$\sum_{i=0}^{k} \alpha_i z_{n-i} = 0 \tag{10.21}$$

If the method is convergent, $y_n/t_n^j \longrightarrow 1$ for fixed t_n, so that

$$\frac{z_n}{n^j} \longrightarrow 1 \qquad \text{as } n \longrightarrow \infty \tag{10.22}$$

For $j = 0$, the limit of (10.21) implies

$$\sum_{i=0}^{k} \alpha_i = 0$$

which implies that $\xi = 1$ is a root of $\rho(\xi)$ by (10.20). Now let us assume that $\xi = 1$ is a j-fold root of ρ, $j \geq 1$. By (10.21),

†This can be shown by the following: If $\xi = 1$ is a $(j + 1)$-fold root of $\rho(\xi)$, then for $0 \leq q \leq j$

$$0 = \left(\frac{d^q}{d\xi^q}(\rho(\xi)\xi^n)\right)_{\xi=1} = \sum_{i=0}^{k} \alpha_i \tau_q(i - n - k)$$

where $\tau_q(x) = -x(-x - 1) \cdots (-x - q + 1)$ is a polynomial of degree q in x. Since the $\tau_q(x)$, $0 \leq q \leq j$, are a basis for the space of jth degree polynomials, we can express any polynomial $p_j(x)$ of degree j as $\sum_{j=0}^{q} \gamma_q \tau_q(x)$. If we set $m = -n - k$, we immediately get

$$\sum_{i=0}^{k} \alpha_i p_j(m + i) = \sum_{q=0}^{j} \gamma_q \sum_{i=0}^{k} \alpha_i \tau_q(i - n - k) = 0$$

Conversely, if (10.20) holds, it holds for $p_j(m + i) = \tau_j(i - k)$, so

$$\left(\frac{d^j}{d\xi^j}\rho(\xi)\right)_{\xi=1} = 0$$

If $\xi = 1$ is a j-fold root of $\rho(\xi) = 0$, then this implies that it is a $(j + 1)$-fold root.

$$
\begin{aligned}
0 &= \sum_{n=k}^{N} p_{j-1}(2k-n) \sum_{i=0}^{k} \alpha_i z_{n-i} \\
&= p_{j-1}(k) \qquad [\alpha_0 z_N + \alpha_1 z_{N-1} + \cdots + \alpha_k z_{N-k}] \\
&\quad + p_{j-1}(k-1) \qquad [\alpha_0 z_{N-1} + \cdots + \alpha_{k-1} z_{N-k} + \alpha_k z_{N-k-1}] \\
&\quad + \cdots \\
&\quad + p_{j-1}(2k-N+1)[\cdots + \alpha_k z_1] \\
&\quad + p_{j-1}(2k-N) \qquad [\cdots + \alpha_{k-1} z_1 + \alpha_k z_0] \\
&= \sum_{n=N-k}^{N} z_n \sum_{i=0}^{N-n} \alpha_i p_{j-1}(n-N+k+i) \\
&\quad + \sum_{n=k}^{N-k-1} z_n \sum_{i=0}^{k} \alpha_i p_{j-1}(n-N+k+i) \\
&\quad + \sum_{n=0}^{k-1} z_n \sum_{i=k-n}^{k} \alpha_i p_{j-1}(n-N+k+i)
\end{aligned}
$$

The second term vanishes by (10.20), while the third term is a polynomial of degree $j-1$ in N. Dividing by N^j, the last term $\longrightarrow 0$ as $N \longrightarrow \infty$, while $z_n/N^j \longrightarrow 1$ for $n \in [N-k, N]$. Hence, we get

$$
\begin{aligned}
0 &= \sum_{n=N-k}^{N} \sum_{i=0}^{N-n} \alpha_i p_{j-1}(n-N+k+i) \\
&= \sum_{i=0}^{k} \alpha_i \sum_{n=N-k+i}^{N} p_{j-1}(n-N+k+i) \\
&= \sum_{i=0}^{k} \alpha_i \sum_{n=0}^{k-i} p_{j-1}(n+2i)
\end{aligned}
$$

Since $p_{j-1}(n)$ is a polynomial of degree $j-1$ in n, $\sum_{n=0}^{k-i} p_{j-1}(n+2i)$ is a polynomial of degree j in i. Hence from (10.20), $\xi = 1$ is a $(j+1)$-fold root of $\rho(\xi)$. Since we only restricted $j < p$, it follows that ξ is a p-fold root of $\rho(\xi)$ and

$$\rho(1) = \rho'(1) = \cdots = \rho^{(p-1)}(1) = 0.$$

To complete the proof, we must show that the last of equations (10.17) is satisfied. To do this, consider the problem $y^{(p)} = p!$, $y^{(q)}(0) = 0$, $0 \le q < p$. The solution is $y = t^p$. The difference equation (10.2) gives

$$\sum_{i=0}^{k} \alpha_i y_{n-i} + h^p \sum_{i=0}^{k} \beta_i = 0 \tag{10.23}$$

Since $\xi = 1$ is a p-fold root of $\rho(\xi)$,

$$\sum_{i=0}^{k} \alpha_i (n-i)^p = \left[\left(\xi \frac{d}{d\xi}\right)^p (\xi^{n-k} \rho(\xi))\right]_{\xi=1} = \rho^{(p)}(1)$$

Since the method is convergent, Theorem 10.3 tells us that $\rho^{(p)}(1) \neq 0$, hence

$$y_n = -(hn)^p \frac{\sigma(1)}{\rho^{(p)}(1)} \tag{10.24}$$

satisfies (10.23). The values of y_i, $0 \leq i < k$, differ from the correct starting values of $(hi)^p$ by at most $0(h^p)$, so a convergent method will be such that $y_n \longrightarrow t_n^p = (hn)^p$. This and (10.24) imply the last of equations (10.17) and so a convergent method has order $\geq p$. Q.E.D.

Although one suspects that the order of a general predictor-corrector multistep or multivalue method must also be $\geq p$, a proof is not known except in the case $p = 1$ for multistep methods.

THEOREM 10.5

The order of a predictor-corrector method for first order equations must be ≥ 1 *if it is convergent.*

Proof: Let the polynomials defining the predictor and corrector be ρ, σ and ρ^*, σ^*, respectively. Since Theorem 10.4 showed that the order of the corrector must be at least one, the only case to consider is a predictor order of -1 and a single correction of order one or greater. Normalize so that $\alpha_0 = \alpha_0^* = -1$. From the definition of order given by Eqs. (10.18) and (10.19), we have

$$\sum_{i=1}^{k} [\alpha_i^* y(t - ih) + h\beta_i^* y'(t - ih)] + h\beta_0^* f\left[\sum_{i=1}^{k} [\alpha_i y(t - ih) + h\beta_i y'(t - ih)]\right] - y(t) = 0\,(h^{r+1}) \tag{10.25}$$

for any differential equation $y' = f(y, t)$ whose solution $y(t) \in C_{r+1}$. We consider the equation $y' = y$. For this we get from (10.25)

$$e^{t-kh}\{\rho^*(e^h) + h\sigma^*(e^h) + h\beta_0^*[\rho(e^h) + h\sigma(\epsilon^h)]\} = 0\,(h^{r+1}) \tag{10.26}$$

If ξ is a root of

$$\rho^*(\xi) + h\sigma^*(\xi) + h\beta_0^*\rho(\xi) + h^2\beta_0^*\sigma(\xi) = 0 \tag{10.27}$$

then $y_n = \xi^n$ is a solution of the predictor-corrector method. We examine (10.27) for a solution of the form $\xi = e^h + \Delta$, where Δ is small. We get by substitution,

$$\Delta\rho^{*\prime}(e^h) + \rho^*(e^h) + h\sigma^*(e^h) + h\beta_0^*\rho(e^h) + h^2\beta_0^*\sigma(e^h) = 0\,(\Delta^2 + h\Delta)$$

The last four terms of the left-hand side are $0\,(h^{r+1})$ by (10.26) and $\rho^{*\prime}(e^h) = \rho^{*\prime}(1) + 0\,(h)$, so

$$\Delta = \frac{1}{\rho^{*\prime}(1)} 0\,(\Delta^2 + h\Delta + h^{r+1})$$

By Theorem 10.4 the corrector has order one, hence $\rho^*(1) = 0$. Since the method converges, it satisfies the root condition (Theorem 10.3) so $\rho^*(\xi)$ does not have a double root at $\xi = 1$. Hence, $\rho^{*\prime}(1) \neq 0$.

Therefore,

$$\Delta = Kh^{r+1} + 0(h^{r+2}) \qquad K \neq 0$$

If $r = 0$,

$$\begin{aligned} y_n = \xi^n &= (e^h + Kh + 0(h^2))^n \\ &= e^{(K+1)t} + 0(h^2) \end{aligned}$$

for $t = nh$.
Thus, the initial value problem $y' = y$, with initial condition $y(0) = 1$ and starting conditions $y_i = (e^h + \Delta)^i$, $0 \leq i < k$, converges to other than the answer e^t if the order of the predictor is -1. Q.E.D.

10.1.3 Consistency and Convergence

DEFINITION 10.9

A method for a pth order equation is called consistent if its order is at least p.

A result that is probably true is that q-stability and consistency are necessary and sufficient conditions for q-convergence for the general equation (10.1).

We have already shown part of this, namely:

1. q-stability $\longrightarrow$ q-root condition.
2. q-convergence $\longrightarrow$ q-root condition.
3. Convergence $\longrightarrow$ consistency of corrector only methods.
4. Convergence $\longrightarrow$ consistency of *PC* methods for first order equations.

In this section we will show that

5. q-root condition and consistency $\Rightarrow$ q-convergence.
6. q-root condition and consistency $\Rightarrow$ q-stability.

We conjecture that

7. Convergence $\longrightarrow$ consistency of multivalue methods for pth order equations.

We start by proving convergence for the simplest case, a single equation of first order.

THEOREM 10.6

A stable consistent multivalue method for first order equations is convergent.

Proof: Although it is possible to show this for problems satisfying only continuity and Lipschitz conditions on $f(y, t)$, we will assume that f

has continuous first derivatives (as usually happens except at a finite number of points which can be easily handled). In that case, $y \in C_2$ and by substituting Taylor's series for the quantities $\tilde{\mathbf{a}}_{(m)}$ and $\tilde{\mathbf{a}}(t)$ about $(t - h)$ in the definition of order given by Eqs. (10.18) and (10.19) we will find that

$$\tilde{\mathbf{a}}(t) - \mathbf{a}(t) = P(y, y', y'', f, f_y, f_t, h)h^2$$

where $P(\ldots)$ is a polynomial in its arguments, and the arguments are evaluated at some points between the solutions of the difference and the differential equations.† Because the arguments can be bounded over the range of the solution, there exists a D such that

$$\|\tilde{\mathbf{a}}(t_n) - \mathbf{a}(t_n)\| = \|\mathbf{d}_n\| \leq Dh^2 \tag{10.28}$$

T is a combination of bounds on the derivatives.

We have shown in Eqs. (9.12) and (9.13) that

$$\mathbf{e}_n = S_n\mathbf{e}_{n-1} + \mathbf{d}_n \tag{10.29}$$

where

$$S_n = \prod_{i=0}^{M-1} \left[I + \mathbf{l}\frac{\partial F}{\partial \mathbf{a}}(\xi_i)\right] A$$

For multivalue methods in the normal form

$$\frac{\partial F}{\partial \mathbf{a}}(\xi_i) = [hf_y(\xi_i), -1, 0, \ldots, 0] = hf_y(\xi_i)\boldsymbol{\delta}_0^T - \boldsymbol{\delta}_1^T$$

Since $|f_y| < L$, we can write

$$S_n = S + h\tilde{S}_n$$

where

†If an unlimited number of corrector iterations are allowed in order to get convergence, this statement must be modified since the polynomial could be of arbitrarily high order, and could not be bounded by simply bounding its arguments. In this case, we note that

$$\tilde{\mathbf{a}}(t) = \mathbf{a}_{(0)} + \omega\mathbf{l}$$

where ω represents the total of all corrections used to get convergence. Thus $\tilde{\mathbf{a}}(t)$ is such that $F(\tilde{\mathbf{a}}(t)) = (\tilde{\mathbf{a}}(t))_1 - hf((\tilde{\mathbf{a}}(t))_0) = 0$. Hence,

$$(\tilde{\mathbf{a}}_{(0)})_1 + \omega l_1 - hf((\tilde{\mathbf{a}}_{(0)})_0 + \omega l_0) = 0$$

If L is an upper bound for f_y, this equation has a unique solution for ω whenever $h < h_0 = l_1/(l_0L)$. This solution is given by

$$\omega = \frac{1}{l_1}[hf((\tilde{\mathbf{a}}_{(0)})_0) - (\tilde{\mathbf{a}}_{(0)})_1][1 - hl_0(f_y/l_1)]^{-1}$$

where f_y is evaluated at a suitable point. If, for fixed $\tilde{h} < h_0$ we keep $h \leq \tilde{h}$, we can write

$$\tilde{\mathbf{a}}(t) = \mathbf{a}_{(0)} + \frac{1}{l_1}[hf((\tilde{\mathbf{a}}_{(0)})_0) - (\tilde{\mathbf{a}}_{(0)})_1][1 + hl_0(f_y/l_1) + Kh^2]$$

where K is bounded. $\tilde{\mathbf{a}}(t) - \mathbf{a}(t)$ can then be expressed as a polynomial with bounded coefficients.

$$S = [I - \mathbf{1}\boldsymbol{\delta}_1^T]^M A$$

and $\tilde{S}_n$ is a matrix whose elements are polynomials in h with bounded coefficients.† Hence,

$$\|\tilde{S}_n\| \leq C_0 \qquad \text{for } h \leq h_0$$

and (10.29) leads to

$$\begin{aligned} \|\mathbf{e}_n\| &= \left\| \sum_{j=1}^{n} S^{n-j}(h\tilde{S}_j \mathbf{e}_{j-1} + \mathbf{d}_j) + S^n \mathbf{e}_0 \right\| \\ &\leq \sum_{j=1}^{n} \|S^{n-j}\| (hC_0 \|\mathbf{e}_{j-1}\| + \|d_j\|) + \|S^n\| \|\mathbf{e}_0\| \end{aligned} \tag{10.30}$$

The condition of stability on S implies that $\|S^m\| \leq C_1$ independent of m. An upper bound for (10.30) can now be found by the usual techniques. It is instructive to see how this is done. Replacing $\|S^m\|$ and $\|d_j\|$ by their bounds, we get

$$\|\mathbf{e}_n\| + \frac{Dh}{C_0} \leq \sum_{j=1}^{n} hC_0C_1\left(\|\mathbf{e}_{j-1}\| + \frac{Dh}{C_0}\right) + \frac{Dh}{C_0} + C_1 \|\mathbf{e}_0\| \tag{10.31}$$

Viewing (10.31) with an equality sign as a recurrence relation for $\|\mathbf{e}_n\| + (Dh/C_0)$, we look for a solution of the form

$$\|\mathbf{e}_N\| + \frac{Dh}{C_0} = K(1 + hC_0C_1)^N \qquad N \geq 1$$

and find

$$K = \frac{Dh}{C_0} + C_1 \|\mathbf{e}_0\|$$

Therefore,

$$\|\mathbf{e}_N\| \leq \left[\frac{Dh}{C_0} + C_1 \|\mathbf{e}_0\|\right](1 + hC_0C_1)^N - \frac{Dh}{C_0} \tag{10.32}$$

Since $(1 + hC_0C_1)^N \leq e^{hNC_0C_1} \leq e^{bC_0C_1} = \tilde{K}$, where $\tilde{K}$ is independent of h, we can bound the error by

$$\|\mathbf{e}_N\| \leq K_1 h + K_2 \|\mathbf{e}_0\| \tag{10.33}$$

Therefore, if $\|\mathbf{e}_0\| \to 0$ as $h \to 0$, $\|e_N\| \to 0$ for all N such that $Nh \leq b$. Q.E.D.

The bound provided by (10.33) is both crude and difficult to apply. C_0 is dependent on the size of $\tilde{S}_n$, which is controlled by the Lipschitz constant, but it is difficult to compute C_0 for a general method. The proof of Theorem 10.6 can be used to show the following result directly:

†If an unlimited number of corrector iterations are allowed, the same trick as used in the previous footnote must be employed to get a polynomial with bounded coefficients for $\tilde{S}_n$.

THEOREM 10.7

If a multivalue method for first order equations has order r and if the starting errors $\|\mathbf{e}_0\|$ *are bounded by* $D'h^{r'}$, *the error at time* $t = Nh$ *is bounded by*

$$\|\mathbf{e}_N\| \leq \frac{Dh^r}{C_0}(e^{C_0C_1t} - 1) + D'h^{r'}C_1e^{C_0C_1t} \tag{10.34}$$

if the solution $y(t)' \in C_{r+1}$.

The local truncation error can be bounded by Dh^{r+1} since $y \in C_{r+1}$. This replaces Dh^2 in the previous proof and the result follows.

The bound provided by Theorem 10.7, although of the right asymptotic order, still suffers from the fact that a lot of information is thrown away when norms are taken. The bound can be improved a little, but further improvements must come by obtaining estimates, as was done in one-step methods. This will be taken up in a later discussion of error.

We will now give a general convergence and error bound theorem for methods for the system of differential equations

$$(y^i)^{(p_i)} = f^i(\{y^j\}, \{y'^j\}, \ldots, t) \qquad i, j = 1, \ldots, s \tag{10.35}$$

each possibly of different order p_i, where it is understood that no higher derivative than the $(p_j - q_j)$th of the jth dependent variable appears on the right-hand side, where $1 \leq q_j \leq p_j$. The ith equation will be handled by the method

$$\begin{aligned} \mathbf{a}^i_{n,(0)} &= A_i\mathbf{a}^i_{n-1} \\ \mathbf{a}_{n,(m+1)} &= \mathbf{a}^i_{n,(m)} + \mathbf{l}_iF^i(\{\mathbf{a}^j_{n,(m)}\}) \\ \mathbf{a}^i_n &= \mathbf{a}^i_{n,(M)} \end{aligned} \tag{10.36}$$

where $\mathbf{a}^i$ is the normal form multivalue method information for the ith variable and

$$F^i(\{\mathbf{a}^j\}) = \frac{h^{p_i}}{p!}f^i\left(\{a^j_0\}, \left\{\frac{a^j_1}{h}\right\}, \left\{2!\frac{a^j_2}{h^2}\right\}, \ldots, t\right) - a^i_p \tag{10.37}$$

with a^i_k being the kth component of $\mathbf{a}^i$.

THEOREM 10.8

If the method given by A_i, $\mathbf{l}_i$ *satisfies the* q_i*-root condition for* p_i*th order equations and its order* $r_i \geq p_i$ *(i.e., it is consistent for* q_i*-differential equations) and if the* f^i *are sufficiently continuously differentiable, then*

$$\|\mathbf{a}^i_n - \mathbf{a}^i(t_n)\|_h^{p_i - q_i + 1} = 0(h^d)$$

where

$$d = \min(r'_i, r_i - p_i + 1) \tag{10.38}$$

provided that the error in the starting values $\|\mathbf{a}_0^i - \mathbf{a}^i(0)\|_h^{p_i}$ *is bounded by* $D'h^{r_i'}$ *for the ith variables and the ith equation is* q_i*-differential.*

The methods can be put into normal form by a transformation. We will assume that this has been done.

Proof: The steps in this proof are:

1. To show that the local truncation errors in the ith equation are bounded by $D^i h^{p_i+d}$.
2. To put a bound on the elements of S^n.
3. To repeat the proof of Theorem 10.6.

Step 1

Since the method given by A_i, $\mathbf{l}_i$ has order r_i for p_ith order equations using M corrector iterations, the predictor has order $\geq r_i - M \geq p_i + d - M$. Therefore, the truncation errors in the predictor step can be bounded by $D^i h^{p_i+d-M}$ (by use of Taylor's series). In the correction step, each $\mathbf{a}^j$ can enter into the calculation of another $\mathbf{a}^i$ via the term $h^{p_i} f^i(\ldots, y^j, \ldots, (y^j)^{(p_j-1)}, \ldots)$ so that the worst errors that can occur are due to the terms $h^{p_i}(y^j)^{(p_j-1)} = h^{p_i} a^j_{p_j-1}(p_j - 1)!/h^{p_j-1}$. The error in these is $0\,(h^{p_i}h^{p_j+d-M}/h^{p_j-1})$ or $0\,(h^{p_i+d-M+1})$. By use of Taylor's series with remainders, this can be made into a bound of the form $D^i h^{p_i+d-M+1}$ for $h \leq h_0$. We see that the order of the bound has been increased by one for one correction. Repeating $M - 1$ more times, we get the required bound.

Step 2

In view of Lemma 10.3 and the fact that the method A_i, $\mathbf{l}_i$ is consistent,

$$S_i = (I - \mathbf{l}_i \boldsymbol{\delta}_{p_i}^T)^M A_i$$

has the form

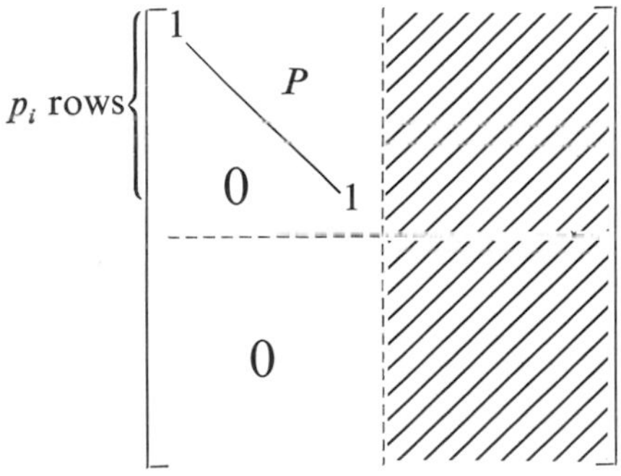

where the shading represents possibly nonzero elements and P is the Pascal triangle matrix, all of whose elements are nonzero. Consequently, the upper left-hand block is a single elementary divisor of order p_i with eigenvalue one. Because S_i satisfies the q_i-root condition, all other eigenvalues are less than one or one with elementary divisors of rank $\leq q_i$. It is then easy to see by induction on n that the sizes of the elements of S_i^n as powers of n are as shown below:

$$\left[\begin{array}{ccccc|ccc}
1 & n & n^2 & \cdots & n^{p_i-1} & n^{p_i-1} & \cdots & n^{p_i-1} \\
 & 1 & n & \cdots & n^{p_i-2} & n^{p_i-2} & \cdots & n^{p_i-2} \\
 & & \ddots & & \cdots & \cdots & & \cdots \\
 & 0 & & \ddots & n^{q_i-1} & n^{q_i-1} & \cdots & n^{q_i-1} \\
 & & & & \cdots & \cdots & & \cdots \\
 & & & & n & & & \\
 & & & & 1 & n^{q_i-1} & \cdots & n^{q_i-1} \\
\hline
 & & 0 & & & & n^{q_i-1} &
\end{array}\right]$$

Step 3

Defining $\mathbf{e}_n^i = \mathbf{a}_n^i - \mathbf{a}^i(t_n)$ and $\mathbf{e}_{n,(m)}^i = \mathbf{a}_{n,(m)}^i - \tilde{\mathbf{a}}_{(m)}^i(t_n)$, we have

$$\mathbf{e}_{n,(0)}^i = A_i \mathbf{a}_{n-1}^i - A_i \mathbf{a}^i(t_n - h) = A_i \mathbf{e}_{n-1}^i \tag{10.39}$$

$$\begin{aligned}\mathbf{e}_{n,(m+1)}^i &= \mathbf{a}_{n,(m)}^i + \mathbf{l}_i F^i(\{\mathbf{a}_{n,(m)}^i\}) - \tilde{\mathbf{a}}_{(m)}^i(t_n) - \mathbf{l}_i F^i(\{\tilde{\mathbf{a}}_{(m)}^j(t_n)\}) \\ &= \mathbf{e}_{n,(m)}^i + \mathbf{l}_i \sum_j \frac{\partial F^i}{\partial \mathbf{a}^j} \mathbf{e}_{n,(m)}^j\end{aligned} \tag{10.40}$$

where $\partial F^i/\partial \mathbf{a}^j$ is evaluated at some point between the true and numerical solution and summation limits have been omitted where obvious.

$$\mathbf{e}_n^i = \mathbf{a}_{n,(M)}^i - \mathbf{a}^i(t_n) = \mathbf{e}_{n,(M)}^i + \mathbf{d}_n^i \tag{10.41}$$

where $\mathbf{d}_n^i$ is the truncation error in the nth step for the ith equation. Now

$$\frac{\partial F^i}{\partial \mathbf{a}^j} = -\delta_j^i \boldsymbol{\delta}_{p_i}^T + \sum_{q=0}^{p_j-1} \frac{h^{p_i-q}}{p_i!} q! \frac{\partial f^i}{\partial y^{j(q)}} \boldsymbol{\delta}_q^T$$

where $\delta_j^i = 0$ if $i \neq j$, $\delta_j^i = 1$ if $i = j$. Substituting this in (10.40), we get

$$\mathbf{e}_{n,(m+1)}^i = (I - \mathbf{l}_i \boldsymbol{\delta}_{p_i}^T)\mathbf{e}_{n,(m)}^i + \sum_j \sum_q h^{p_i-q} \frac{q!}{p_i!} \frac{\partial f^i}{\partial y^{j(q)}} \mathbf{l}_i \boldsymbol{\delta}_q^T \mathbf{e}_{n,(m)}^j \tag{10.42}$$

Hence, from (10.41), (10.42), and (10.39),

$$\begin{aligned}\mathbf{e}_n^i &= S_i \mathbf{e}_{n-1}^i + \mathbf{d}_n^i \\ &\quad + \sum_{m=0}^{M-1} (I - \mathbf{l}_i \boldsymbol{\delta}_{p_i}^T)^{M-1-m} \sum_j \sum_q h^{p_i-q} \frac{q!}{p_i!} \frac{\partial f^i}{\partial y^{j(q)}} \mathbf{l}_i \boldsymbol{\delta}_q^T \mathbf{e}_{n,(m)}^j \\ &= S_i \mathbf{e}_{n-1}^i + \mathbf{d}_n^i + \mathbf{U}_n^i\end{aligned} \tag{10.43}$$

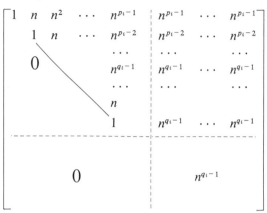

where the triple summation term has been called $\mathbf{U}_n^i$. (Note that the partial derivatives of f^i that appear in $\mathbf{U}_n^i$ are usually evaluated at different points for different values of m. Fortunately, this will not affect us, as we only need rough bounds on these terms.) Applying (10.43) repeatedly, we get

$$\mathbf{e}_N^i = \sum_{n=1}^{N} S_i^{N-n}(\mathbf{d}_n^i + \mathbf{U}_n^i) + S_i^N \mathbf{e}_0 \tag{10.44}$$

We complete the proof by showing inductively that for $N \geq 1$,

$$\| \mathbf{e}_N^i \|_h^{s_i} \leq h^d[(1 + hk)^N(k_1 + k_2) - k_2] \tag{10.45}$$

and

$$\| \mathbf{e}_{N,(m)}^i \|_h^{s_i} \leq h^d[(1 + hk)^{N-1}(k_{1,m} + k_{2,m}) - k_{2,m}] \tag{10.46}$$

for some constants $k, k_1, k_2, k_{1,m}, k_{2,m}$, where $s_i = p_i - q_i + 1$.

Inequality (10.45) is true for $N = 0$ if we choose $k_1 \geq D'$. We proceed by double induction, on N and then on $m \leq M$. Assume (10.45) true for all $0 \leq N \leq n - 1$. From (10.39), $k_{1,0}$ and $k_{2,0}$ can be expressed in terms of k_1 and k_2 to satisfy (10.46) for $m = 0$, $1 \leq N \leq n$. Now assume (10.46) is true for $\mathbf{e}_{N,(m)}^i$, all $1 \leq N \leq n$, $0 \leq m \leq m'$. From (10.46) for $q < s_j$,

$$| \boldsymbol{\delta}_q^T \mathbf{e}_{n,(m)}^j | \leq h^{d+q}[(1 + hk)^{n-1}(k_{1,m} + k_{2,m}) - k_{2,m}] \tag{10.47}$$

so Eq. (10.42) shows that, by using bounds on the partial derivatives occurring, $k_{1,m+1}$ and $k_{2,m+1}$ can be chosen such that (10.46) is satisfied for $m = m' + 1$. Hence, we conclude that (10.45) for $0 \leq N \leq n - 1$ implies (10.46) for $0 \leq N \leq n$. Substituting (10.46) into the definition of $\mathbf{U}_n^i$ following (10.43) and using (10.47), we get

$$\| \mathbf{U}_n^i \| \leq h^{p_i+d}[(1 + hk)^{n-1}(k_{1U} + k_{2U}) - K_{2U}]$$

for some constants k_{1U} and k_{2U}, while from Step 1 we have

$$\| \mathbf{d}_n^i \| \leq Dh^{p_i+d}$$

Let $\boldsymbol{\epsilon}$ be a column vector of all ones. From the form of S_i^m found in Step 2, we have

$$\| S_i^n \boldsymbol{\epsilon} \|_h^{s_i} \leq k_4 h^{-p_i+1} \qquad \text{for } nh \leq b \tag{10.48}$$

Hence from (10.44) we get

$$\| \mathbf{e}_N^i \|_h^{s_i} \leq h^d k_4 \left[\sum_{n=1}^{N} h(D + (1 + hk)^{n-1}(k_{1U} + k_{2U}) - k_{2U}) + D' \right]$$

Replacing k_{2U} by $k_5 = \max(D, k_{2U})$,

$$\| \mathbf{e}_N^i \|_h^{s_i} \leq h^d k_4 \left[\frac{(1 + hk)^N - 1}{k}(k_{1U} + k_5) + D' \right] \tag{10.49}$$

Note that k_4 is fixed by (10.48) and not dependent on other k's. In particular, $k_4 \geq 1$, so we can set $k_1 = k_4 D'$. If k_2 is now chosen, k_5 and k_{1U} are determined, and we can then choose k such that

$$\frac{k_4}{k}(k_{1U} + k_5) = k_1 + k_2$$

When these are substituted into (10.49), it reduces to (10.45), which shows that the method converges as $0(h^d)$. Q.E.D.

If in Theorem 10.8 we consider $\mathbf{e}_n$ to be the difference between two numerical solutions $\mathbf{a}_n^*$ and $\mathbf{a}_n$ of a single pth order equation starting from different initial values $\mathbf{a}_0$ and $\mathbf{a}_0^*$, we can apply Steps 2 and 3 as they stand with $\mathbf{d}_n$ set to 0. Consequently, $D = 0$. We can choose $k_2 = k_{2,m} = k_{2U} = k_5 = 0$ throughout the proof. Note that $k_1 = k_4 D'$, so if we set $\| \mathbf{a}_0^* - \mathbf{a}_0 \|_h^p = D'$, we have $r' = 0$, $d = 0$, and (10.45) implies that

$$\| \mathbf{a}_N^* - \mathbf{a}_N \|_h^s = \| \mathbf{e}_N \|_h^s \leq k_4(1 + hk)^N D'$$
$$\leq K \| \mathbf{a}_0^* - \mathbf{a}_0 \|_h^p \qquad \text{for } 0 \leq hN \leq b$$

where K is independent of $\| \mathbf{a}_0^* - \mathbf{a}_0 \|_h^p$. Set $s = p - q + 1$ and we have proved:

THEOREM 10.9

If a consistent multivalue method satisfies the q-root condition, it is q-stable.

10.2 THE MAXIMUM ORDER OF A STABLE MULTISTEP METHOD

We now turn to a discussion of a fundamental result, due to Dahlquist, which puts a bound on the maximum order of a stable multistep method for first order equations. Earlier we saw that it was possible to choose ρ and σ so that order $2k$ could be achieved. We will now show how the important requirement of stability limits the order to $k + 1$ if k is odd and $k + 2$ if k is even. We have already seen that a σ can be chosen for any ρ to make the order $k + 1$. Since ρ can be chosen to be stable, an order of $k + 1$ can be achieved easily. We will see that if and only if k is even and ρ is chosen so that all of its roots are on the unit circle can σ be chosen to make the order $k + 2$.

To discuss the stability, which is equivalent to the condition that the roots of $\rho(\xi)$ are inside or on the unit circle, it is easier to make a transformation such that

$$\xi = \frac{1 + z}{1 - z} \quad \text{or} \quad z = \frac{\xi - 1}{\xi + 1}$$

This maps the interior of the unit circle into the left half plane, as can be seen by setting $\xi = re^{i\theta} = rc + irs$, where $c = \cos\theta$ and $s = \sin\theta$. We have

$$z = \frac{rc + irs - 1}{rc + irs + 1} = \frac{[(rc+1) - irs]\,[(rc-1) + irs]}{[(rc+1) - irs]\,[(rc+1) + irs]}$$
$$= \frac{(r^2 - 1) + 2irs}{(rc+1)^2 + r^2s^2}$$

The denominator is > 0, the numerator is in the left half plane if $r^2 < 1$, on the imaginary axis if $r^2 = 1$, and in the right half plane if $r^2 > 1$. Hence, the shaded regions in Figure 10.2 correspond.

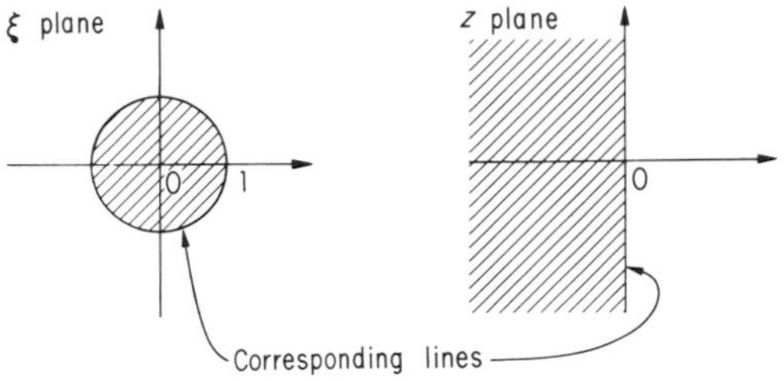

Fig. 10.2

We define

$$R(z) = \left(\frac{1-z}{2}\right)^k \rho\left(\frac{1+z}{1-z}\right)$$

and

$$S(z) = \left(\frac{1-z}{2}\right)^k \sigma\left(\frac{1+z}{1-z}\right)$$

These are both polynomials of degree k in z. The roots of $\rho(\xi) = 0$ for which $z \neq 1$ correspond to roots of $R(z) = 0$. Our hypothesis is that ρ is stable, hence ξ such that $z = 1$ is not a root of $\rho(\xi) = 0$. Consequently, all roots of R and ρ correspond except for roots of ρ at -1. These correspond to roots of R at infinity, that is, to decreases in the degree of R. Assume that the order of the method is r. The steps in the proof are as follows:

We map the condition that the method be of order r [see Eq. (8.10)], namely,

$$\frac{\rho(\zeta)}{\log(\xi)} + \sigma(\xi) = 0((\xi - 1)^r)$$

into

$$\frac{R(z)}{\log\{(1+z)/(1-z)\}} + S(z) = 0(z^r)$$

by substituting for ξ and multiplying by $((1-z)/2)^k$. We expand $R(z)/\log\{(1+z)/(1-z)\}$ as a power series in z, say $\sum_{n=0}^{\infty} r_n z^n$, and show that $r_n < 0$ for $n = k + 1$ or $k + 2$ as k is odd or even. Since $S(z)$ is of maximum degree k, we then see that $r \leq k + 1$ or $k + 2$ as k is odd or even.

First we consider $R(z) = a_0 + a_1z + \cdots + a_kz^k$. Since $\xi = 1$ is a simple root of $\rho(\xi) = 0$, $z = 0$ is a simple root of $R(z) = 0$. Therefore, $a_0 = 0$ and $a_1 \neq 0$. We can take $a_1 > 0$ without loss of generality. We will show that stability $\Rightarrow a_i \geq 0$, $i \geq 2$. ρ is a real polynomial; therefore R is also real. Consequently, the roots of R are either real, x_μ, or occur in conjugate pairs, $x_\nu \pm iy_\nu$, where x_μ and $x_\nu \leq 0$, because roots of ρ which may not be outside of the unit circle in the ξ-plane map into roots of R in the left half z-plane. We can write $R(z)$ as

$$\begin{aligned} &a \prod_\mu (z - x_\mu) \prod_\nu (z - x_\nu - iy_\nu)(z - x_\nu + iy_\nu) \\ &= a \prod_\mu (z + |x_\mu|) \prod_\nu (z^2 + 2z\,|x_\nu| + x_\nu^2 + y_\nu^2) \end{aligned} \tag{10.50}$$

Therefore, every coefficient has the sign of a, but $a_1 > 0$, therefore $a_i \geq 0$, $i \geq 2$.

We next consider

$$\begin{aligned} &\frac{z}{\log[(1+z)/(1-z)]} \\ &= \sum_{\mu=0}^{\infty} c_{2\mu} z^{2\mu} \\ &= \frac{z}{\log(1+z) - \log(1-z)} \\ &= \frac{z}{z - (z^2/2) + (z^3/3) - \cdots + z + (z^2/2) + (z^3/3) + \cdots} \\ &= \frac{1}{2 + (2z^2/3) + (2z^4/5) + \cdots} \end{aligned}$$

Therefore $c_0 = \frac{1}{2}$. We will show that $c_{2\mu} < 0$, $\mu \geq 1$. This is a particular case of the following lemma due to Kaluza (1928):

LEMMA 10.5

If

$$\left(\sum_{n=0}^{\infty} a_n x^n\right)\left(\sum_{n=0}^{\infty} b_n x^n\right) = 1$$

if $a_n > 0$, $n \geq 0$, *and if* $a_{n-1} \cdot a_{n+1} > a_n^2$, $n \geq 1$, *then* $b_n < 0$ *for* $n \geq 1$.

Proof: We set $a_0 = 1$ without loss of generality. Then $b_0 = 1$. If we examine the term in x^n, we have

$$0 = a_n + \sum_{k=1}^{n} a_{n-k} b_k \tag{10.51}$$

By looking at the term in x^{n+1}, we get

$$b_{n+1} = -a_{n+1} - \sum_{k=1}^{n} a_{n+1-k} b_k \tag{10.52}$$

Multiplying (10.51) by a_{n+1}, (10.52) by a_n, and adding, we get

$$a_n b_{n+1} = \sum_{k=1}^{n} b_k (a_{n+1} a_{n-k} - a_n a_{n+1-k}) \tag{10.53}$$

We proceed by induction. If $b_1, \ldots, b_n < 0$, (10.53) shows that $b_{n+1} < 0$ since $a_n > 0$ and $a_{n+1}a_{n-k} - a_n a_{n+1-k} > 0$. (This can be seen by considering $a_{n+1}a_n a_{n-1}^2 \cdots a_{n-k+2}^2 a_{n-k+1}a_{n-k} = (a_{n+1}a_{n-1})(a_n a_{n-2})(a_{n-1}a_{n-3}) \cdots (a_{n-k+2})(a_{n-k}) > a_n^2 a_{n-1}^2 \cdots a_{n-k+1}^2$ and dividing by $a_n a_{n-1}^2 \cdots a_{n-k+2}^2 a_{n-k+1}$.) Since $a_0 b_1 + b_0 a_1 = b_1 + a_1 = 0$, $b_1 = -a_1 < 0$, so the result is true for $n = 1$. Q.E.D.

By taking $z^2 = x$ and $a_n = 2/(2n+1)$ in Lemma 10.4, we immediately have $c_{2\mu} = b_\mu < 0$ for $\mu \geq 1$, since $a_n > 0$ and $a_{n-1}a_{n+1} = 4/(2n-1)(2n+3) = 4/[(2n+1)^2 - 4] \geq a_n^2$.

Returning to our original goal, we now consider

$$\frac{R(z)}{z} \frac{z}{\log[(1+z)/(1-z)]} = \sum_{m=1}^{k} a_m z^{m-1} \sum_{\mu=0}^{\infty} c_{2\mu} z^{2\mu}$$

Since this is $\sum_{n=0}^{\infty} r_n z^n$, we have

$$r_{k+1} = c_2 a_k + c_4 a_{k-2} + \cdots + c_{k+1} a_1 \qquad \text{for odd } k \tag{10.54}$$

and

$$r_{k+1} = c_2 a_k + c_4 a_{k-2} + \cdots + c_k a_2 \qquad \text{for even } k \tag{10.55}$$

Since $a_1 > 0$, $a_j \geq 0$, and $c_{2\mu} < 0$, (10.54) shows that $r_{k+1} < 0$ for odd k, showing that $k+1$ is the maximum stable order. If k is even, (10.55) can give $r_{k+1} = 0$ only if $a_2 = a_4 = \cdots = a_k = 0$. In this case,

$$r_{k+2} = c_4 a_{k-1} + c_6 a_{k-3} + \cdots + c_{k+2} a_1 < 0,$$

so the maximum order is $k+2$, and it can be achieved by making $r_{k+1} = 0$ which requires that $a_2 = a_4 = \cdots = 0$ from (10.55). Since a_0 is also zero, $R(z)$ is an odd polynomial in z. Therefore, $R(-z) = -R(z)$. Consequently, roots of $R(z)$ are also roots of $R(-z)$, so that if $x_\nu + iy_\nu$ is a root of $R(z)$ so is $-x_\nu - iy_\nu$. But, because of stability, both must be in the left half plane implying that $x_\nu = 0$, so that all roots of $R(z)$ are on the imaginary axis, which means that roots of $\rho(\xi)$ are on the unit circle. Conversely, if the roots of $\rho(\xi)$ are on the unit circle, those of $R(z)$ are on the imaginary axis, implying that $R(z)$ is odd and that $a_2 = a_4 = \cdots = 0$, so that the order is $k+2$ if the polynomial $\sigma(\xi)$ is chosen appropriately. We have thus shown:

THEOREM 10.10

A stable k-step method for first order equations has maximum order

$k+1$ *if k is odd*

$k+2$ *if k is even*

An order of k + 2 can only occur if all roots of $\rho(\xi)$ *are on the unit circle.*

An order of $k+2$ implies that the method is weakly stable, so such methods are of limited practical importance.

Similar results have been obtained for methods for higher order equations [e.g., Dahlquist (1959)], but in view of the results of the next section for multivalue methods, these results have little practical effect.

10.3 EXISTENCE OF STABLE MULTIVALUE METHODS

In this section we will show that there exist k-value methods of maximum order that are stable. Exactly what that order is will be discussed further in the following section. We have seen, however, that the $(k - 1)$-step Adams-Bashforth-Moulton predictor-corrector method is of order k and equivalent to a k-value method, so this indicates that the order can be k for first order equations.

In a method for pth order equations we must ensure that

$$S = (I - \mathbf{l}\boldsymbol{\delta}_p^T)^M A$$

satisfies the stability condition. p eigenvalues of S will be one (the principal roots). We will show that $\mathbf{l}$ can be chosen such that the remaining eigenvalues can assume any desired value for maximum order methods. We use the normal form representation with k values in $\mathbf{a}$ so if the predictor has the maximum order of $k - 1$, A is the Pascal triangle matrix.

THEOREM 10.11

If A is the $k \times k$ Pascal matrix $A_{ij} = \binom{j}{i}$ $0 \le i, j \le k - 1$, then for any set of $k - p$ numbers $\{\lambda_i\}$, a column vector $\mathbf{l}$ exists such that the matrix

$$(I - \mathbf{l}\boldsymbol{\delta}_p^T)^M A$$

has p eigenvalues equal to one and $k - p$ eigenvalues equal to the set $\{\lambda_i\}$. The eigenvalues are independent of the first p elements of $\mathbf{l}$ and uniquely determine the last $k - p$ elements of $\mathbf{l}$.

Proof: The independence of the first p elements of $\mathbf{l}$ is obvious from the form of the matrix. Also, this form implies that the first p eigenvalues are one. Therefore, we can take $p = 0$ and A to be the $(k - p) \times (k - p)$ lower principal minor of A. The only properties that are required of A are that it be upper triangular with equal nonzero diagonal elements and nonzero first off-diagonal elements $A_{i,i+1}$. This is true for all principal minors of Pascal matrixes. First note that

$$(I - \mathbf{l}\boldsymbol{\delta}_0^T)^M = I - c\mathbf{l}\boldsymbol{\delta}_0^T$$

where

$$c = \begin{cases} -\dfrac{(1 - l_0)^M - 1}{l_0} & l_0 \neq 0 \\ M & l_0 = 0 \end{cases}$$

Hence it is sufficient to consider $M = 1$. Let

$$T = A(I - \mathbf{l}\boldsymbol{\delta}_0^T)$$

It has the same eigenvalues as

$$S = (I - \mathbf{l}\boldsymbol{\delta}_0^T)A$$

since A is nonsingular and $A^{-1}TA = S$. $T = A - \mathbf{u}\boldsymbol{\delta}_0^T$, where $\mathbf{u} = A\mathbf{l}$. We will show in the lemma below that a similarity transformation Q of A exists that puts A in a Jordan form with nonzero elements immediately above the diagonal, and in which Q and Q^{-1} have the triangular form

$$\begin{bmatrix} d_0 & 0 & 0 & \cdots & 0 \\ 0 & d_1 & x & \cdots & x \\ 0 & 0 & d_2 & \cdots & x \\ & \cdots & & \ddots & \cdot \\ 0 & 0 & 0 & \cdots & d_{k-1} \end{bmatrix}$$

where the x's represent nonzero elements.

It then follows that

$$Q^{-1}TQ = Q^{-1}AQ - Q^{-1}\mathbf{u}\boldsymbol{\delta}_0^T Q = Q^{-1}AQ - \mathbf{v}\boldsymbol{\delta}_0^T$$

where $\mathbf{v} = Q^{-1}\mathbf{u}d_0$. Apart from the ones on the diagonal, $Q^{-1}TQ$ is in the form of a companion matrix for a polynomial whose coefficients are the elements of $\mathbf{v}$. Therefore $\mathbf{v}$ can be chosen so as to make the roots equal to $\{\lambda_i\} - 1$, in which case $\mathbf{l} = A^{-1}Q\mathbf{v}/d_0$ will cause S to have the desired eigenvalues. This uniquely determines $\mathbf{l}$.

It remains to be shown that the following lemma is true.

LEMMA 10.6

If A is an upper triangular matrix such that $A_{ii} = c \neq 0$ (i not summed) and if $A_{i,i+1} \neq 0$, $i = 0, 1, \ldots, k-1$, then there exists a similarity transformation Q with the above form that takes A into a Jordan form without changing its diagonal elements.

Proof: By construction, let Q_i be the elementary transformation

$$Q_i = \begin{bmatrix} 1 & 0 & & & \cdots & 0 \\ 0 & 1 & & & \cdots & 0 \\ \cdots & & & & \cdots & \\ 0 & \cdots & 1 & q_{i+1,i+2} & \cdots & q_{i+1,k-1} \\ \cdots & & & & \cdots & \\ 0 & 0 & & & \cdots & 1 \end{bmatrix}$$

and let $A_{i+1} = Q_i^{-1}A_iQ_i$, $i = 0, 1, \ldots, k-3$, where $A_0 = A$. The

$q_{i+1,j}$ can be chosen to annihilate the last $k - i - 2$ elements of the ith row of A_{i+1} without changing the diagonal or first off-diagonal elements. Thus, $\tilde{Q} = Q_0 Q_1 \cdots Q_{k-3}$ makes A_{k-2} similar to A with unchanged diagonal or first off-diagonal elements. Finally, a diagonal similarity transformation D can be used to make the first off-diagonal elements of A_{k-2} equal to one without changing the diagonal elements. Thus, $Q = \tilde{Q}D$ is the required transformation which puts A into Jordan form. Q.E.D.

If $\mathbf{l}$ is picked to achieve a given set of eigenvalues for a particular value of p, the components of $\mathbf{l}$ which give the same eigenvalues for other values of p can be easily found as follows:

A_{pk}, the $(k - p)$ by $(k - p)$ lower principal minor of the k by k Pascal matrix A is similar to $A_{0,k-p}$ under the diagonal transformation

$$D_{pk} = \operatorname{diag}\left[d_i = \binom{p+i}{i}\right]$$

If the last $k - p$ elements of $\mathbf{l}$ for some p are called $\mathbf{l}_{pk}$, we can find $\mathbf{l}_{pk}$ from $\mathbf{l}_{0,k-p}$ by a linear transformation. Write $S_{pk} = (I - \mathbf{l}_{pk}\boldsymbol{\delta}_0^T)A_{pk}$. We have

$$D_{pk}S_{pk}D_{pk}^{-1} = (I - D_{pk}\mathbf{l}_{pk}\boldsymbol{\delta}_0^T D_{pk}^{-1})D_{pk}A_{pk}D_{pk}^{-1} = (I - D_{pk}\mathbf{l}_{pk}\boldsymbol{\delta}_0^T)A_{0,k-p}$$

so that $\mathbf{l}_{pk} = D_{pk}^{-1}\mathbf{l}_{0,k-p}$.

p elements of $\mathbf{l}$ remain to be chosen. For any choice, the method will remain stable since they do not affect the eigenvalues. In the next section we will see that they can be chosen to improve the order of the method. The importance of the result of this section is that it shows that the limitation on the order of stable multistep methods can be overcome in multivalue methods.

10.4 IMPROVED ORDER FOR NORMAL FORM MULTIVALUE METHODS

We will consider multivalue methods in which A is a $k \times k$ Pascal triangle matrix. The last $k - p$ elements of $\mathbf{l}$ will be assumed to have been selected to give the desired stability properties. The remaining p elements of $\mathbf{l}$ can be chosen so as to get the highest order possible. To do this, we will first redefine truncation error. In the definition of order by Eqs. (10.18) and (10.19), $\mathbf{a}$ was assumed to contain the elements $h^q y^{(q)}/q!$ for the normal form method. Errors will then be no smaller than $0(h^k)$, the first neglected term. It will be seen below that a higher order is possible if another definition is adopted.

Define the $(k' - k)$-dimensional vector $\mathbf{r}(t)$ to be the transpose of

$$\left[\frac{h^k y^{(k)}}{k!}, \ldots, \frac{h^{k'-1}y^{(k'-1)}}{(k'-1)!}\right]$$

and let $\mathbf{a}^C(t)$ be the transpose of the vector $[y, hy', \ldots, h^{k-1}y^{(k-1)}/(k-1)!]$ evaluated on the solution of the differential equation at t. Suppose that $\mathbf{a}(t)$, the vector we are trying to compute, is given by

$$\mathbf{a}(t) = \mathbf{a}^C(t) + E\mathbf{r}(t) \tag{10.56}$$

where E is a $k \times (k' - k)$ matrix of constants. If the method calculates an $\tilde{\mathbf{a}}(t_n)$ from $\mathbf{a}(t_n - h)$ which is related to the correct value by

$$\tilde{\mathbf{a}}(t_n) = \mathbf{a}(t_n) + \mathbf{d}_n h^{k'} + 0(h^{k'+1}) \tag{10.57}$$

for equations with solutions in $C_{k'+1}$, then the truncation error is defined as $\mathbf{d}_n h^{k'} + 0(h^{k'+1})$ and the method has order $k' - 1$.

This means that the vector $\mathbf{a}$ that we calculate contains components of the derivatives of order k and greater, and thus differs from the previously assumed values by $0(h^k)$. In practice, this definition means that starting and step-changing errors are of a lower degree than truncation errors. It is unlikely that the user will calculate starting values with these components included, so errors from start-up of $0(h^k)$ can be expected. Also, the simple method for changing step size, namely premultiplying by a diagonal matrix, will not preserve E, and therefore also introduces errors of the same order.

If this new definition of $\mathbf{a}(t)$ is used in Theorem 10.8, its proof is unchanged, so, provided that the starting errors are of the right order, the global convergence rate of the method will be $0(h^{k'-p})$.

Let the $(k' + 1) \times (k' + 1)$ Pascal matrix be partitioned as follows:

$$\begin{array}{c} k \\ k'-k \\ 1 \end{array} \left[\begin{array}{c|c|c} A & D & \mathbf{c}_1 \\ \hline 0 & B & \mathbf{c}_2 \\ \hline 0 & 0 & 1 \end{array}\right]$$

If the solution is in $C_{k'+1}$, then

$$\mathbf{a}^C(t) = A\mathbf{a}^C(t - h) + D\mathbf{r}(t - h) + \mathbf{c}_1 a_{k'} + 0(h^{k'+1}) \tag{10.58}$$

and

$$\mathbf{r}(t) = B\mathbf{r}(t - h) + \mathbf{c}_2 a_{k'} + 0(h^{k'+1})$$

by Taylor's series, where $a_{k'} = h^{k'} y^{k'}(t - h)/k'!$ To simplify the arithmetic, consider one corrector iteration only so that

$$\tilde{\mathbf{a}}(t) = A\mathbf{a}(t - h) + \mathbf{1}F(A\mathbf{a}(t - h)) \tag{10.59}$$

Substitute (10.56) and (10.57) into (10.59) to get

$$\begin{aligned} \mathbf{a}^C(t_n) + E\mathbf{r}(t_n) + \mathbf{d}_n h^{k'} + 0(h^{k'+1}) &= A(\mathbf{a}^C(t_n - h) \\ &+ E\mathbf{r}(t_n - h)) + \mathbf{1}F(A\mathbf{a}^C(t_n - h) + AE\mathbf{r}(t_n - h)) \end{aligned} \tag{10.60}$$

Substitute (10.58) into (10.60) and use $F(\mathbf{a}^C) = 0$ to get

$$\begin{aligned} \mathbf{d}_n h^{k'} &= \left[\left(I + \mathbf{1}\frac{\partial F}{\partial \mathbf{a}}\right)(AE - D) - EB\right]\mathbf{r}(t_n - h) \\ &\quad - \left[E\mathbf{c}_2 + \left(I + \mathbf{1}\frac{\partial F}{\partial \mathbf{a}}\right)\mathbf{c}_1\right]a_{k'} + 0(h^{k'+1}) \end{aligned} \tag{10.61}$$

But

$$\frac{\partial F}{\partial \mathbf{a}} = -\boldsymbol{\delta}_p^T + \sum_{q=1}^{p} \frac{(p-q)!\,h^q f_{y^{(p-q)}}}{p!} \boldsymbol{\delta}_{p-q}^T$$

In order to satisfy (10.61), terms in h^q, $q < k'$, must vanish, while $\mathbf{d}_n$ is given by the terms in $h^{k'}$. Hence,

$$R = (I - \mathbf{l}\boldsymbol{\delta}_p^T)(AE - D) - EB = 0 \tag{10.62}$$

$$d_n = -\frac{[Ec_2 + (I - \mathbf{l}\boldsymbol{\delta}_p^T)\mathbf{c}_1]y^{(k')}}{(k')!} + \mathbf{l}\sum_{q=1}^{\tilde{p}} \frac{(p-q)!\,f_{y^{(p-q)}}}{p!(k'-q)!} \boldsymbol{\delta}_{p-q}^T (AE - D)\boldsymbol{\delta}_{k'-k-q} y^{(k'-q)} \tag{10.63}$$

where $\tilde{p} = \min(p, k'-k)$, and

$$\sum_{q=1}^{k'-k-i-1} \frac{(p-q)!\,f_{y^{(p-q)}}}{(k'-q)!} \boldsymbol{\delta}_{p-q}^T (AE - D)\boldsymbol{\delta}_{k'-k-q-i-1} y^{(k'-q-i-1)}) = 0 \qquad i = 0, 1, \ldots, k'-k-2 \tag{10.64}$$

If we write

$$\mathbf{u}^T = \boldsymbol{\delta}_p^T(AE - D), \tag{10.65}$$

Eq. (10.62) becomes

$$R = AE - EB - \mathbf{l}\mathbf{u}^T - D = 0 \tag{10.66}$$

Noting that A and B are upper triangular matrices with ones on the diagonal, we see that the first column of R gives

$$\sum_{j=0}^{k-1} (A_{ij} - \delta_{ij})E_{j0} - l_i u_0 = D_{i0} \qquad 0 \le i \le k-1 \tag{10.67}$$

A *strongly stable* method has no extraneous roots on the unit circle so $l_{k-1} \neq 0$. Consequently, (10.67) provides solutions for u_0 when $i = k-1$ and for $E_{i+1,0}$ when $p \le i \le k-2$. Similarly, the last $k-p$ elements of the mth column of R can be solved for u_m and $E_{i+1,m}$, $p \le i \le k-2$ in the order $m = 1, 2, \ldots, k-k'-1$. Now $\mathbf{u}^T$ is known, (10.65) provides a nonsingular equation for $E_{p,i}$, $0 \le i \le k'-k-1$.

Since u_0 is nonzero, the first p rows of (10.66) form equations that can be successively solved by working up diagonals starting from the leftmost, i.e.,

$$R_{p-1,0}, R_{p-1,1}, R_{p-2,0}, R_{p-1,2}, \ldots$$

In this process, the first p elements of the last column of E can be picked without restriction. *In particular, if $k'-k \le p$, they can be picked to make the top row of E zero* so that at least the function value does not contain additional derivatives. This process also determines the values of l_i for $i = p-k'+k, \ldots, p-1$; the remaining components l_i are determined by the values of

$$E_{k'-k-1,i+k'-k} \qquad i = 0, \ldots, p-k'+k-1$$

The dependence of elements of E and $\mathbf{l}$ on other elements can be seen from the diagram below. The element in the (i, j)th position is the one found by equating $R_{ij} = 0$. These elements are treated in the order shown by the arrows, and each time a new diagonal is started from the right column, a free parameter $E_{i, k'-k-1}$ is introduced.

$$
\begin{bmatrix}
l_0 & E_{0,0} & E_{0,1} & \cdots & E_{0,k'-k-2} \\
l_1 & E_{1,0} & E_{1,1} & \cdots & E_{1,k'-k-2} \\
 & & & \cdots & \\
l_{p-2} & E_{p-2,0} & E_{p-2,1} & \cdots & E_{p-2,k'-k-2} \\
l_{p-1} & E_{p-1,0} & E_{p-1,1} & \cdots & E_{p-1,k'-k-2}
\end{bmatrix}
\quad
\begin{matrix}
E_{0,k'-k-1} \\
E_{1,k'-k-1} \\
\\
E_{p-2,k'-k-1} \\
E_{p-1,k'-k-1}
\end{matrix}
$$

$\uparrow$ Start
   evaluation
                                    Free parameters

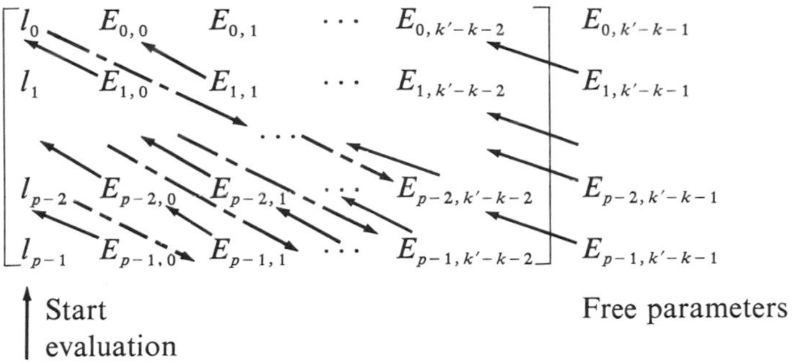

In general, (10.64) can only be satisfied by an appropriate lower triangular part of $AE - D$ being zero or by the appropriate partial derivatives of f vanishing everywhere. The former condition is independent of those parameters (the last column of E) that are still free, so that it is unlikely to be satisfied.

Therefore we have shown:

THEOREM 10.12

> If the smallest value of j for which $\partial f/\partial y^{(p-j)} \neq 0$ is q, there exist strongly stable k-value methods of order $k + q - 1$.

In particular, there are methods of order $k + p - 1$ for the special pth order equation $y^{(p)} = f(y, t)$. Notice that for $p = 1$, the $(k - 1)$-step Adams-Bashforth-Moulton method expressed as a k-value method has an order $k - 1$ predictor [which is all that is possible if terms beyond $y^{(k-1)}$ are not carried] but is a kth order method in the sense that E is a $k \times 1$ matrix. Therefore, the $\mathbf{a}$ calculated by this method includes a fixed multiple of $h^k y^{(k)}$ except in position 0. If the initial values of $\mathbf{a}_0$ are set to $h^q y^{(q)}/q!$, $0 \leq q < k$, there is a starting error of $0(h^k)$ (Theorem 10.7). Similarly a *fixed* number of step changes will only perturb the error by $0(h^k)$, but if the number of step changes $= 0(h^{-1})$, a global error of $0(h^{k-1})$ will result.

If M corrector iterations are used, Eq. (10.61) is modified by replacing $(I + \mathbf{l}(\partial F/\partial \mathbf{a}))$ by

$$
\prod_{i=1}^{M} \left[I + \mathbf{l}\left(\frac{\partial F}{\partial \mathbf{a}}\right)_i \right]
$$

throughout, where the subscript i refers to the fact that the partial derivatives are evaluated at different points for different iterations i. The conclusions of this section remain unchanged if $\mathbf{l}$ is replaced by $\tilde{\mathbf{l}} = \mathbf{l}(1 - (1 - l_p)^M)/l_p$ if $l_p \neq 0$ or by $\mathbf{l}M$ if $l_p = 0$.

10.5 ASYMPTOTIC BEHAVIOR OF THE ERROR

With one-step methods we were able to express the error as

$$e_n = h^r\delta(t_n) + 0(h^{r+1})$$

In this section we are going to examine similar results for multistep and multivalue methods. We start by considering corrector only multistep methods for first order equations. The technique is identical to the one used on the Euler method in Theorem 1.3. We have

$$\sum_{i=0}^{k} (\alpha_i y_{n-i} + h\beta_i f_{n-i}) = 0 \tag{10.68}$$

and

$$\sum_{i=0}^{k} (\alpha_i y(t - ih) + h\beta_i f(y(t - ih))) = C_{r+1}h^{r+1}y^{(r+1)}(t) + O(h^{r+2}) \tag{10.69}$$

Subtract to get

$$\sum_{i=0}^{k} \left[\alpha_i e_{n-i} + h\beta_i f_y(\xi_{n-i})e_{n-i} + \frac{C_{r+1}}{\sum_{j=0}^{k} \beta_j}\beta_i h^{r+1}y^{(r+1)}(t_n)\right] = 0(h^{r+2})$$

Set $e_n = h^r\delta_n$ and note that

$$f_y(\xi_{n-i}) = f_y(y(t_{n-i})) + 0(h^r)$$
$$\beta_i y^{(r+1)}(t_n) = \beta_i y^{(r+1)}(t_{n-i}) + 0(h)$$

to get

$$\sum_{i=0}^{k} \left[\alpha_i\delta_{n-i} + h\beta_i\left(f_y(y(t_{n-i}))\delta_{n-i} + \frac{C_{r+1}}{\sum \beta_j}y^{(r+1)}(t_{n-i}) + 0(h)\right)\right] = 0 \tag{10.70}$$

This is the solution, by the multistep method we are analyzing, of an equation which is $0(h)$ different from

$$\delta'(t) = g(t)\delta(t) + \frac{C_{r+1}}{\sum \beta_j}y^{(r+1)}(t) \tag{10.71}$$

where $g(t) = f_y(y(t))$. If the starting values are exact, the solution of (10.70) converges to that of (10.71) by Theorem 10.6 and the fact that (10.71) is well-posed, so we conclude that

THEOREM 10.13

For a convergent multistep method for first order equations

$$e_n = h^r\delta(t_n) + 0(h^{r+1})$$

where $\delta(t)$ satisfies (10.71) *with* $\delta(0) = 0$ *if the starting values are exact.*

We again see the reason for normalizing so that $\sum \beta_j = 1$.

In general, our starting values will be inexact. We saw in Theorem 10.7 that there was an additional error of $0(h^{r'})$ if the starting errors were $0(h^{r'})$. We would like to know what asymptotic effect these have on the error, particularly in the case $r' \leq r$. We must assume that the actual starting errors are given as $\tilde{\delta}_i h^{r'} + 0(h^{r'+1})$, $0 \leq i < k$, where the $\tilde{\delta}_i$ are constants. Consequently, we wish to examine the solution of (10.70) with initial conditions $\delta_i = \tilde{\delta}_i h^{r'-r}$. If $r' > r$, the asymptotic error is as given by Theorem 10.13. If $r' < r$, then the error will be of the form

$$e_n = \delta_n h^{r'} + 0(h^{r'+1})$$

where δ_n is given by (10.70) with $C_{r+1} = 0$ and starting values $\delta_i = \tilde{\delta}_i$. If $r' = r$, then we wish to solve (10.70) as is with starting values $\delta_i = \tilde{\delta}_i$.

We state the following theorem:

THEOREM 10.14

Let the roots of $\rho(\xi) = 0$ be $\xi_1 = 1, \xi_2, \ldots, \xi_k$, and the numbers z_{ij}, $0 \leq j < k$, $1 \leq i \leq k$, be defined by

$$z_{ij} = \begin{cases} j^m \xi_i^j & \text{if } \xi_i \neq 0 \\ \delta_{ij} + 1 & \text{if } \xi_i = 0 \end{cases}$$

where m is 0 *for the first appearance of the root ξ_i,* 1 *for its second appearance, etc. If u_i is defined by*

$$\tilde{\delta}_j = \sum_{i=1}^{k} u_i z_{ij} \qquad 0 \leq j < k$$

(*this is always possible as $\{z_{ij}\}$ is nonsingular*), *then if $r' = r$ the error e_n for a strongly stable convergent method satisfies*

$$e_n = h^{r'}\delta(t_n) + 0(h^{r'+1})$$

where $\delta(t)$ is the solution of (10.71) *with $\delta(0) = u_1$. If $r' < r$, $\delta(t)$ is given by the same problem with $C_{r+1} = 0$.*

The proof can be found in Henrici (1962), Section 5.3, where a stronger result for the behavior in the presence of additional roots on the unit circle is given. The effect of these is to introduce other components $h^{r'}u_i\delta_i(t_n)$, where $\delta_i(t_n)$ is the solution of a related differential equation. Since weakly stable methods are not recommended, the details are not given here.

We now examine the asymptotic form of the error for multivalue methods for pth order equations. Since the arithmetic becomes unnecessarily complex otherwise, only one corrector iteration will be considered. The result we will get is the natural extension of Theorem 10.13, namely:

THEOREM 10.15

If the truncation error of a convergent multivalue method is expressed as

$$\left[\mathbf{d}_2 y^{(k')} + \mathbf{l}\sum_{q=1}^{p}\frac{\partial f}{\partial y^{(p-q)}} y^{(k'-q)} d_{1q}\right] h^{k'} + 0(h^{k'+1}) \tag{10.72}$$

[*cf. Eq.* (10.63)] *and if* $\{\gamma_m\}$ *are defined by*

$$\mathbf{d}_2 = \sum_{m=0}^{k-1} \gamma_m S^m \mathbf{l} \tag{10.73}$$

where

$$S = (I - \mathbf{l}\boldsymbol{\delta}_p^T)A$$

then the error in integrating from exact initial conditions has components due to an error $h^{k'-p}\delta(t_n) + 0(h^{k'-p+1})$ *in* y_n, *where* $\delta(t)$ *is the solution of*

$$\delta^{(p)}(t) = \sum_{q=1}^{p}\frac{\partial f}{\partial y^{(p-q)}}(\delta^{(p-q)} + d_{1q} y^{(k'-q)}(t)) + \sum_{m=0}^{k-1}\gamma_m y^{(k')}(t) \tag{10.74}$$

with $\partial(0) = \partial^{(1)}(0) = \cdots = \partial^{(p-1)}(0) = 0$. *If the method is strongly stable,* γ_m *can be found to satisfy* (10.73).

Proof: If we look at $\boldsymbol{\delta}_n = h^{-k'+p}\mathbf{e}_n$ where $\mathbf{e}_n = \mathbf{a}_n - \mathbf{a}^C(t_n) - E\mathbf{r}(t_n)$ [see Eq. (10.56)],

$$\boldsymbol{\delta}_{n,(0)} = A\boldsymbol{\delta}_{n-1} \tag{10.75}$$

$$\begin{aligned}\boldsymbol{\delta}_n = \boldsymbol{\delta}_{n,(0)} + \mathbf{l}\frac{\partial F}{\partial \mathbf{a}}\boldsymbol{\delta}_{n,(0)} + \mathbf{l}h^p \sum_{q=1}^{p}\frac{\partial f}{\partial y^{(p-q)}} y^{(k'-q)}(t_n) d_{1q} \\ + h^p \sum_{m=0}^{k-1}\gamma_m S^m \mathbf{l} y^{(k')}(t_n) + 0(h^{p+1})\end{aligned} \tag{10.76}$$

Instead of adding $h^p\gamma_m S^m \mathbf{l} y^{(k')}(t_n)$ to $\boldsymbol{\delta}_n$, we could have added $h^p\gamma_m \mathbf{l} y^{(k')}(t_n - mh) + 0(h)^{p+1}$ to $\boldsymbol{\delta}_{n-m}$ with the same effect. If this is done, subsequent steps will also "cast back" their γ_i, so the solution to (10.75) and (10.76) within $0(h)$ is the same as if the S^m had been dropped in (10.76). (The changes in the starting values due to the "casting back" to negative n, and in the value at t_n due to the casting back from larger n are of order p smaller than the solution, so they can be ignored.) After this change, (10.75) and (10.76) are the difference equations that would result if (10.74) were to be solved by the method under discussion. The first part of the result follows.

The second part follows from the fact that the $S^m\mathbf{l}$, $m = 0, 1, \ldots, k-1$ are linearly independent. This can be seen by noting that, since $l_{k-1} \neq 0$ for a strongly stable method, $\mathbf{l}$ is a principal vector of rank k corresponding to the eigenvalue one of A. [That is, $(A - I)^m\mathbf{l} = 0 \Rightarrow$

$m \geq k$.] By expansion, we find that

$$S^m \mathbf{l} = A^m \mathbf{l} + w_{m1} A^{m-1} \mathbf{l} + \cdots + w_{mm} \mathbf{l}$$

Each new $A^m \mathbf{l}$ introduces a new principal vector component into $S^m \mathbf{l}$ $m = 0, \ldots, k-1$, hence the $S^m \mathbf{l}$ are linearly independent.

PROBLEMS

1. Suppose we use the predictor formulas given by (10.2) and (10.3) with $\alpha_0 = -1$ and $\beta_0 = \beta_{q0} = 0$, and then correct using (10.4) and

$$\frac{h^q y_{n,(m+1)}^{(q)}}{q!} = \sum_{i=1}^{k} \left(\alpha_{qi}^* y_{n-i} + \beta_{qi}^* \frac{h^p}{p!} f_{n-i}\right) + \beta_{q0}^* \frac{h^p}{p!} f(y_{n,(m)}, \ldots, y_{n,(m)}^{(p-1)}, t_n)$$

Give an expression for the coefficients α_{qi} and β_{qi} in terms of α_{qi}^*, β_{qi}^*, α_i^*, β_i^*, α_i, and β_i which makes it possible to phrase the method in a matrix form using the vector

$$\mathbf{y}_n = \left[y_n, y_{n-1}, \ldots y_{n-k+1}, hy_n', \frac{h^2 y_n''}{2}, \ldots, \frac{h^p y_n^{(p)}}{p!}, \ldots, \frac{h^p y_{n-k+1}^{(p)}}{p!} \right]$$

2. Prove that the matrix which determines the a_{ij} in Lemma 10.1 is nonsingular. [*Hint:* Find what factors of the form $(\xi_i^p - \xi_i^q)$ can occur in its determinant.]

3. If an rth order predictor and an sth order corrector are used for the equation

$$y''' = f(y, y', t)$$

what is the order of the method for this equation after M corrector iterations?

4. Prove Theorem 10.3.

5. By considering

$$y^{(p)} = k! \frac{t^{k-p}}{(k-p)!}$$

prove Lemma 10.4.

6. By considering $y^{(p)} = y^{(p-1)}$, show that the order order of the predictor plus M (the number of corrector iterations) cannot be less than the order r of the method for pth order equations.

7. Find $\tilde{S}_n$ as defined in the proof of Theorem 10.5 for the Adams-Bashforth-Moulton scheme of second order in both predictor and corrector with one corrector iteration. What difference do additional iterations make?

8. Consider the following method.

$$y_{n+\frac{1}{2}} = y_n + \frac{5}{24} hf(y_n) + \frac{h}{3} f(y_{n+\frac{1}{2}}) - \frac{h}{24} f(y_{n+1})$$

$$y_{n+1} = y_n + \frac{h}{6} f(y_n) + \frac{2h}{3} f(y_{n+\frac{1}{2}}) + \frac{h}{6} f(y_{n+1})$$

(a) $f(y_{n+\frac{1}{2}})$ and $f(y_{n+1})$ are estimated using the Euler method to determine $y_{n+\frac{1}{2}}$ and y_{n+1}.

(b) The equations are solved exactly by some technique (it doesn't matter how). What is the order of the method in each of the cases?

9. A strongly stable method for pth order equations satisfies the strict root condition. Consider the special pth order equation and the result in Theorem 10.12. A remark in the proof of that theorem indicated that E could be selected so that $E_{0,i} = 0$, $0 \leq i \leq k' - k - 1$. Show how the proofs of Theorems 10.8 and 10.9 can be adapted to show that if $\mathbf{a}_0$ is set to the exact values of $h^q y^{(q)}/q!$, $0 \leq q < k'$, a global error of $0(h^k)$ is possible.

10. (Very complex!)
(a) Develop the equations preceding Theorem 10.12 for an M-iteration method. If a one-iteration method achieves order $k + q - 1$, is that order maintained during subsequent iterations? What is the truncation error?
(b) Prove the extension of Theorem 10.15 to M corrector iterations.

11. Prove that the maximum order of a stable explicit k-step method is k. (*Hint:* Use the fact that $\sum_{\nu=0}^{\mu} C_\nu > 0$ for $\mu \geq 0$. Try to prove this also.)

11 SPECIAL METHODS FOR SPECIAL PROBLEMS

The bulk of this chapter deals with stiff differential equations. These arise in many computer aided design techniques, particularly network analysis and simulation. They also arise in almost all chemical kinetic studies. We will also discuss a problem that turns out to be closely related, that of solving sets of nonlinear algebraic equations simultaneously with the differential equations. These occur frequently in network analysis and often in simulation. Finally, the problem of finding values for unknown parameters when the solution of the differential equation is known at a number of points will be studied briefly. This is the problem that often faces researchers performing integrations of chemical kinetic equations.

11.1 STIFF EQUATIONS

Stiff differential equations frequently arise in physical equations due to the existence of greatly differing time constants. *Time constant* is the term used by engineers and physicists to refer to the rate of decay. For example, the equation $y' = \lambda y$ has the solution $ce^{\lambda t}$. If λ is negative, then y decays by a factor of e^{-1} in time $-1/\lambda$. This is the time constant. The more negative λ, the shorter the time constant. Physical systems frequently behave, at least locally, in an exponential fashion (capacitors discharging, chemical reactions proceeding to equilibrium, etc.). In a system, different components will be decaying at different rates. For the system

$$\mathbf{y}' = \mathbf{f}(\mathbf{y})$$

the decay rates may be related locally to the eigenvalues of $\partial f/\partial y$. If some of the reactions are slow and others are fast, the fast ones will control the stability of the method, although the components may have decayed to insignificant levels so that the truncation error is determined by the slow components. Consider, for example, the "system"

$$\begin{aligned} y' &= -y, \qquad & y(0) &= 1 \\ z' &= -1000z, \qquad & z(0) &= 1 \end{aligned} \tag{11.1}$$

In fact, these equations are independent of each other, so we can analyze the behavior of each separately. If we do this for most of the methods discussed so far, we will find that stability requirements will force $1000h$ to be bounded. Euler's method will limit it to two in order that $|1 + \lambda h| < 1$, while the fourth order Runge-Kutta method will restrict it to about 2.8. From Figure 8.2 we see that the Adams-Moulton method restricts λh even more for higher order methods. If Eqs. (11.1) are integrated by one of these methods, steps of not much more than the time constant of z can be taken. After a few steps, the value of z will be so small that it will be negligible compared to y. From this point on very small steps must be used because of the second component, although only the first component contains any significant information. This, then, is the problem of stiff equations.

In the particular example above the two components can be separated and different methods or step sizes could be applied to each. A number of proposals have been made along these lines, and they can be effective in the region in which the rapidly varying component is changing. However, it is not possible to separate general equations into two or more components that simply.

Consider, for example,

$$\begin{aligned} u' &= 998u + 1998v, \qquad & u(0) &= 1 \\ v' &= -999u - 1999v, \qquad & v(0) &= 0 \end{aligned} \tag{11.2}$$

These are derived from (11.1) by the transformation

$$\begin{bmatrix} u \\ v \end{bmatrix} = \begin{bmatrix} 2 & -1 \\ -1 & 1 \end{bmatrix} \begin{bmatrix} y \\ z \end{bmatrix}$$

and therefore behave similarly. Their solution is

$$u = 2e^{-t} - e^{-1000t},$$
$$v = -e^{-t} + e^{-1000t}$$

so that both dependent variables contain both fast and slow components. Their solution is shown in Figure 11.1.

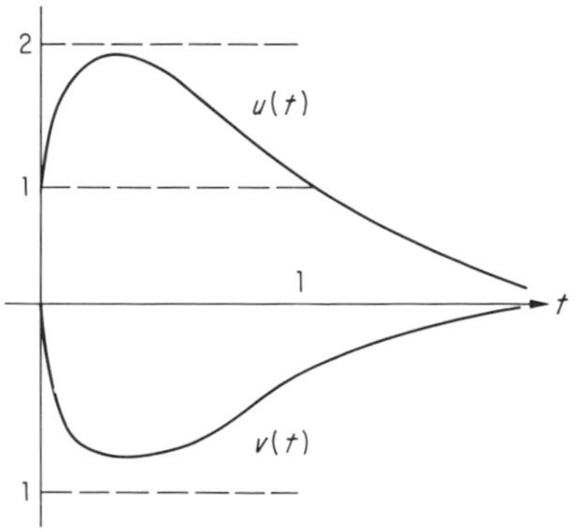

Fig. 11.1 A stiff problem.

Equations (11.2) cause the same restrictions on h even when the rapidly changing component is insignificant. If the inverse transformation to return (11.2) to (11.1) could be found, the components could be treated separately, but this is a nontrivial task for large systems of equations or if the transformation depends on t.

It is not necessary to consider a system of equations to observe this problem. The simple example

$$y' = \lambda(y - F(t)) + F'(t) \tag{11.3}$$

with $\lambda \ll 0$ has a similar behavior if $F(t)$ is a smooth, slowly varying function. The solution of (11.3) is

$$y = (y_0 - F(0))e^{\lambda t} + F(t) \tag{11.4}$$

Even if $y_0 - F(0) \neq 0$, λt will soon be sufficiently negative that the first component will be insignificant compared to the second. If we examine the error equation for (11.3) using any of the one-step or multistep methods discussed, we will see that the local truncation error is determined by h and a derivative of F by the time that $\lambda t \ll 0$, whereas the stability is dependent on the value of $h\lambda$.

In systems of equations such as

$$\mathbf{y}' = A(\mathbf{y} - \mathbf{F}(t)) + \mathbf{F}'(t) \tag{11.5}$$

the eigenvalues of A play the part of λ. In this case, we must consider complex λ. If all of the eigenvalues of A have negative real parts, the solution of (11.5) converges to $F(t)$ as t tends to infinity.

While it is true that the numerical approximation of (11.3) by any one of the techniques discussed so far converges to the solution as $h \to 0$, h has to be intolerably small before acceptable accuracy is obtained in practice, so small in fact, that round-off errors and computation time become critical. The problem, then, is to develop methods that do not restrict the step size for stability reasons.

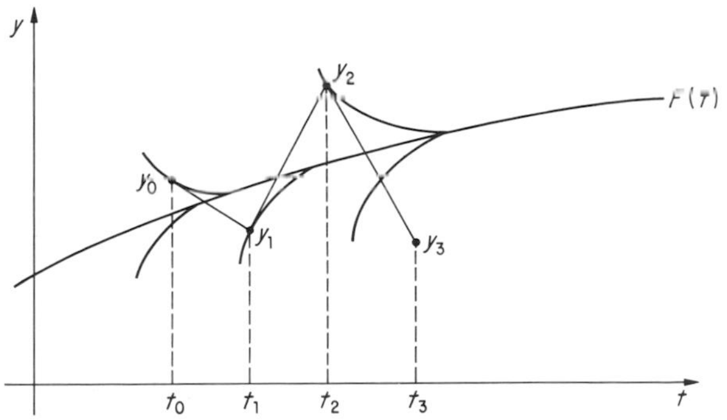

Fig. 11.2 Euler's method for stiff problems.

When the components such as $ce^{\lambda t}$ are insignificant compared to other components in the solution, and they are continuing to decay, it is not necessary to approximate $e^{\lambda h}$ closely, it is only necessary that the approximation be bounded by one so that the terms are not amplified. Figure 11.2 shows the effect of Euler's method on Eq. (11.3). If $|1 + h\lambda| > 1$, the error is amplified at each stage.

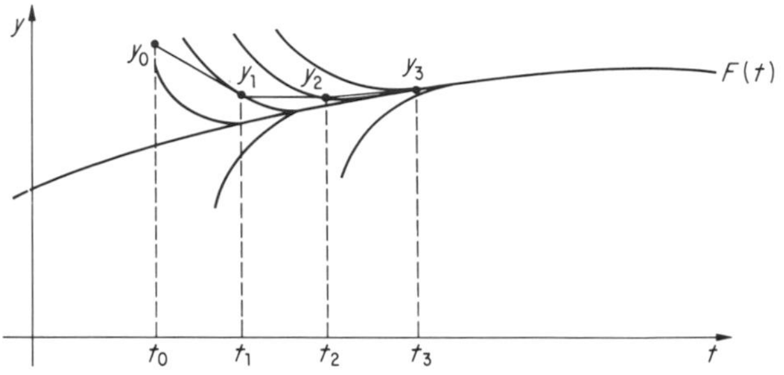

Fig. 11.3 The backward Euler method for stiff problems.

Figure 11.3 shows the same problem solved by the *backward Euler method* $y_{n+1} = y_n + hf(y_{n+1}, t_{n+1})$—an implicit method that can be directly solved for equations linear in y. As can be seen from that figure, it is stable, although the accuracy of representation of the $e^{\lambda t}$ term is poor. By analysis it can be seen that the error is amplified by $(1 - h\lambda)^{-1}$ at each step. If $\text{Re}(\lambda) < 0$, this is less than one. Obviously this is a desirable property, and has been formalized by Dahlquist (1963) as A-stability:

DEFINITION 11.1

> A method is said to be A-stable if all numerical approximations tend to zero as $n \longrightarrow \infty$ when it is applied to the differential equation $y' = \lambda y$ with a fixed positive h and a (complex) constant λ with a negative real part.

Throughout this section we will talk about first order equations only, although it is evident that higher order equations can exhibit similar properties. Some of these will be examined in Section 11.2. A comprehensive survey of application areas that give rise to stiff problems and methods that have been used for their solution is given in Bjurel et al. (1970).

11.1.1 Multistep Methods

Once again Dahlquist (1963) provided one of the first important contributions in the area. He proved that a multistep method that is A-stable cannot have order greater than two, and that the method of order two with the smallest error constant C_3 is the trapezoidal method for which $C_3 = \frac{1}{12}$.

This is a restrictive result which indicates that if multistep methods are to be used, the requirement of A-stability has to be relaxed. If multistep methods are applied to the linear problem $\mathbf{y}' = A\mathbf{y}$, the stability is determined by the roots of

$$\rho(\xi) + h\lambda_i\sigma(\xi) = 0 \qquad i = 1, \ldots, s$$

where λ_i are the eigenvalues of A. Since these are fixed, we do not need stability in the negative half plane, only for those regions occupied by the $h\lambda_i$. For slowly varying A, these regions will be a series of wedges in the $h\lambda$-plane, as shown in Figure 11.4. Widlund (1967) defined $A(\alpha)$-*stability* as follows:

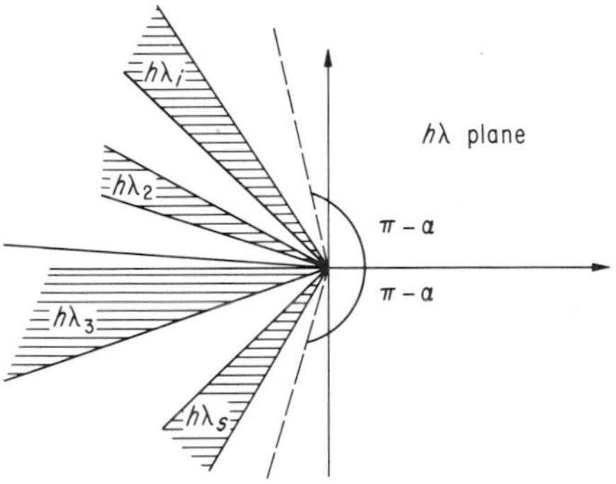

Fig. 11.4 $A(\alpha)$-stability region.

DEFINITION 11.2

A method is $A(\alpha)$-stable, $\alpha \in (0, \pi/2)$, if all numerical approximations to $y' = \lambda y$ converge to 0 as $n \longrightarrow \infty$ with h fixed for all $|\arg(-\lambda)| < \alpha$, $|\lambda| \neq 0$.

This means that the region of absolute stability includes the wedge to the left of the dashed line in Figure 11.4. Widlund then showed that there exist r-step methods of order r which are $A(\alpha)$-stable for any $\alpha < \pi/2$ and $r \leq 4$. Thus, for a *given* linear problem whose solution is *asymptotically stable* (i.e., the eigenvalues of A are strictly in the negative half plane), there exist multistep methods of order four or less which are stable for any (positive) h. Since $A(\pi/2)$ stable methods are A-stable, the truncation errors of $A(\alpha)$-stable methods of order $r \geq 3$ increase unboundedly as $\alpha \longrightarrow \pi/2$.

An alternative weakening of A-stability is defined as *Stiff stability* [Gear (1969)]. Figure 11.5 defines regions in the $h\lambda$-plane.

DEFINITION 11.3

A method is stiffly stable if in the region $R_1(\text{Re}\,(h\lambda) \leq D)$ it is absolutely stable, and in $R_2(D < \text{Re}\,(h\lambda) < \alpha, |\text{Im}\,(h\lambda)| < \theta)$ it is accurate.

The rationale for this definition is as follows. $e^{h\lambda}$ is the change in a component in one step due to an eigenvalue λ. If $h\lambda = u + iv$, then the change in magnitude is e^u. If $u < D < 0$, then the component is reduced by at least e^D in one step. We are not interested in the accuracy of components that are very small, so for some D we are willing to ignore all components in R_1.

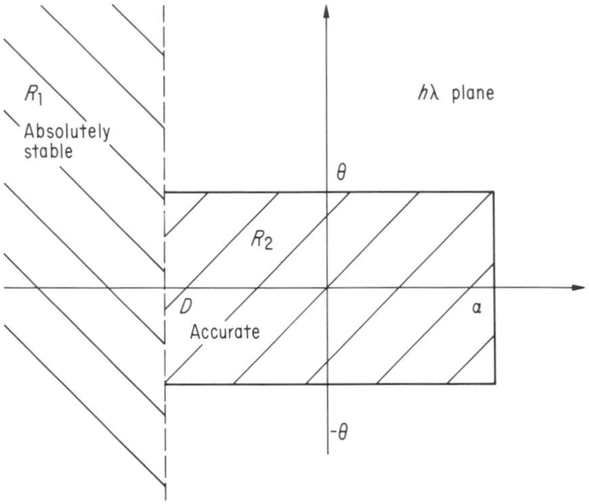

Fig. 11.5 Stiff stability.

We just require that the method be absolutely stable. Around the origin we are interested in accuracy, for which relative or absolute stability is necessary. If $u > \alpha > 0$, a component is increasing by at least e^{α} in one step. We must limit this in order that our mesh be fine enough to follow this change, hence we never use the region $u > \alpha$. If $|v| > \theta$, there are at least $\theta/2\pi$ complete cycles of oscillation in one step. Except in R_1, where we are not interested in the decaying components, and $u > \alpha$, which is not used, we must follow these components. It is well-known that for band limited signals at least two samples must be taken at the highest frequency present in order to represent the signal. In practice, about five times that number is necessary for numerical accuracy, so θ certainly is less than $\pi/5$.

Naturally we ask about the existence of stiffly stable methods of order greater than two. In Gear (1969) the k-step methods of order k with $\sigma(\xi) = \xi^k$ are shown to be stiffly stable for $k \leq 6$ for some D, α, and θ. The result is obtained by first computing $\rho(\xi)$ from $\sigma(\xi)$ so as to get an order k method. (See Section 8.1.1.) The locus in the μ-plane for which a root of $\rho(\xi) + \mu\sigma(\xi) = 0$ has magnitude one can then be plotted by plotting $\mu = -\rho(e^{i\theta})/\sigma(e^{i\theta})$, $\theta \in [0, 2\pi]$. These loci are shown in Figure 11.6 for $k = 1, 2, 3$ and Figure 11.7 for $k = 4, 5, 6$. At $\mu = \infty$, the roots of $\rho(\xi) + \mu\sigma(\xi) = 0$ are those of $\sigma(\xi) = 0$, or all zero. All points outside of the closed locus are connected by a continuous arc to $\mu = \infty$. Since the roots are continuous functions of μ, all roots for μ outside of the locus are less than one in magnitude. Thus the absolute stability region is the exterior of the closed curve. For $k = 7(1)15$, these methods are not stiffly stable. These methods are given in Henrici (1962), Section 5.1.4, where they are called "methods based on numerical differentiation." The usefulness of the first two ($k = 1, 2$) for

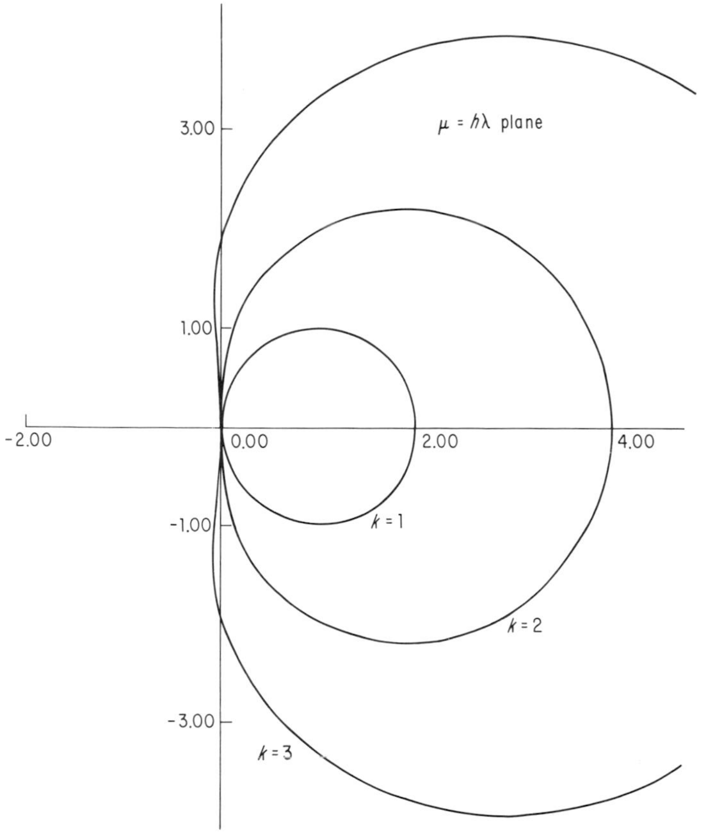

Fig. 11.6 Regions of absolute stability for stiffly stable methods of orders one through three. Methods are stable outside of closed contours.

stiff equations was pointed out by Curtiss and Hirschfelder (1952). For $k = 1$ we get the backward Euler method. Recently, stiffly stable multistep methods of orders seven and eight [Dill (1969)] and up to eleven [Jain (1970)] have been found but no tests of their utility have been made as yet.

Solving the Corrector Equation

For a stiffly stable method, $\sigma(\xi)$ must have at least as great a degree as $\rho(\xi)$. Otherwise, one root at $\mu = \infty$ is ∞. This means that the methods are implicit. Consequently, we have to solve the corrector equation. Our previous iteration was

$$y_n^{(m+1)} = \text{linear sum} + h\beta_0 f(y_n^{(m)})$$

which converged if

$$\left\| h\beta_0 \frac{\partial f}{\partial y} \right\| < 1$$

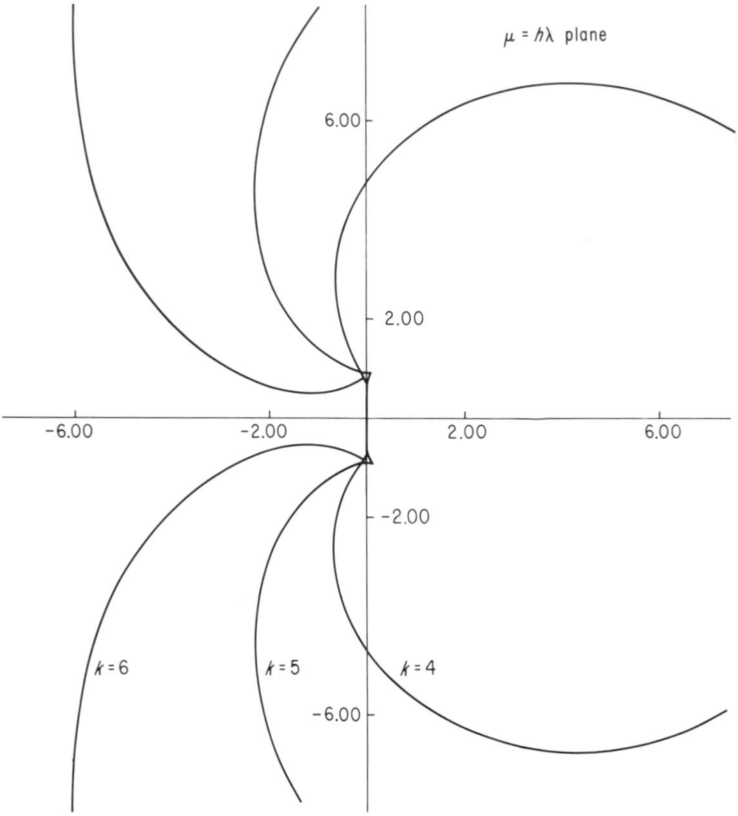

Fig. 11.7 Regions of absolute stability for stiffly stable methods of orders four through six.

Unfortunately in this case $h(\partial f/\partial y)$ can be very negative, so we must use a different iteration. A Newton solution can be expensive if the evaluation of $\partial f/\partial y$ is expensive, but $\partial f/\partial y$ need not be re-evaluated at each iteration if it does not change much. If a Newton type method in which $\partial f/\partial y$ is not re-evaluated at each step converges, it does to a solution of the equation. When the method is expressed in a $(k+1)$-value normal form, we usually perform the corrector iteration

$$\mathbf{a}_{n,(m+1)} = \mathbf{a}_{n,(m)} + \mathbf{l}F(\mathbf{a}_{n,(m)}) \tag{11.6}$$

Hence, if the method converged, it would converge to

$$\mathbf{a}_n = \mathbf{a}_{n,(0)} + \mathbf{l}\omega \tag{11.7}$$

where ω is a scalar such that

$$F(\mathbf{a}_n) = F(\mathbf{a}_{n,(0)} + \mathbf{l}\omega) = 0 \tag{11.8}$$

Let us solve (11.8) by a Newton iteration. We will have

$$\omega_{(m+1)} = \omega_{(m)} - \left[\frac{\partial F}{\partial \mathbf{a}} \cdot \mathbf{l}\right]^{-1} F(\mathbf{a}_{n,(0)} + \mathbf{l}\omega_{(m)}) \tag{11.9}$$

If we write $\mathbf{a}_{n,(m)} = \mathbf{a}_{n,(0)} + \mathbf{l}\omega_{(m)}$, (11.9) becomes

$$\mathbf{a}_{n,(m+1)} = \mathbf{a}_{n,(m)} - \mathbf{l}\left[\frac{\partial F}{\partial \mathbf{a}} \cdot \mathbf{l}\right]^{-1} F(\mathbf{a}_{n,(m)}) \tag{11.10}$$

Since $F(\mathbf{a}) = hf(a_0) - a_1$, we have

$$W = \left[\frac{\partial F}{\partial \mathbf{a}} \cdot \mathbf{l}\right]^{-1} = \left[-l_1 + hl_0 \frac{\partial f}{\partial y}\right]^{-1} \tag{11.11}$$

If the multistep method is initially written as

$$y_n = \sum_{i=1}^{k} \alpha_i y_{n-i} + h\beta_0 f_n$$

we have $l_1 = 1$, $l_0 = \beta_0$. We see that W depends on the order of the method (via β_0), h, and $\partial f/\partial y$. If $\partial f/\partial y$ is slowly varying (as frequently happens in practice), W will not change much during the iteration of (11.10) for a single step or over several steps in which the step and order do not change. In the program given in Chapter 9, this fact is used. If a stiff method is requested, the $\mathbf{l}$ corresponding to $\sigma(\xi) = \xi^k$, $1 \leq k \leq 6$, is selected. The α_i, β_0 and $\mathbf{l}$ for these methods are given in Tables 11.1 and 11.2.

Table 11.1 COEFFICIENTS OF STIFFLY STABLE METHODS

k	2	3	4	5	6
β_0	$\frac{2}{3}$	$\frac{6}{11}$	$\frac{12}{25}$	$\frac{60}{137}$	$\frac{60}{147}$
α_1	$\frac{4}{3}$	$\frac{18}{11}$	$\frac{48}{25}$	$\frac{300}{137}$	$\frac{360}{147}$
α_2	$-\frac{1}{3}$	$-\frac{9}{11}$	$-\frac{36}{25}$	$-\frac{300}{137}$	$-\frac{450}{147}$
α_3		$\frac{2}{11}$	$\frac{16}{25}$	$\frac{200}{137}$	$\frac{400}{147}$
α_4			$-\frac{3}{25}$	$-\frac{75}{137}$	$-\frac{225}{147}$
α_5				$\frac{12}{137}$	$\frac{72}{147}$
α_6					$-\frac{10}{147}$

Table 11.2 COEFFICIENTS OF STIFFLY STABLE METHODS IN NORMAL FORM

k	2	3	4	5	6
l_0	$\frac{2}{3}$	$\frac{6}{11}$	$\frac{24}{50}$	$\frac{120}{274}$	$\frac{720}{1764}$
l_1	$\frac{3}{3}$	$\frac{11}{11}$	$\frac{50}{50}$	$\frac{274}{274}$	$\frac{1764}{1764}$
l_2	$\frac{1}{3}$	$\frac{6}{11}$	$\frac{35}{50}$	$\frac{225}{274}$	$\frac{1624}{1764}$
l_3		$\frac{1}{11}$	$\frac{10}{50}$	$\frac{85}{274}$	$\frac{735}{1764}$
l_4			$\frac{1}{50}$	$\frac{15}{274}$	$\frac{175}{1764}$
l_5				$\frac{1}{274}$	$\frac{21}{1764}$
l_6					$\frac{1}{1764}$

The matirix W is re-evaluated only if the order is changed or if the corrector fails to converge in the sense that the corrections $WF(\mathbf{a}_{n,(m)})$ are not small by the third iteration.

A Test Example

Krogh [private communication] has proposed the following example to test programs for stiff equations.

Define

$$(z^i)' = -\beta_i z^i + (z^i)^2 \qquad i = 1, 2, 3, 4$$

where β_i are nonzero constants. The solution is

$$z^i = \frac{\beta_i}{1 + c_i e^{\beta_i t}} \tag{11.12}$$

If the initial value is $z^i(0) = -1$, $c_i = -(1 + \beta_i)$. Define the unitary matrix U as

$$U = \frac{1}{2}\begin{bmatrix} -1 & 1 & 1 & 1 \\ 1 & -1 & 1 & 1 \\ 1 & 1 & -1 & 1 \\ 1 & 1 & 1 & -1 \end{bmatrix}$$

and define $\mathbf{y} = U\mathbf{z}$, where $\mathbf{z} = [z^1, z^2, z^3, z^4]^T$. The differential equation for $\mathbf{y}$ is

$$\mathbf{y}' = -B\mathbf{y} + U\mathbf{w} \tag{11.13}$$

where $B = U \operatorname{diag}[\beta_1, \beta_2, \beta_3, \beta_4]U$, $\mathbf{w} = [(z^1)^2, (z^2)^2, (z^3)^2, (z^4)^2]^T$. The solution is given by (11.12) and $\mathbf{y} = U\mathbf{z}$. The Jacobian matrix $\partial f/\partial y$ for (11.13) is

$$J = U \operatorname{diag}[-\beta_i + 2z^i]U$$

and hence the eigenvalues are $2z^i - \beta_i$ since $U^{-1} = U$. With initial values $\mathbf{y}(0) = [-1, -1, -1, -1]^T = \mathbf{z}(0)$, we see from (11.12) that

$$z^i \longrightarrow \left\{ \begin{array}{ll} \beta_i & \text{if } \beta_i < 0 \\ 0 & \text{if } \beta_i > 0 \end{array} \right\} \qquad \text{as } t \longrightarrow \infty$$

Hence the eigenvalues $\longrightarrow -|\beta_i|$. As long as either $\beta_i > 0$ and $c_i < -1$, or $\beta_i < 0$ and $c_i > -1$, z^i is finite and negative.

Following Krogh's suggestion, the problem was integrated with $\beta_1 = 1000$, $\beta_2 = 800$, $\beta_3 = -10$ and $\beta_4 = 0.001$. Initially, the eigenvalues are -1002, -802, 8 and -2.001. When $0.001t \gg 1$, they are -1000, -800, -10 and -0.001. In the initial periods of integration the step will be limited by truncation errors in the terms involving e^{-1002t} and e^{-800t}. By the time that $t = 0.01$ these will be less important. However, if a method restricts h to ensure stability, h will have to remain about 10^{-3}, whereas a method that is absolutely stable in most of the negative half plane can allow h to increase. (Stability along the negative real axis is sufficient for this example.)

The effect of stiffness is demonstrated in Table 11.3 and 11.4. Table 11.3 shows the maximum error in the components of $\mathbf{y}$ at the first mesh point after $t = 10^i$, $i = -2, -1, \ldots, 3$, followed by the number of steps, derivative evaluations, and matrix inversions. The requested error was set to 10^{-6} in the program. The average step size used is also shown, although this gives a pessimistic indication of the size of the step in Table 11.3. We can see that the last 21 steps had an average size of nearly 30. In fact, it can be seen that the step is increasing by a factor of about 10 for each decade in t. Table 11.4 shows Adams' method used for the same problem. The step cannot be increased because of stability, so it takes about 30 times the number of steps to reach $t = 10$. It would have taken about 1.5×10^6 steps to reach $t = 1000$, but the program was stopped when the number of derivative evaluations exceeded 10^5 at $t = 16.8$.

Table 11.3 STIFFLY STABLE METHOD—EXPERIMENTAL RESULTS FOR STIFF PROBLEMS

Present error	*Steps*	*Evaluations*	*Inversions*	*Average step*	*Current time*
0.9100D-07	70	179	7	0.1463D-03	0.0102436701
0.2667D-05	110	262	12	0.9535D-03	0.1048869068
0.2208D-05	168	405	15	0.6025D-02	1.0122667324
0.2870D-05	216	523	20	0.4635D-01	10.0110785897
0.2984D-05	252	616	25	0.4067D 00	102.4771283917
0.1199D-05	283	693	29	0.3625D 01	1025.7769259724

Table 11.4 ADAMS' METHOD—EXPERIMENTAL RESULTS FOR STIFF PROBLEMS

Present error	*Steps*	*Evaluations*	*Inversions*	*Average step*	*Current time*
0.1863D-07	60	178	0	0.1698D-03	0.0101889615
0.4525D-06	182	548	0	0.5519D-03	0.1004531167
0.3001D-07	1796	5421	0	0.5569D-03	1.0001087924
0.4639D-05	15285	59246	0	0.6543D-03	10.0004462820

In favor of Adams' method, however, we should note that for $t < 0.01$, where stiffness is not a problem, Adams' method uses fewer steps and is more accurate than the stiff method. [The truncation error of these kth order stiff methods is $1/k$ compared to γ_k^* (see Table 7.4) for the kth order Adams-Moulton methods.]

Figure 11.8 shows the maximum error in the components of $\mathbf{y}$ versus the number of function evaluations for the derivative plus the number for the partial derivatives (which is four times the number of matrix inversions) in order to reach $t = 100$. (The one point that is strangely out of line is probably due to a fortunate cancellation of errors in different components.)

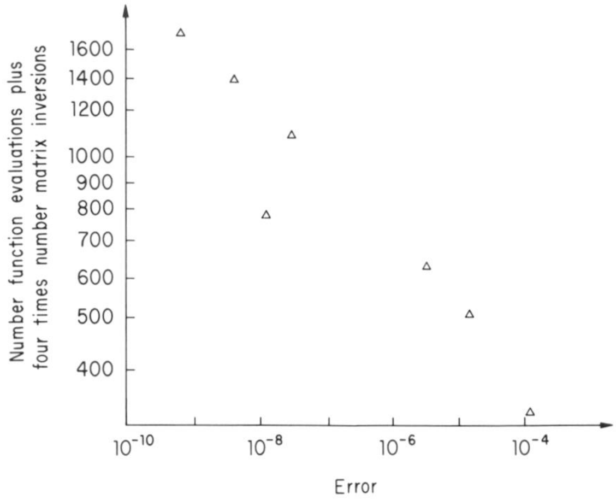

Fig. 11.8 Error versus function evaluations.

There results were obtained using the program given in Section 9.3 with the parameter MF set to two.

The reader should note that this test example is not well-posed in the large; errors greater than about 10^{-3} near equilibrium can cause the perturbed solution to be unbounded.

11.1.2 A-Stable Methods

Since multistep methods are not A-stable if their order is greater than two, we naturally ask about the existence of classes of A-stable methods. In this section we will examine some classes of methods that do not require a knowledge of $\partial f/\partial y$ (except to solve implicit equations). In subsequent sections we will look at some specialized methods that have been proposed that require values or reasonable approximations to $\partial f/\partial y$.

One class of A-stable methods was introduced in Chapter 2. The q-stage implicit Runge-Kutta method of order $2q$ is A-stable. This was pointed out by Ehle (1968). Substituting $y' = \lambda y$ into Eqs. (2.23), we get for a q-stage method,

$$y_{n+1} = \frac{P_q(h\lambda)}{Q_q(h\lambda)} y_n$$

where $P_q(\mu)$ and $Q_q(\mu)$ are polynomials of degree q in μ. Since the method is of order $2q$,

$$R_q(h\lambda) = \frac{P_q(h\lambda)}{Q_q(h\lambda)} = e^{h\lambda} + 0(h^{2q+1})$$

Hence $R_q(\mu)$ is the nth diagonal Pade approximation to e^{μ}, and it is known

[Birkhoff and Varga (1965)] that these are such that

$$|R_q(\mu)| < 1 \qquad \text{if } \operatorname{Re}(\mu) < 0$$

Thus the methods are *A*-stable.

Axelsson (1969) has defined methods to be *stiffly A-stable* if for the equation $y' = \lambda y$, $y_{n-1}/y_n \longrightarrow 0$ as $\operatorname{Re}(h\lambda) \longrightarrow -\infty$. Such methods would cause rapidly decaying components to also decay rapidly in the numerical approximation, so that it would not be necessary to use small steps even when the components were still present—at least in linear systems. The q-stage $2q$th order Runge-Kutta methods are not stiffly *A*-stable since $R_q(\mu) \longrightarrow (-1)^{-q}$ as $\mu \longrightarrow -\infty$. If the order is reduced by one and the coefficients restricted so that the last k_i evaluated is an estimate of the derivative at the end of the interval, $e^{h\lambda}$ is approximated by

$$e^{h\lambda} = \frac{P_{q-1}(h\lambda)}{Q_q(h\lambda)} + 0(h^{2q})$$

This approximation is shown to be stiffly *A*-stable by Axelsson (1969).

Another class of *A*-stable methods is given by an extension of Taylor's series methods. If derivatives at both t_n and t_{n+1} can be computed, formulas of the type

$$y_{n+1} = y_n + \sum_{j=1}^{q} h^q(\beta_{qj1} y_n^{(j)} + \beta_{qj0} y_{n+1}^{(j)}) \tag{11.14}$$

can be considered. If the β_{qji} are picked to match the first $2q + 1$ terms in the Taylor's series of y_{n+1}, we find

$$\beta_{qj} = \beta_{qj1} = (-1)^{j+1}\beta_{qj0}$$

For $q = 1$, we have the trapezoidal rule $\beta_{11} = \frac{1}{2}$. For $q = 2$, we get

$$y_{n+1} = y_n + \frac{h}{2}(y'_n + y'_{n+1}) + \frac{h^2}{12}(y''_n - y''_{n+1})$$

If (11.14) is used to integrate $y' = \lambda y$, we get

$$y_{n+1} = \frac{1 + \sum_{j=1}^{q} \beta_{qj1}(h\lambda)^j}{1 - \sum_{j=1}^{q} \beta_{qj0}(h\lambda)^j} y_n = R_q(h\lambda) y_n$$

Since the order is $2q$, $R_q(h\lambda)$ is again the diagonal Pade approximation to $e^{h\lambda}$ and is *A*-stable. By decreasing the order by one and setting $\beta_{qq1} = 0$, we get stiffly *A*-stable methods.

Due to the amount of work involved in solving the implicit equations (2.23) or the amount of work both in evaluating the derivatives and solving the implicit equation (11.14), neither of these methods has been used extensively except in the lowest order case, which is the trapezoidal rule for the

Taylor's series methods and the implicit midpoint rule for the Runge-Kutta methods. The latter is given by

$$k_1 = hf\left(y_n + \frac{k_1}{2}\right)$$

$$y_{n+1} = y_n + k_1$$

11.1.3 Methods Based on a Knowledge of $\partial f/\partial y$

A variety of methods has been proposed that make use of $A = \partial f/\partial y$ explicitly. If the eigenvalues of A are known, then methods can be derived which are exact for exponentials with those decay rates, just as we choose the multistep methods to be exact for polynomials [Pope (1963)]. Some methods proposed in the literature require a knowledge of those eigenvalues or of an approximation to A [Lawson (1967) and Dahlquist (1969)].

Linear Equations

If we have the system of equations

$$\mathbf{y}' = Ay$$

where A is a constant matrix, and we can find the similarity transformation such that QAQ^{-1} is in Jordan form, then we can solve the equations

$$\mathbf{z}' = Jz$$

explicitly, where $\mathbf{z} = Q\mathbf{y}$, $J = QAQ^{-1}$. If we now make the transformation

$$\mathbf{z}(t) = e^{Dt}\boldsymbol{\eta}(t)$$

where D is the diagonal part of J, and e^{Dt} is a diagonal matrix each of whose entries is the exponential of the equivalent diagonal entry of Dt, we get the equations

$$\begin{aligned}\boldsymbol{\eta}'(t) &= e^{-Dt}(J - D)e^{DT}\boldsymbol{\eta}(t) \\ &= (J - D)\boldsymbol{\eta}(t)\end{aligned}$$

Since $J - D$ is a matrix with at most ones on the upper diagonal, $\boldsymbol{\eta}(t)$ is a polynomial and we get

$$\mathbf{y} = Q^{-1}\mathbf{z} = Q^{-1}e^{Dt}\boldsymbol{\eta}(t)$$

Generally, A is not constant, but if such a transformation is made for a given value $A(t_0)$, the resulting differential equations, although much more complicated, can be handled by standard techniques without stability problems, since the eigenvalues are now small. Unfortunately, this method requires that the matrix Q be found, which involves a lot of work. If $A(t)$ changes at all rapidly, the new Q must be re-evaluated frequently, so the method is prohibitively expensive.

Extensions of Runge-Kutta Methods

Rosenbrock (1963) proposed an extension of the explicit Runge-Kutta process involving $\partial f/\partial y$. The most general form is

$$\begin{aligned} k_1 &= hf(y_n) + hb_1 A(y_n)k_1 \\ k_2 &= hf(y_n + \beta_{21}k_1) + hb_2 A(y_n + \eta_{21}k_1)k_2 \\ &\cdots \\ k_q &= hf\Big(y_n + \sum_{i=1}^{q-1} \beta_{qi}k_i\Big) + hb_q A\Big(y_n + \sum_{i=1}^{q-1} \eta_{qi}k_i\Big)k_q \\ y_{n+1} &= y_n + \sum_{i=1}^{q} \gamma_i k_i \end{aligned} \tag{11.15}$$

where $A(y) = \partial f(y)/\partial y$. A number of specific examples of these methods are given in the literature [Calahan (1968) and Allen (1969)]. Since they use $\partial f/\partial y$, they are, in a sense, implicit methods. The type of equation that must be solved in implicit multistep or Runge-Kutta methods for the problem $y' = A(t)y$ is similar to (11.15). An example of this method is Calahan's (1968) third order method given by

$$\begin{aligned} k_1 &= hf(y_n) + hb_1 A(y_n)k_1 \\ k_2 &= hf(y_n + \beta_{21}k_1) + hb_1 A(y_n)k_2 \\ y_{n+1} &= y_n + \gamma_1 k_1 + \gamma_2 k_2 \end{aligned} \tag{11.16}$$

where

$$\begin{aligned} b_1 &= \tfrac{1}{2}(1 + \sqrt{\tfrac{1}{3}}) = 0.788675 \\ \beta_{21} &= -2\sqrt{\tfrac{1}{3}} = -1.154701 \\ \gamma_1 &= 0.75, \qquad \gamma_2 = 0.25 \end{aligned}$$

When this class of methods is applied to the equation $y' = \lambda y$, it leads to a relation of the form

$$y_{n+1} = R(h\lambda)y_n$$

where $R(h\lambda)$ is a rational polynomial in $h\lambda$. If it can be shown that $|R(h\lambda)| \leq 1$ for Re $(h\lambda) < 0$, the method will be A-stable. To show this, it is sufficient to show that $\lim |R(\mu)| \leq 1$ as $\mu \to \infty$, $|R(i\omega)| \leq 1$, $-\infty < \omega < \infty$, with inequality holding somewhere, and that $R(\mu)$ is regular in Re $(\mu) \leq 0$.

11.2 ALGEBRAIC AND SINGULAR EQUATIONS

There are two classes of problems that are related to stiff equations. The first is called *singular perturbations*. In this problem systems of equations are given for two vectors $\mathbf{y}$ and $\mathbf{z}$ (not necessarily of the same length) in the form

$$\mathbf{y}' = \mathbf{f}(\mathbf{y}, \mathbf{z}, t) \tag{11.17a}$$

$$\epsilon\mathbf{z}' = \mathbf{g}(\mathbf{y}, \mathbf{z}, t) \tag{11.17b}$$

where ϵ is a very small constant. Within a very small region $[0, t_\epsilon]$ the solution displays a rapidly changing behavior due to (11.17b) (this is often called a *boundary layer* since the problem occurs in hydrodynamic flow problems near boundary walls); thereafter the second equation can effectively be replaced by

$$0 = \mathbf{g}(\mathbf{y}, \mathbf{z}, t) \tag{11.18}$$

The second problem occurs when the higher order implicit equation

$$F(y, y', \ldots, y^{(p)}, t) = 0 \tag{11.19}$$

is such that as t increases, $\partial F/\partial y^{(p)} \longrightarrow 0$ (or gets small). It is evident that when $\partial F/\partial y^{(p)} = 0$, (11.19) becomes an equation of order $p - 1$ or less. Consider, for example, the linear second order equation

$$\epsilon(t)y'' + ay' + by = g(t)$$

where $\epsilon(t) \longrightarrow 0$ as t increases. If we write $z = y'$, we get the system

$$y' = z \tag{11.20a}$$

$$\epsilon(t)z' = -az - by + g(t) \tag{11.20b}$$

If these equations are such that the transients due to a small ϵ damp out, there will be a boundary layer, followed by a region in which (11.20b) can be replaced by

$$0 = -az - by + g(t)$$

Substituting this into (11.20a), we get the first order equation

$$y' = -\frac{by - g(t)}{a}$$

In another class of problems we are initially given the set of differential and algebraic equations (11.17a) and (11.18). We see that in all three cases there are regions in which just this problem has to be solved. In the first two cases there are regions in which a completely differential problem must first be solved and then the additional problem of deciding when to switch to the other technique.

In this section we will examine a method in which both stages of the process can be handled uniformly, so that, for example, it is not necessary to solve (11.19) explictly for $y^{(p)}$ or $y^{(p-1)}$, or to decide which region is appropriate.

Consider a normal form multivalue method for the solution of (11.19). We wish to find an ω such that

$$\mathbf{a}_n = A\mathbf{a}_{n-1} + \mathbf{l}\omega$$

satisfies (11.19). Let $F(\mathbf{a}) = F(a_0, a_1/h, 2!\,a_2/h^2, \ldots, p!\,a_p/h^p, t)$. If F has continuous derivatives, there exists an $\boldsymbol{\xi}$ such that

$$\mathbf{a}_n = A\mathbf{a}_{n-1} - \mathbf{l}\left[\frac{\partial F}{\partial \mathbf{a}}(\xi)\cdot\mathbf{l}\right]^{-1} F(A\mathbf{a}_{n-1}) \tag{11.21}$$

(Numerically we can find this solution by an iterative method such as the quasi-Newton method

$$\mathbf{a}_{n,(0)} = A\mathbf{a}_{n-1}$$
$$\mathbf{a}_{n,(m+1)} = \mathbf{a}_{n,(m)} - \mathbf{l}\left[\frac{\partial F}{\partial \mathbf{a}} \cdot \mathbf{l}\right]^{-1} F(\mathbf{a}_{n,(m)})$$

where $\partial F/\partial \mathbf{a}$ is evaluated at some conveient value of its arguments as for stiff methods.) If we examine the stability of (11.21) to numerical perturbations $\mathbf{e}_n$, we find that

$$\mathbf{e}_n = S_n \mathbf{e}_{n-1} + 0(\mathbf{e}_n + \mathbf{p}_n)^2$$

where $\mathbf{p}_n$ is of the order of the truncation error in the predictor (this last term vanishes if F is linear in $\mathbf{a}$),

$$S_n = \left[I - \left[\frac{\partial F}{\partial \mathbf{a}}(\xi) \cdot \mathbf{l}\right]^{-1} \mathbf{l} \frac{\partial F}{\partial \mathbf{a}}(\xi_1)\right] A$$

and ξ_1 is a point between the two numerical solutions. Suppose that $\partial F/\partial y^{(p)}$ is nonzero. For small h, we find that

$$\begin{aligned} S_n &= \left(I - \frac{\mathbf{l}\boldsymbol{\delta}_p^T}{l_p}\right) A + 0(h) \\ &= S + 0(h) \end{aligned} \tag{11.22}$$

Notice that this differs from the error amplification matrix S that we obtained in Chapter 9 only by the term $1/l_p$. (In fact, the Adams' type methods for pth order equations given in Table 9.1 had $l_p = 1$, so there would be no difference.)

In Chapter 9 we found $\mathbf{l}$ that depended on the order of the equation to be solved. If we could find $\mathbf{l}$ independent of p such that S given by (11.22) is stable for several values of p, then the resulting method could be used for the implicit equation (11.19) without knowledge of its actual order, as long as it was known only to take on values for which S is stable. *An interesting fact is that the* $\mathbf{l}$ *given in Table 11.2 are such that* S *is stable for all* $0 \leq p \leq k$. Thus, these methods are applicable to up to kth order implicit equations for $k < 6$. If the order is constant, that is, $\partial F/\partial y^{(p)}$ is bounded away from zero, the methods converge as $h \leftarrow 0$ regardless of p in the range $0 \leq p \leq k$.

However, in the case that $\partial F/\partial y^{(p)}$ tends to zero at some point in the interval, problems can arise. These can be seen by considering the equation

$$F(y, y', t) = g(t) + yf(t) - \epsilon(t)y' = 0 \tag{11.23}$$

For this problem, S_n is given by

$$S_n = (I - \mathbf{l}[l_0 hf(t_n) - l_1 \epsilon(t_n)]^{-1}[hf(t_n)\boldsymbol{\delta}_0^T - \epsilon(t_n)\boldsymbol{\delta}_1^T])A$$

If $\epsilon(t)$ tends to zero as t increases but $f(t) \neq 0$, we can find an h in the range $0 < h \leq h_0$ such that for some t, $\mu = hf(t)/\epsilon(t)$ takes on any value in the negative half plane. (Points in the positive half plane need not be considered as the differential equation would then be increasingly unstable as t increased.) Consequently, we must be concerned with the stability of

$$S_n = \left[\frac{I - \mathbf{l}[\mu\boldsymbol{\delta}_0^T - \boldsymbol{\delta}_1^T]}{l_0\mu - l_1}\right]A$$

for $\text{Re}(\mu) \leq 0$. This is precisely the problem of stiff stability discussed in the last section, and we have seen that S_n can be made stable in the whole of $\text{Re}(\mu) \leq 0$ with order two methods, and in most of $\text{Re}(\mu) \leq 0$ by methods of order up to at least eleven. For pth order equations we are interested in the regions of stability of

$$S_n = \left[\frac{I - \mathbf{l}\left[\sum_{i=0}^{p} \mu_i \boldsymbol{\delta}_i^T\right]}{\sum_{i=0}^{p} \mu_i l_i}\right]A$$

for any equation of the form

$$\mu_0 y + \mu_1 h y' + \cdots + \mu_p \frac{h^p y^{(p)}}{p!} = 0$$

which is stable. No results are known to the writer at this time.

Algebraic Equations

When $p = 0$ in (11.19), we have an algebraic equation to solve. In this sense, we view an algebraic equation as a differential equation of order zero. If the technique proposed above is used to solve such an equation, we arrive at the error amplification matrix S given by (11.22) with $p = 0$. We have pointed out that it is stable for the $\mathbf{l}$ given by Table 11.2 for stiff equations.

In fact, all of the eigenvalues of S are zero for this case. This can be seen by looking at the limit of the equation

$$\epsilon y' = f(y, t)$$

as $\epsilon \longrightarrow 0+$. If $\partial f/\partial y < 0$, this is a stiff equation for small positive ϵ. The eigenvalues of S are the zeros of $\rho(\xi) + (1/\epsilon)(\partial f/\partial y)\sigma(\xi)$. As $\epsilon \longrightarrow 0$, these become the zeros of $\sigma(\xi)$, which are zero since $\sigma(\xi) = \xi^k$. When $\epsilon = 0$, the differential equation becomes an algebraic equation.

When this technique is used for algebraic equations, we are predicting the solution of $F(y, t) = 0$ at t_n by polynomial extrapolation from values of y at earlier t_i's, then correcting by the technique used to solve (10.19). If the Newton corrector iteration is used, it is the same as solving $F(y, t) = 0$ by Newton's method.

We could have arrived at this type of method directly by proposing to find an initial guess for y_n by an extrapolation formula through $y_{n-1}, y_{n-2}, \ldots, y_{n-k-1}$, and then using Newton's method. The error amplification matrix for this method would have all zero eigenvalues since no error is propogated beyond $k + 1$ steps if the corrector is iterated to convergence. Hence this method is equivalent to the previous method, and will have the same $\mathbf{l}$ when put in normal form.

The fact that the one method works for both algebraic and differential equations is important for the types of equations that arise in network analysis and simulation. These are frequently implicit systems of equations of the form

$$\mathbf{F}(\mathbf{y}, \mathbf{y}', t) = 0$$

where $\mathbf{F}$ and $\mathbf{y}$ are vectors which represent the s equations and the s dependent variables. In this case the corrections for each dependent variable are the components of

$$\left[\frac{\partial \mathbf{F}}{\partial \mathbf{y}} l_0 + \frac{\partial \mathbf{F}}{\partial \mathbf{y}'} \frac{l_1}{h}\right]^{-1} \mathbf{F}(\mathbf{y}, \mathbf{y}', t)$$

It is not necessary to solve $\mathbf{F}$ for $\mathbf{y}'$ explicitly or to determine which equations are differential. The matrix whose inverse has to be multiplied by $\mathbf{F}$ is usually sparse and in this formulation sparse techniques can be used profitably. [See Tewardson (1967), Tinney and Walker (1967), and Willoughby (1969).]

If some of the dependent variables appear only linearly and not in derivative form (as frequently happens), we can rename those $\mathbf{v}$ and write the equation as

$$\mathbf{F}(\mathbf{y}, \mathbf{y}', t) + P\mathbf{v} = 0$$

where the components of $\mathbf{v}$ have been removed from $\mathbf{y}$ and P is a constant matrix. It can then be seen that the changes to $\mathbf{y}$ in the corrector iteration are independent of the predicted values of $\mathbf{v}$. It therefore is only necessary to store the values of $\mathbf{v}$, not its derivatives, and no prediction process need be used on the $\mathbf{v}$. More details on this approach can be found in Gear (1971), Calahan (1969), and Hachtel et al. (1971).

11.3 PARAMETER ESTIMATION

Frequently, the experimentalist will have hypothesized a behavior for a system given by a set of differential equations which depend on a number of parameters $\mathbf{p} = \{p^i\}$. Experimental measurements at various times give approximate values of the solution; the job is to find the value of the parameters. Suppose the system of equations is

$$\mathbf{y}' = \mathbf{f}(\mathbf{y}, t, \mathbf{p}), \qquad \mathbf{y}(0) = \mathbf{y}_0(\mathbf{p}) \tag{11.24}$$

and values of $\mathbf{z}_i = \mathbf{y}(\tau_i)$ are given, $i = 1, \ldots, m$. If there are just as many values of $\mathbf{y}(\tau_i)$ given as there are parameters and equations, we have a type of boundary value problem to solve. Usually there are more values of $\mathbf{y}(\tau_i)$ than can be satisfied by the number of parameters and initial values, so a least squares approach to satisfying the conditions is appropriate. The solution of (11.24) can be expressed as

$$\mathbf{y}(t) = \mathbf{F}(t, \mathbf{p}) \tag{11.25}$$

A least squares fit will require that a function of the form

$$\sum_{i=1}^{m} \| \mathbf{y}(\tau_i) - \mathbf{z}_i \|_i^2 \, \omega_i^2 \tag{11.26}$$

be minimized where $\| \cdot \|_i$ is a norm which may exclude some components of the vector if they are not specified. We will not discuss the subject of minimization here; the reader can refer to the many papers and books in the field, for example, Kowalik and Osborne (1968).

Some of the methods require that the partials of (11.26) with respect to the parameters be computed. This can be done numerically by differencing, requiring $q + 1$ integrations of s equations if there are q parameters and s components of $\mathbf{y}$. Alternatively, we can get differential equations for $\partial y^i/\partial p^j$. We have from (11.24)

$$\frac{d}{dt}\left(\frac{\partial \mathbf{y}}{\partial \mathbf{p}}\right) = \frac{\partial \mathbf{f}}{\partial \mathbf{p}} + \frac{\partial \mathbf{f}}{\partial \mathbf{y}} \frac{\partial \mathbf{y}}{\partial \mathbf{p}} \tag{11.27}$$

(11.24) and (11.27) are a system of $(q + 1)s$ equations which must be integrated once. If $\partial \mathbf{f}/\partial \mathbf{p}$ is complicated, this is likely to be more costly than numerical differentiation, but otherwise direct differentiation may have significant advantages. In particular, if the original problem is stiff as frequently happens, the matrix $\partial \mathbf{f}/\partial \mathbf{y}$ has to be approximated anyway. System (11.27) is then also stiff because

$$\frac{\partial\left[\dfrac{d}{dt}\left(\dfrac{\partial \mathbf{y}}{\partial p^j}\right)\right]}{\partial\left(\dfrac{\partial \mathbf{y}}{\partial p^j}\right)} = \frac{\partial \mathbf{f}}{\partial \mathbf{y}}$$

so it has the same eigenvalues. The corrector can first be applied to (11.24) to compute $\mathbf{y}_{n+1}$, then (11.27) can be handled. For each p^j, the vector $\partial \mathbf{y}/\partial p^j$ uses the same matrix in the Newton iteration of the corrector as was used for $\mathbf{y}$, so it need not be recalculated.

If the initial values are unknown, they can be counted as parameters. If there are errors in the times τ_i at which the $\mathbf{z}_i$ are measured, they can be included in the unknowns used to minimize (11.26) since

$$\left.\frac{\partial \mathbf{y}(\tau_i)}{\partial \tau_i}\right|_{\mathbf{p}\text{ fixed}} = \mathbf{f}(\mathbf{y}(\tau_i), \tau_i, \mathbf{p})$$

is known.

PROBLEMS

1. Show that the test example given in Section 11.1.1 is ill-posed for perturbations in y in excess of 0.005 in each component (with appropriate signs) after the system has reached equilibrium.

2. (a) Express y_{n+1} in terms of y_n, λ, and h for the equation $y' = \lambda y$ using the method

$$q_1 = y_n + \beta_{11}hf(q_1) + \beta_{12}hf(y_{n+1})$$
$$y_{n+1} = y_n + \beta_{21}f(q_1) + \beta_{22}hf(y_{n+1})$$

(b) What is the maximum order of this method?
(c) What are the values of the β_{ij} for this order?

3. Suppose the trapezoidal rule is used to integrate from t_n to $t_n + h$ in one step to get $\tilde{y}_{n+1}$, and then is used to integrate in two steps of size $h/2$ to get $\hat{y}_{n+1}$. It can be shown that the local truncation error of

$$y_{n+1} = \frac{4\hat{y}_{n+1} - \tilde{y}_{n+1}}{3}$$

is $0(h^5)$. Is the method A-stable?

4. Prove that the method given by (11.16) is A-stable.

5. If you were integrating the equation

$$y' = 10(e^t - y) + e^t, \qquad y(0) = 1$$

by the Euler method, at what accuracies would you consider stiffness a problem?

6. Consider the single first order equation

$$y' = f(y, t), \qquad y(0) = y_0, \qquad t \in [0, b]$$

in which f has continuous bounded derivatives. You also know that for this particular problem

$$\left|\frac{\partial f}{\partial y} - L\right| < u$$

where L and u are constants such that $0 < u < -L$. A fixed step size multivalue method is used to integrate this problem, and the corrector is iterated to convergence by a Newton method. You are told that its local truncation error can be bounded by Th^{r+1} for $h \leq h_0$ and that the matrix

$$S(h) = \left[\frac{I - \mathbf{l}(\boldsymbol{\delta}_1^T - hL\boldsymbol{\delta}_0^T)}{l_1 - hLl_0}\right]A$$

has all eigenvalues less than one for $\delta \leq h \leq h_0$. The method is also known to be stable [$S(0)$ satisfies the root condition]. Derive an error bound that is independent of L for $\delta \leq h \leq h_0$.

7. A chemist is studying some reactions and he knows that the concentrations of two

components in his experiment obey the ordinary differential equations

$$\frac{dy}{dt} = -k_1 y + k_2(b - 2y - z)z$$

$$\frac{dz}{dt} = -k_3 z + k_4(b - 2y - z)(a - y - z) - \frac{dy}{dt}$$

where the k_i are unknown positive constants, and a and b are known positive constants.

The following experiment was made:

(a) At $t = 0$, the values of a, b, $y(0)$, and $z(0)$ were set to

$$a = 1.0$$
$$b = 2.0$$
$$y(0) = 0.25$$
$$z(0) = 0.50$$

(b) The values of $y(t)$ and $z(t)$ were sampled at various times. The following data was gathered.

t	$y(t)$	$z(t)$
0	0.250	0.500
0.333	0.301	0.403
0.672	0.324	0.362
1.012	0.335	0.345
∞	0.345	0.332

By $t = \infty$ the chemist means a sufficiently long time that the system was stable (i.e., $dy/dt = dz/dt = 0$). In this case it was certainly stable by $t = 100$.

From physical considerations the chemist knows that the k_i should be close to one. What values of k_i can you calculate for him? Estimate the error in these answers.

8. Prove the statement at the end of Section 11.2 to the effect that the predicted values of $\mathbf{v}$ do not affect the corrector iteration (ignoring round-off errors).

12 CHOOSING A METHOD

In this short chapter we will attempt to give some guidelines concerning the choice of a method for a given problem. Necessarily these comments are based on the present state of the art. Consequently, the areas in which future developments seem likely to change these guidelines will also be discussed.

The criteria for choosing the step size and order for a given class of methods were discussed in Chapter 5 for one-step methods. Those criteria are equally applicable to other methods if we continue to assume that the objective is to get the maximum step size per function evaluation. The choice between different representations of equivalent multistep methods was covered in Chapter 9. Thus, the decisions that remain to be made are between the various types of methods (Taylor's series, Runge-Kutta, multivalue, Bulirsch-Stoer, and some of the techniques that assume a greater knowledge of the equation) and, in the case of Runge-Kutta and multivalue methods, between the various values that the constants of the method can take (these affect the truncation error in Runge-Kutta methods and both the truncation error and stability regions in the case of multivalue methods).

Trivial Problems

Many equations that are solved on digital computers can be classified as trivial by the fact that even with an inefficient method of solution, little computer time is used. Economics then dictates that the best method is the one that minimizes the human time of preparation of the program. The classical Runge-Kutta method with a fixed step size is then one of the most convenient. The step can be reduced by a factor of two to decide if the accuracy is sufficient. However, if the computer library contains other methods in canned programs that require no special starting procedures, these can just as well be used. If the problem is simple (for example, linear in y) but stiff (as evidenced by Runge-Kutta methods requiring very small step sizes) then the trapezoidal rule can be effective. However, an implicit equation must be solved. This can be done by direct methods in the case of linear equations. Library subroutines are usually available for this.

Smooth Non-Stiff Problems

Either the Bulirsch-Stoer or multivalue methods are good for these problems. For many problems the Bulirsch-Stoer method is one of the fastest, but currently good criteria for deciding on the best order and step size are not known. Multivalue methods, such as the automatic one given in Chapter 9, use considerable effort in starting, but afterwards are fairly competitive. Consequently, if the differential equation is to be integrated over a large interval (in the sense that many steps are needed for the accuracy requested) and the relative sizes of the derivatives do not change rapidly, multivalue methods may be best. If the problem is known not to have any components that decay at a rate close to the rate of growth of increasing components, the smaller truncation errors possible with methods that are weakly stable or nearly so are worth considering. Otherwise, the Adams' methods are more desirable.

Problems with Discontinuities

Many problems have discontinuous derivatives at a number of points. A rocket trajectory, for example, will have a discontinuous second derivative when the engine is turned on or off. Methods which assume a Taylor's series expansion over such points must be limited in order. The rocket trajectory, for example, cannot use higher than a first order method in the step in which a discontinuity occurs. A k-step method would be limited to first order as long as the discontinuity was interior to the interval $[t_{n-k}, t_n]$. In this case, a one-step method whose step sizes are such that the discontinuities only occur at mesh points is superior because its order is not limited. For these problems either a Runge-Kutta method or Bulirsch-Stoer method is desirable.

Higher Order Equations

Tests [Gear (1967)] and preliminary results [Rutishauser (1960)] on direct methods for higher order equations indicate that in some cases it is better to transform to a system of lower order equations, in other cases to deal directly with the high order equation. If the latter is done, the multivalue extensions of Adams' method are among the most convenient, since the program for higher order equations is essentially identical to the first order program given in Chapter 9.

Stiff Equations

General nonlinear equations that are known to be stiff can be handled by the multistep method discussed in Chapter 11, and embodied in the program in Chapter 9. Other methods [Allen (1969), Calahan (1968), Pope (1963), and Rosenbrock (1963)] may be better for some cases, but are more difficult to cast in a general program applicable to many cases. The implicit Runge-Kutta methods look promising for those cases where discontinuities can occur, particularly those of Axelsson (1969) which are stiffly A-stable. However, they require a considerable number of function evaluations and probably more work to converge to the solution of the q-stage implicit equations than the multivalue methods because the system is q times as large.

If the equations are linear, or the stiff parts arise from linear terms, the methods of Lawson (1967) and Dahlquist (1969) may be superior, although it is necessary to find appropriate ways of controlling the error in these. A situation similar to stiff equations occurs when $\partial f/\partial y$ has eigenvalues close to the imaginary axis. If these are relatively fixed, it is known that the solution includes components oscillating with given frequencies. Gautschi (1916) has proposed methods that are exact for low order trigonometric polynomials much as conventional multistep methods are exact for low order ordinary polynomials.

Comparison of Five Methods

As a guide to the relative efficiency of different methods presented in earlier chapters, the three problems given at the end of Section 6.2 (negative exponential, Euler's equation, and J16) were integrated using the automatic Runge-Kutta program in Chapter 5 (RK), the polynomial and rational extrapolation program in Chapter 6 (POLY and RAT), and the multivalue method using Adams or stiff formulas (ADAM and STIFF). The parameters MAXORD and MAXPTS for the extrapolation algorithm were set to 6 and 8, respectively, giving an $0(h^{14})$ error per step. The maximum order for ADAM and STIFF was set to 7 and 6, respectively. The initial step size tried was set to EPS for RK, ADAM, and STIFF, and to $(\text{EPS})^{.25}$ for POLY and RAT. (The latter two methods are very sensitive to initial step size selection, indicating that the step control mechanism needs improvement.)

The relation between actual relative error and number of function evaluations is shown in Figures 12.1 to 12.3. To achieve this, EPS was varied from 10^{-2} to 10^{-11} in the case of exp (-20) and Euler, and 10^{-6} to 10^{-11} in the case of J16. We see that the steeper slope of the Runge-Kutta graph indicates a superiority for low accuracies. The extrapolation methods are generally better than the multistep methods on these tests. This is in part offset by the fact that the initial step had to be selected carefully for the extrapolation

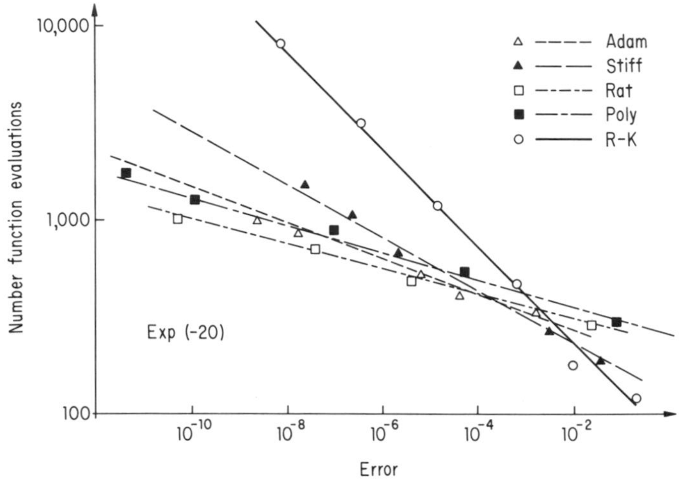

Fig. 12.1 Errors in exp (−20).

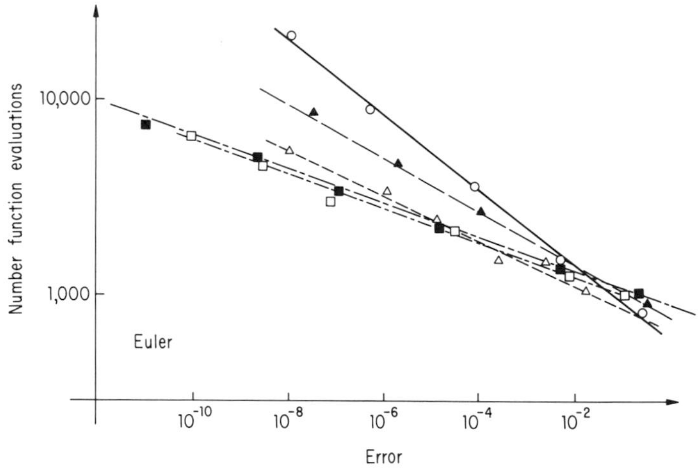

Fig. 12.2 Errors in Euler equations.

methods—a bad choice would lead to poorer results than given by the multi-value methods—and that only a seventh order Adams' method was used. If a higher maximum order had been programmed, these smooth problems would have been handled in fewer steps.

Table 12.1 shows the times used by the subroutines if they are assumed to be of the form $A + B \times N$, where N is the number of equations in the system. (IBM 360 FORTRAN H, OPT = 2 was used on a 360/91 system.)

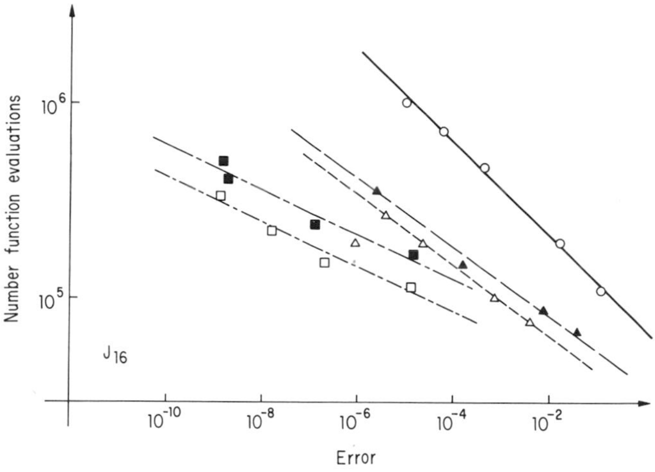

Fig. 12.3 Errors in J16 equations.

As can be seen, Runge-Kutta methods execute significantly faster than the other methods per function evaluation. For problems in which the function evaluation time is small, this overhead may be significant, and hence the Runge-Kutta method is to be preferred even for medium accuracies. If the function evaluation time is large, only the number of function evaluations should be taken into account.

Table 12.1 IBM 360/91 EXECUTION TIMES PER FUNCTION EVALUATIONS (ALL TIMES IN 10^{-6} SECONDS)

Method	*Adam*	*Stiff*	*Rk*	*Rat*	*Poly*
Time for no equations	112.7	92.5	49.7	33.3	28.8
Time per additional equation	17.9	45.6	1.2	5.8	5.0

12.1 EFFECT OF FUTURE DEVELOPMENTS

Any comments on this topic are necessarily in the nature of forecasting, and destined to be proved wrong, in part or whole, within a short time. The justification for including them as a concluding section is to suggest some of the more obvious areas which need further analysis. The areas are

1. Methods for linear and quasi-linear equations of forms
   (a) $y' = Ay + f(t)$.
   (b) $y' = A(t)y + f(t)$.

2. Criteria to indicate when higher order equations should be handled directly as opposed to being transformed to a lower order system.
3. Extrapolation methods for stiff equations.
4. Adequate theory for stiff methods.
5. Extension of theory to variable step multivalue methods.

While it seems likely that present methods are effective for general nonlinear equations in which nothing is known a priori about the solution or the eigenvalues of $\partial f/\partial y$, it seems probable that restricted equations, particularly linear and quasi-linear equations, can be handled far more efficiently than possible at present. Many large systems, arising, for example, in networks, contain largely linear components, but this property has yet to be exploited effectively in all but simple cases.

It is unusual to encounter higher than second order equations in practice. Except for the case when the second derivative is absent, it is not apparent when to use direct methods. It seems reasonable to expect an advantage up to a speed factor of two for the most appropriate direct method because reduction to a first order system is almost doubling the amount of information to be processed.

The tests reported above on extrapolation methods show a possible favorable behavior when compared with other methods for non-stiff equations. Dahlquist (1969) has indicated that extrapolation applied to the trapezoidal rule may be effective for stiff equations. Extrapolation methods seem to offer the most flexible approach to order and step control and may replace other methods for general nonlinear problems.

The theoretical treatment of stiff problems has not yet progressed very far. The greatest need is error bounds which are relatively independent of the negativeness of the eigenvalues of the Jacobian and an idea of what are the minimum additional constraints on the system to achieve such a result. Dahlquist (1963) discusses the trapezoidal method for a nonlinear equation.

No discussion has been given on a number of methods, reflecting this writer's belief that they do not currently compete with the best of the three principal methods emphasized in this book. However many of these methods have not received the attention of the Runge-Kutta, Adams, or extrapolation methods, so it must be conceded that perhaps great future strides will be taken in another direction. Among the methods not mentioned earlier are the combinations of Runge-Kutta and multistep methods [Butcher (1965B), Gragg and Stetter (1964), and Gear (1965)] in which information from previous points is used as well as evaluations of the function at nonmesh points in the interval of integration. Preliminary tests by this writer indicate that their range of absolute stability is not large, and that more function evaluations may be taken than with a simple Adams' method when problems of error estimation and step selection are taken into account.

BIBLIOGRAPHY

The following bibliography includes books, papers, and reports referred to in the text plus those that have come to the attention of the writer and have been published since 1961. An extensive bibliography up to that date can be found in Henrici (1962).

Allen, B. T. (1960), "A New Method of Solving Second Order Differential Equations When the First Derivative is Present," *Comp. J.*, **8**, pp. 392–394.

Allen, R. H. (1969), "Numerically Stable Explicit Integration Techniques using a Linearized Runge-Kutta Extension," *Boeing Scientific Laboratories Document dl*-82-0929.

Anderson, N. H., Ball, R. B., and Voss, J. R. (1960), "A Numerical Method for Solving Control Differential Equations on Digital Computers," *JACM*, **7**, pp. 61–68.

Axelsson, O. (1969), "A Class of *A*-Stable Methods," *BIT*, **9**, pp. 185–199.

Bard, A., Ceschino, F., Kuntzmann, J., and Laurent, P. (1961), "Formules de base de la méthode de Runge-Kutta," *Chiffres*, **4**, pp. 31–37.

Bashforth, F. and Adams, J.C. (1883), *Theories of Capillary Action.*, Cambridge U. P., New York.

Birkhoff, G. and Varga, R. S. (1965), "Discretization Errors for Well Set Cauchy Problems," *Jour. Math. and Phys.*, **44**, pp. 1–23.

Bjurel, G. (1969), "Preliminary Report on Modified Linear Multistep Methods for a Class of Stiff Ordinary Differential Equations," *Dept. of Information Processing*, The Royal Institute of Technology, Stockholm, Report # NA 69.02.

Bjurel, G., Dahlquist, G., Lindberg, B., Linde, S., and Odén, L. (1970), "Survey of

Stiff Ordinary Differential Equations, *Dept. of Information Processing*, Royal Institute of Technology, Stockholm, Report # NA 70.11.

Blum, E. K. (1962), "A Modification of the Runge-Kutta Fourth Order Method," *Math. Comp.*, **16**, pp. 176–187.

Brayton, R. K., Gustavson, F. G., and Liniger, W. (1966), "A Numerical Analysis of the Transient Behavior of a Transistor Circuit,", *IBM Jour.* **10**, pp. 292–299.

Brock, P. and Murray, F. J. (1952), "The Use of Exponential Sums in Step by Step Integration," *MTAC*, **6**, pp. 63–78, 138–150.

Brown, R. R., Riley, J. D. and Bennett, M. M. (1965), "Stability Properties of Adams-Moulton Type Methods," *Math. Comp.*, **19**, pp. 90–96.

Brush, D. G., Kohfeld, J. J. and Thomson, G. T. (1967), "Solution of Ordinary Differential Equations Using Two Off-Step Points," *JACM*, **14**, pp. 769–784.

Bulirsch, R. and Stoer, J. (1964), "Fehlerabschatzungen und Extrapolation mit rationalen Funktioner bei Verfahren von Richardson-typus," *Num. Math.*, **6**, pp. 413–427.

Bulirsch, R. and Stoer, J. (1966), "Numerical Treatment of Ordinary Differential Equations by Extrapolation Methods," *Num. Math.*, **8**, pp. 1–13.

Bulirsch, R. and Stoer, J. (1966B), "Asymptotic Upper and Lower Bounds for Results of Extrapolation Methods," *Num. Math.*, **8**, pp. 93–104.

Butcher, J. C. (1963), "Coefficients for the Study of Runge-Kutta Integration Processes," *Jour. Australian Math. Society*, **3**, pp. 185–201.

Butcher, J. C. (1964), "Implicit Runge-Kutta Processes,", *Math, Comp*, **18**, pp. 50–64.

Butcher, J. C. (1964B), "On Runge-Kutta Processes of High Order," *Jour. Australian Math. Society*, **4**, pp. 179–194.

Butcher. J. C. (1964C), "Integration Processes Based on Radau Quadrature Formulas," *Math. Comp.*, **18**, pp. 233–244.

Butcher, J. C. (1965), "On the Attainable Order of Runge-Kutta Methods," *Math, Comp.*, **19**, pp. 408–417.

Butcher, J. C. (1965B), "A Modified Multistep Method for the Numerical Integration of Ordinary Differential Equations," *JACM*, **12**, pp. 124–135.

Butcher, J. C. (1966), "On the Convergence of Numerical Solutions to Ordinary Differential Equations," *Math. Comp.*, **20**, pp. 1–10.

Butcher, J. G. (1967), "A Multistep Generalization of Runge-Kutta with Four or Five Stages," *JACM*, **14**, pp. 84–99.

Byrne, G. D. (1967) "Paramameters for Pseudo Runge-Kutta Methods," *Comm ACM* **10**, No. 2, pp. 102–104

Byrne, G. D. and Lambert, R. J. (1966), "Pseudo Runge-Kutta Methods Involving Two Points," *JACM*, **13**, pp. 114–123.

Calahan D. (1969), "Numerical Considerations in the Transient Analysis and Optimal Design of Nonlinear Circuits," *Digest Record of Joint Conference on Mathematical and Computer Aids to Design*, ACM/SIAM/IEEE, Anaheim, Cal., pp. 129–145.

Calahan, D. A. (1967), "Numerical Solution of Linear Systems with Widely Separated Time Constants," *Proc. IEEE*, **55**, pp. 2016–2017.

Calahan, D. A. (1968), "A Stable Accurate Method of Numerical Integration for Non-Linear Systems," *Proc. IEEE*, **56**. p. 744.

Calahan, D. A. and Abbott, N. E. (1970), "Stability Analysis of Numerical Integration," *Proc. of the Tenth Midwest Symposium on Circuit Theory*, pp. I-2-1 to I-2-20.

Calahan, D. A. and Gear, C. W. (1969), "An Ill-Conditioning Problem with Implicit Integration," *Proc. IEEE*, **57**, pp. 1775–1776.

Case, J. (1969), "A Note on the Stability of Predictor Corrector Techniques," *Math. Comp.*, **23**, pp. 741–750.

Casity, C. R. (1966), "Solutions of the Fifth Order Runge-Kutta Equations," *SINUM*, **3**, pp. 598–606.

Cassity, C. R. (1969), "The Complete Solution of the Fifth Order Runge-Kutta Equations," *SINUM*, **6**, No. 3, pp. 432–436.

Certaine, J. (1960), "The Solution of Ordinary Differential Equations with Large Time Constants," in *Mathematical Methods for Digital Computers*, ed. A. Ralston and H. S. Wilf. Wiley, New York, pp. 128–132.

Ceschino, F. (1961), "Modification de la longueur du pas dan l'intégration numérique par les méthods à pas liés," *Chiffres*, **2**, pp. 101–106.

Ceschino, F. (1961B), "Une méthode de mise en oeuvre des formules d'Obrechkoff pour l'intégration des équations différentielles," *Chiffres*, **2**, pp. 49–54.

Ceschino, F. and Kuntzmann, J. (1963), *Problèmes différentiels de conditions initials*. Dunod, Paris. Translation by D. Boyanovitch, as *Numerical Solution of Initial Value Problems*. Prentice-Hall, Englewood Cliffs, N. J., 1966.

Chase, P. E. (1962), "Stability Properties of Predictor-Corrector Methods for Ordinary Differential Equations," *JACM*, **9**, pp. 457–468.

Christiansen, J. (1970), "Handbook Series Numerical Integration, Numerical Solution of Ordinary Simultaneous Differential Equations of the First Order Using a Method for Automatic Step Change," *Num. Math*, **14**, No. 4, pp. 317–324.

Clark, N. A. (1966), "Program Description for Library Subroutine ANL D250 DIFSUB," Argonne National Laboratory, Argonne, Ill.

Clark, N. A. (1968), "A Study of Some Numerical Methods for the Integration of Systems of First Order Ordinary Differential Equations," *Argonne National Lab Report No. 7428*.

Clenshaw, C. W. (1957), "The Numerical Solution of Linear Differential Equations in Chebyshev Series," Proc. Cambridge Phil. Society, **53**, pp. 134–149.

Clenshaw, C. W. (1960), "The Numerical Solution of Ordinary Differential Equations in Chebyshev Series," in *PICC Symposium, Rome, on Differential and Integral Equations*. Birkhäuser, Basel, pp. 222–227.

Clenshaw, C. W. and Curtiss, A. R. (1960), "A Method for Numerical Integration on an Automatic Computer," *Num. Math.*, **2**, pp. 197–205.

Cohen, C. J. and Hubbard, E. C. (1960), "An Algorithm Applicable to Numerical Integration of Orbits in Multiple Revolution Steps," *Astron. J.*, **65**, pp. 454–456.

Collatz, L. (1960), *The Numerical Treatment of Differential Equations*, 3rd ed. Springer, Berlin.

Cooper, G. J. (1967), "A Class of Single Step Methods for Systems of Nonlinear Differential Equations," *Math. Comp.*, **21**, pp. 597–610.

Cooper, G. J. (1968), "Interpolation and Quadrature Methods for Ordinary Differential Equations," *Math. Comp.*, **22**, 69–76.

Cooper, G. J. and Gal, E. (1967), "Single Step Methods for Linear Differential Equations," *Num. Math.*, **10**, pp. 307–315.

Courant, R. (1936), *Differential and Integral Calculus*, Vol. 2. Interscience, New York.

Cowell, P. H. and Crommelin, A. C. D (1910), "Investigation of the Motion of Halley's Comet from 1759 to 1910," appendix to *Greenwich Observations for 1909*. Edinburgh. p. 84.

Crane, R. L. and Klopfenstein, R. W. (1965), "A Predictor-Corrector Algorithm with Increased Range of Absolute Stability," *JACM*, pp. 227–241.

Crane, R. L. and Lambert, R. J. (1962), "Stability of a Generalized Corrector Formula," *JACM*, **9**, No. 1. pp. 104–117.

Curtiss, C. F. and Hirschfelder, J. O. (1952), "Integration of Stiff Equations," *Proc. Nat. Acad. Science, U. S.*, **38**, pp. 235–243.

Dahlquist, G. (1956), "Numerical Integration of Ordinary Differential Equations," *Math. Scandinavica*, **4**, pp. 33–50.

Dahlquist, G. (1959), "Stability and Error Bounds in the Numerical Integration of Ordinary Differential Equations," *Trans. Roy. Inst. Tech., Stockholm*, No. 130.

Dahlquist, G. (1963), "A Special Stability Problem for Linear Multistep Methods," *BIT*, **3**, pp. 27–43.

Dahlquist, G. (1963B), "Stability Questions for Some Numerical Methods for Ordinary Differential Equations," *Proc. Symposium for Applied Math.*, **15**, pp. 147–158.

Dahlquist, G. (1966), "On Rigorous Error Bounds in the Numerical Solution of Ordinary Differential Equations," in *The Numerical Solution of Nonlinear Differential Equations*, ed. D. Greenspan. John Wiley and Sons, New York, pp. 89–96.

Dahlquist, G. (1969), "A Numerical Method for some Ordinary Differential Equations with Large Lipshitz Constants," in *Information Processing 68*, ed. A. J. H. Morrell. North Holland Publishing Co., Amsterdam, pp. 183–186.

Danchick, R. (1968), "Further Results on Generalized Predictor-Corrector Methods," *Jour. Comp. and Sys. Sciences*, **2**, No. 2, pp. 203–218.

Davison, E. (1967), "A High Order Crank-Nicholson Technique for Solving Differential Equations," *Comp. J.*, **10**, pp. 195–197.

Day, J. T. (1964), "A One-Step Method for the Numerical Solution of Second Order Linear Ordinary Differential Equations," *Math. Comp.*, **18**, p. 664.

Day, J. T. (1965), "A Runge-Kutta Method for the Numerical Integration of the Differential Equation $y'' = f(x, y)$," *ZAMM*, **5**, pp. 354–356.

Day, J. T. (1965B), "A One-Step Method for the Numerical Integration of the Differential Equation $y'' = f(x)y + g(x)$," *Comp. J.*, **7**, p. 314.

Decell, H. P., Jr., Guseman, L. F. and Lea, R. N. (1966), "Concerning the Numerical Solution of Differential Equations," *Math. Comp.*, **20**, No. 95, pp. 431–434.

DeGroat, J. J. and Abbett, M. J. (1965), "A Computation of One-Dimensional Combustion of Methane," AIAA Jour., **3**, pp. 381–383.

Dejon, B. (1966), "Stronger than Uniform Convergence of Multistep Difference Methods," *Num. Math.*, **8**, pp. 29–41.

Dejon, B. (1967), "Numerical Stability of Difference Equations with Matrix Coefficients," *SINUM*, **4**, No. 1, pp. 119–128.

Dennis, S. C. R. (1960), "The Numerical Integration of Ordinary Differential Equations Possessing Exponential Type Solutions." *Proc. Cambridge Phil. Society*, **56**, pp. 240–246.

Dennis, S. C. R. (1962), "Step by Step Integration of Ordinary Differential Equations," *Applied Math. Quarterly*, **20**, pp. 359–372.

Descloux, J. (1963), "Note on a Paper by A. Nordsieck," *Department of Computer Science Report No.* 131, University of Illinois, Urbana, Ill.

Dill, C. (1969), "A Computer Graphic Technique for Finding Numerical Methods for Ordinary Differential Equations," *Department of Computer Science Report No.* 295, University of Illinois, Urbana, Ill.

Dyer, J. (1968), "Generalized Multistep Methods in Satellite Orbit Computation," *JACM*, **15**, No. 4, pp. 712–719.

Ehle, B. L. (1968), "High Order *A*-stable Methods for the Numerical Solution of Systems of Differential Equations," *BIT*, **8**, pp. 276–278.

Emanuel, G. (1964), "Numerical Analysis of Stiff Equations," *Aerospace Report No.* TDR-269 (4230-20)-3.

Engeli, M. E. (1969), "Achievements and Problems in Formula Manipulation," in *Information Processing, 68*, ed. A. J. H. Morrell. North Holland Publishing Co., Amsterdam. pp. 24–32.

Fehlberg, E. (1960), "Neue genauere Runge-Kutta Formeln fur Differentialgleichungen *n*-ter Ordung," *ZAMM*, **40**, pp. 449–455.

Fehlberg, E. (1966), "New High Order Runge-Kutta Formulas with an Arbitrary Small Truncation Error," ZAMM, **46**, pp. 1–16.

Feldstein, M. A. and Stetter, H. J. (1963), "Simplified Predictor-Corrector Methods," *Proceedings 18th ACM National Conference.*

Forrington, C. V. D. (1961), "Extensions of the predictor-corrector method for the solution of systems of ordinary differential equations," *Comp. J.*, **4**, pp. 80–84.

Forsythe, G. and Moler, C. B. (1967), *Computer Solution of Linear Algebraic Systems.* Prentice-Hall, Inc., Englewood Cliffs, N. J.

Fowler, M. E. and Warten, R. M. (1967), "A Numerical Integration Technique for Ordinary Differential Equations with Widely Separated Eigenvalues," IBM Jour., **11**, pp. 537–543.

Fox, L. (1962), "Chebyshev Methods for Ordinary Differential Equations," *Comp. J.*, **4**, pp. 318–331.

Fox, L. (1962), *Numerical Solution of Ordinary and Partial Differential Equations.* Pergamon, New York.

Froese, C. (1961), "An evaluation of Runge-Kutta Type Methods for Higher Order Differential Equations," *JACM*, **8**, pp. 637–644.

Fyfe, D. J. (1966), "Economical Evaluation of Runge-Kutta Formulas," *Math. Comp.*, **20**, No. 95, pp. 392–398.

Gabel, G. (1968), "Predictor Corrector Methods using Divided Differences," Master's thesis, University of Toronto.

Gates, L. D., Jr. (1964), "Numerical Solution of Differential Equations by Repeated Quadratures," *SIAM Review*, **6**, pp. 134–147.

Gautschi, W. (1961), "Numerical Integration of Ordinary Differential Equations Based on Trigonometric Polynomials," *Num. Math.*, **3**, pp. 381–397.

Gear, C. W. (1965), "Hybrid Methods for Initial Value Problems in Ordinary Differential Equations," *SINUM*, **2**, p. 69.

Gear, C. W. (1966), "The Numerical Integration of Ordinary Differential Equations of Various Orders," *Argonne National Lab Report*, #ANL 7126.

Gear, C. W. (1967), "The Numerical Integration of Ordinary Differential Equations," *Math. Comp.*, **21**, pp. 146–156.

Gear, C. W. (1969), "The Automatic Integration of Stiff Ordinary Differential Equations," in *Information Processing*, *68*, ed. A. J. H. Morrel. North Holland Publishing Company, Amsterdam. pp. 187–193.

Gear, C. W. (1970), "Rational Approximations by Implicit Runge-Kutta Schemes," *BIT*, **10**, pp. 20–22.

Gear C. W. (1971) "The Simulataneous Numerical Solution of Differential-Algebraic Systems," *IEEE Transactions on Circuit Theory*, **18**, No. 1, pp. 89–95.

Giese, C. (1967), "State Variable Difference Methods for Digital Simulation, *IEEE Transactions on Computers*, **8**, pp. 263–271.

Giloi, W. and Grebe, H (1968), "Construction of Multistep Integration Formulas for Simulation Purposes," *IEEE*, **17**, No. 12, pp. 1121–1131.

Gragg, W. (1963), "Repeated Extrapolation to the Limit in the Numerical Solution of Ordinary Differential Equations," Doctoral dissertation, UCLA.

Gragg, W. (1965), "On Extrapolation Algorithms for Ordinary Initial Value Problems," *SINUM*, **2**, pp. 384–403.

Gragg, W. and Stetter, H. (1964), "Generalized Multistep Predictor-Corrector Methods," *JACM*, **11**, No. 2, pp. 188–209.

Greenspan, H., Hafner, W., and Ribaric, M. (1965), "On Varying Step Sizes in Numerical Integration of First Order Differential Equations," *Num. Math.*, **7**, pp. 286–291.

Hachtel, G., Brayton, R., and Gustavson, F. (1971), "The Sparse Tableux Approach to Network Analysis and Design," *IEEE Transactions on Circuit Theory*, **18**, No. 1.

Hafner, P. A. (1969), "Stability Charts of Various Numerical Methods for Solving Systems of Ordinary Differential Equations," *Weapons Research Establishment Technical Note* WSD 112, Salisbury, South Australia, November, 1969.

Hain, K. and Hertweck, F. (1960), "Numerical Integration of Ordinary Differential Equations by Difference Methods with Automatic Determination of Steplength," in *PICC Symposium, Rome, Differential and Integral Equations.* Birkhauser, Basel.

Haines, C. F. (1969), "Implicit Integration Processes with Error Estimate for the Numerical Solution of Differential Equations," *Comp. J.*, **12**, No. 2, pp. 183–187.

Hansen, K. F., Koen, B. V., and Little, W. W. (1966), "Stable Numerical Solutions of the Reactor Kinetics Equations," *Nuc. Science Eng.*, **25**, pp. 183–188.

Henrici, P. (1962), *Discrete Variable Methods for Ordinary Differential Equations.* John Wiley and Sons, New York.

Henrici, P. (1963), *Error Propagation for Difference Methods.* John Wiley and Sons, New York.

Henrici, P. (1964), *Elements of Numerical Analysis.* Wiley, New York.

Heun, K. (1900), "Neue Methode zur approximativen Integration der Differentialgleichungen einer unabhängigen Veränderlichen," *Z. Math. u. Phys.*, **45**, pp. 23–38.

Hildebrand, F. B. (1956), *Introduction to Numerical Analysis.* McGraw-Hill Book Co., New York.

Hildbrand, F. B. (1968), *Finite Difference Equations and Simulation.* Prentice-Hall, Inc., Englewood Cliffs, N. J.

Hull, T. E. (1962), "Corrector Formulas for Multistep Integration Methods," *SIAM Jour.*, **10**, pp. 351–369.

Hull, T. E. (1967), "A Search for Optimum Methods for the Numerical Integration of Ordinary Differential Equations," *SIAM Review*, **9**, 647–654.

Hull, T. E. (1968), "The Effectiveness of Numerical Methods for Ordinary Differential Equations," *Studics in Numerical Analysis*, **2**, pp. 114–121.

Hull, T. E. (1969), "The Numerical Integration of Ordinary Differential Equations," in *Information Processing 68*, ed. A. J. H. Morrell. North Holland Publishing Company, Amsterdam.

Hull, T. E. and Creemer, A. L. (1963), "Efficiency of Predictor-Corrector Schemes," *JACM*, **10**, pp. 291–301.

Hull, T. E. and Johnston, R. L. (1964), "Optimum Runge-Kutta Methods," *Math. Comp.*, **18**, pp. 306–310.

Hull, T. E. and Swenson, J. R. (1966), "Tests of Probabilistic Models for Propogation of Round-Off Error,", CACM, **9**, pp. 108–113.

Imhof, J. P. (1963), "On the Method for Numerical Integration of Clenshaw and Curtis," *Num. Math.*, **5**, pp. 138–141.

Ince, E. L. (1956), *Ordinary Differential Equations.* Dover, New York.

Ira, M. (1964), "A Stabilizing Device for Unstable Solutions of Ordinary Differential Equations, Design and Application of a Filter," *Information Processing in Japan*, **4**, pp. 65–73.

Isaacson, E. and Keller H. B. (1966), *Analysis of Numerical Methods.* John Wiley and Sons, Inc., New York.

Jain, M. K. and Srivastava, V. K. (1970), "High Order Stiffly Stable Methods for Ordinary Differential Equations," *Department of Computer Science Report No.* 394, University of Illinois, Urbana, Ill.

Kahan, W. (1966), "A Computable Error Bound for Systems of Ordinary Differential Equations," *SIAM Review (Abstract)*, **8**, pp. 568–569.

Kaluza, T. (1928), "Uber die Koefficienten reziproker Petenzreihen," *Math. Zs.*, **28**, pp. 161–170.

Karim, A. I. A. (1966), "The Stability of the Fourth Order Runge-Kutta Method for the Solution of Systems of Differential Equations," *CACM*, **9**, pp. 113–116.

Karim, A. I. A. (1968), "A Theorem for the Stability of General Methods for the Solution of Differential Equations," *JACM*, **15**, No. 4, pp. 706–711.

Keller, H. B. (1968), *Numerical Methods for Two-point Boundary Value Problems.* Blaisdell, Waltham, Mass.

King, R. (1966), "Runge-Kutta Methods with Constrained Minimum Error Bounds," *Math. Comp.*, **20**, No. 95, pp. 386–391.

Klopfenstein, R. W. and Millman, R. S. (1968), "Numerical Stability of One-Evaluation Predictor-Corrector Methods," *Math. Comp.*, **22**, No. 103, pp. 557–564.

Kohfeld, J. J. and Thompson, G. T. (1967), "Multistep Methods with Modified Predictors and Correctors," *JACM*, **14**, pp. 155–166.

Kohfeld, J. J. and Thompson, G. T. (1968), "A Modification of Nordsieck's Method using an Off Step Point," *JACM*, **15**, No. 3, pp. 390–401.

Konen, H. P. and Luther, H. A. (1967), "Some Singular Explicit Fifth Order Runge-Kutta Solutions," *SINUM*, **4**, pp. 607–619.

Kopal, Z. (1961), *Numerical Analysis*, 2nd ed. Wiley, New York.

Kowalik J. and Osborne M. R. (1968), *Methods for Unconstrained Optimization Problems.* American Elsevier Co., New York.

Krogh, F. T. (1966), "Predictor-Corrector Methods of High Order with Improved Stability Characteristics," *JACM*, **13**, pp. 374–385.

Krogh, F. T. (1967), "A Note on the Effect of Conditionally Stable Correctors," *Math. Comp.*, **21**, No. 100, pp. 717–719.

Krogh, F. T. (1967B), "A Test for Instability in the Numerical Solution of Ordinary Differential Equations," *JACM*, **14**, pp. 351–354.

Krogh, F. T. (1967C), "On Methods of Adams' Type for the Numerical Solution of Ordinary Differential Equations," *TRW Report* No. 67.3122.2.317.

Krogh, F. T. (1969), "A Variable Step, Variable Order Multistep Method for the Numerical Solution of Ordinary Differential Equations," in *Information Processing 68*, Vol. I, ed. A. J. H. Morrell. North Holland Publishing Company, Amsterdam, pp. 194–199.

Krogh, F. T. (1969B), "On Testing a Subroutine for the Numerical Integration of Ordinary Differential Equations," *Jet Propulsion Lab Tech. Report #217.*

Kruckberger, F. and Unger, H. (1960), "On the Numerical Integration of Ordinary Differential Equations and the Determination of Error Bounds," in *PICC*

Symposium, Rome, on Differential and Integral Equations. Birkhauser, Basel. pp. 369–379.

Kuntzmann, J. (1961), "Neuere Entwickelungen der Methode von Runge und Kutta," *ZAMM*, **41**, pp. 28–31.

Kuntzmann, J. (1962), "Nouvelle méthode pour l'intégration approchée des équations différentielles," in *Information Processing 62*, ed. C. Popplewell, North Holland Publishing Company, Amsterdam.

Lambert, R. J. (1967), "An Analysis of the Numerical Stability of Predictor-Corrector Solutions of Nonlinear Ordinary Differential Equations," *SINUM*, **4**, No. 4, pp. 597–606.

Lambert, J. and Mitchell, A. (1962), "On the Solution of $y' = f(x, y)$ by a Class of High Accuracy Difference Formulas of Low Order," *Z. Angew. Math. Phys.*, **13**, pp. 223–232.

Lambert, J. and Shaw, B. (1965), "Numerical Solution of $y' = f(x, y)$ by a Class of Formulas Based on Rational Approximation," *Math. Comp.*, **19**, No. 91, pp. 456–462.

Lambert, J. D. and Shaw, B. (1966), "A Generalization of Multistep Methods for Ordinary Differential Equations," *Num. Math.*, **8**, pp. 250–263.

Lambert, J. D. and Shaw, B. (1966B), "A Method for the Numerical Solution of $y' = f(x, y)$ Based on a Self-Adjusting Nonpolynomial Interpolant," *Math. Comp.*, **20**, pp. 11–20.

Lanczos, C. (1960), "Solution of Ordinary Differential Equations by Trigonometric Interpolation," in *PICC Symposium, Rome, on Differential and Integral Equations.* Birkhäuser, Basel. pp. 22–32.

Laurent, P. J. (1961), "Méthodes spéciales du type de Runge-Kutta," *Premier Congress AFCAL*, pp. 27–36.

Lawson, J. D. (1966), "An Order Five Runge-Kutta Process with Extended Region of Stability," *SINUM*, **3**, pp. 593–597.

Lawson J. D. (1967), "Generalized Runge-Kutta Processes for Stable Systems with with Large Lipschitz Constants," *SINUM*, **4**, pp. 372–380.

Lawson, J. D. (1967B), "An Order Six Runge-Kutta Process with Extended Region of Stability," *SINUM*, **4**, pp. 620–625.

Lee, H. B. (1967), "Matrix Filtering as an Aid to Numerical Integration," *Proc. IEEE*, **55**, pp. 1826–1831.

Lether, F. G. (1966), "The Use of Richardson Extrapolation in One-Step Methods with Variable Step Size," *Math. Comp.*, **20**, No. 95, pp. 379–385.

Lewis, H. R. and Stovall, E. J., Jr. (1967), "Comments on a Floating-Point Version of Nordsieck's Scheme for the Numerical Integration of Differential Equations," *Math. Comp.*, **21**, pp. 157–161.

Liniger, W. (1968), "Optimization of a Numerical Method for Stiff Systems of Ordinary Differential Equations," *IBM Research Report No.* RC-2198.

Liniger, W. (1968), "A Criteria for *A*-Stability of Linear Multistep Integration Formulae," *Computing*, **3**, pp. 280–285.

Liniger, W. (1969), "Global Accuracy and *A*-Stability of One- and Two-Step Integration Formulae for Stiff Ordinary Differential Equations," *IBM Research Report No.* RC-2396.

Liniger, W. and Willoughby, R. (1970), "Efficient Integration Method for Stiff Systems of Ordinary Differential Equations," *SINUM*, **7**, No. 1, pp. 47–66.

Liou, M. L. (1966), "A Novel Method of Evaluating Transient Response," *Proc. IEEE*, **54**, No. 1, pp. 20–23.

Lomax, H. (1968), "On the Construction of Highly Stable Explicit Numerical Methods for Integration of Coupled Ordinary Differential Equations with Parasitic Eigenvalues," *NASA Tech. Report No. TN* 4547.

Lomax, H. (1968B), "Stable Implicit and Explicit Numerical Methods for Integrating Quasi-Linear Differential Equations with Parasitic-Stiff and Parasitic-Saddle Eigenvalues," *NASA Tech. Note NASA, No. TN* D-4703 Ames Research Center, Moffett Field, Calif.

Lomax, H. and Bailey, H. E. (1967), "A Critical Analysis of Various Numerical Integration Methods for Computing the Flow of a Gas in Chemical Non-Equilibrium," *NASA Tech. Report No. TN* D-4109.

Loscalzo, F. R. (1969), "An Introduction to the use of Spline Functions in Ordinary Differential Equations," in *Theory and Applications of Spline Functions*, ed. T. N. E. Greville. Academic Press, New York, pp. 37–64.

Loscalzo, F. R. and Talbot, T. D. (1967), "Spline Function Approximations for Solution of Ordinary Differential Equations," *SINUM*, **4**, No. 3, pp. 433–445.

Lotkin, M. (1951), "On the Accuracy of Runge-Kutta's Method," *MTAC*, **5**, pp. 128–132.

Luther, H. A. (1966), "Further Explicit Fifth Order Runge-Kutta Formulas," *SIAM Review*, **8**, pp. 374–380.

Luther, H. A (1968), "An Explicit Sixth Order Runge-Kutta Formula," *Math. Comp.*, **22**, No. 102, pp. 344–346.

Luther, H. A. and Konen, H. P. (1965), "Some Fifth Order Classical Runge-Kutta Formulas," *SIAM Review*, **7**, pp. 551–558.

Magnus, D. and Schecter, H., "Analysis and Application of the Pade Approximation for the Integration of Chemical Kinetic Equations," *General Applied Science Labs Tech. Report No.* 642, (Project 5810).

Makinson, G. J. (1968), "Stable High Order Implicit Methods for the Numerical Solution of Systems of Differential Equations," *Comp. J.*, **11**, pp. 305–310.

Merson, R. H. (1957), "An Operational Method for the study of Integration Processes," *Proceedings of a Symposium on Data Processing*, Weapons Research Establishment, Salisbury, South Australia. A description of the Runge-Kutta-Merson method can be found in Fox (1962).

Miller, J. C. P. (1966), "The Numerical Solution of Ordinary Differential Equa*tions*," in *Numerical Analysis: An Introduction*, ed. J. Walsh. Academic Press, London. pp. 63–98.

Milne, W. E. and Reynolds, R. R. (1962), "Fifth-Order Methods for the Numerical Solution of Ordinary Differential Equations," *JACM*, **9**, pp. 64–70.

Miranker, W. L. (1968), "Difference Schemes for the Integration of Stiff Systems of Ordinary Differential Equations," *IBM Research Report No.* RC-1977.

Miranker, W. L. and Liniger, W. (1967), "Parallel Methods for the Numerical Integration of Ordinary Differential Equations," *Math. Comp.*, **21**, pp. 303–320.

Moore, R. E. (1965), "Automatic Local Coordinate Transformations to Reduce the Growth of Error Bounds in the Interval Computation of Solutions of Ordinary Differential Equations," in *Error in Digital Computation*, Vol. 2, ed. L. B. Ball. Wiley, New York.

Moore, R. E. (1966), *Interval Analysis*. Prentice-Hall, Englewood Cliffs, N. J.

Moretti, G. (1965), "A New Technique for the Numerical Analysis of Nonequilibrium Flows," *AIAA Jour.*, **3**, pp. 381–383.

Morrison, D. (1962), "Optimal Mesh Size in the Numerical Integration of an Ordinary Differential Equation," *JACM*, **9**, pp. 98–103.

Moulton, F. R. (1926), *New Methods in Exterior Ballistics*. U. of Chicago, Chicago.

Mysovskikh, I. P. (1969), *Lectures on Numerical Methods*. Translation by L. B. Ball. Wolters-Noordhoff Publ. Co., Groningen, Netherlands.

Newberry, A. C. R. (1963), "Multistep Integration Formulas," *Math. Comp*, **17**, pp. 452–455.

Newberry, A. C. R. (1967), "Convergence of Successive Substitution Starting Procedures, *Math. Comp.*, **21**, No. 99, pp. 489–491.

Nordsieck, A. (1962), "On the Numerical Integration of Ordinary Differential Equations," *Math. Comp.*, **16**, pp. 22–49.

Norsett, S. P. (1969), "A Criterion for $A(\alpha)$ Stability of Linear Multistep Methods," *BIT*, **9**, pp. 259–263.

Nugeyre, J. B. (1961), "Un procédé mixte (Runge-Kutta, pas liés) d'intégration des systèmes différentiels du type $x'' = X(x, t)$," *Chiffres*, **4**, pp. 55–68.

Osborne, M. R. (1964), "A Method for Finite Difference Approximation to Ordinary Differential Equations," *Comp. J.*, **7**, pp. 58–65.

Osborne, M. R. (1967), "Minimizing Truncation Error in Finite Difference Approximations to Ordinary Differential Equations," *Math. Comp.*, **21**, No. 98, pp. 133–145.

Osborne, M. R. (1969), "A New Method for the Integration of Stiff Systems of Ordinary Differential Equations," in *Information Processing 68*, ed. A. J. H. Morrel. North Holland Publishing Company, Amsterdam, pp. 200–204.

Pope, D. A. (1963), "An Exponential Method of Numerical Integration of Ordinary Differential Equations," *Comm. ACM*, **6**, pp. 491–493.

Rahme, H. S. (1969), "A New Look at the Numerical Integration of Ordinary Differential Equations," *JACM*, **16**, No. 3, pp. 496–506.

Ralston, A. (1961), "Some Theoretical and Computational Matters Relating to Predictor Corrector Methods of Numerical Integration," *Comp. J.*, **4**, pp. 64–67.

Ralston, A. (1962), "Runge-Kutta with Minimum Error Bounds," *Math. Comp.*, **16**, pp. 431–437 (1962); *Math Comp.*, **17**, p. 488 (1963).

Ralston, A. (1965), *A First Course in Numerical Analysis*. McGraw-Hill, New York.

Ralston, A. (1965B), "Relative Stability in the Numerical Solution of Ordinary Differential Equations," *SIAM Review*, **7**, pp. 114–125.

Reimer, M. (1968), "Finite Difference Forms Containing Derivatives of Higher Order," *SINUM*, **5**, No. 4, pp. 725–738.

Richards, P. I., Lanning, W. D., and Torrey, M. D. (1965), "Numerical Integration of Large, Highly-Damped Nonlinear Systems," *SIAM Review*, **7**, No. 3, pp. 376–380.

Richardson, L. F. (1927), "The Deferred Approach to the Limit, I—Single Lattice," *Trans. Roy. Soc., London*, **226**, pp. 299–349.

Robertson, H. H. (1966), "The Solution of a Set of Reaction Rate Equations," in *Numerical Analysis: An Introduction*, ed. J. Walsh. Academic Press, London, pp. 178–182.

Roe, G. M. (1967), "Experiments with a New Integration Algorithm," G. E. Report *No.* 67-C-037, Schenectady, N. Y.

Rosenbrock, H. H. (1963), "Some General Implicit Processes for the Numerical Solution of Differential Equations," *Comp.* J. **5** pp. 329–330.

Rosser, J. B. (1967), "A Runge-Kutta for all Seasons," *SIAM Review*, **9**, No. 3, pp. 417–452.

Rutishauser, H. (1960), "Bemerkungen zur numerischen Integration gewohnlicher Differentialgleichungen *n*-ter Ordnung," *Num. Math.*, **2**, pp. 263–279.

Sachnoff, L. (1960), "Integration of Simultaneous Differential Equations Using Multiple Stepsizes," *15th ACM National Conference*.

Sandberg, I. W. (1967), "Some Properties of a Class of Numerical Integration Formula," *Bell System Tech. Jour.*, **46**, No. 9, pp. 2061–2080.

Sandberg, I. W. (1967B), "Two Theorems on the Accuracy of Numerical Solutions of Systems of Ordinary Differential Equations," *Bell System Tech. Jour.* **46**, No. 6, pp. 1243–1266.

Sandberg, I. W. and Schichman, H. (1968), "Numerical Integration of Systems of Stiff Nonlinear Differential Equations," *Bell System Tech. Jour.*, **47**, No. 4, pp. 511–528.

Sarafyan, D. (1965), "Multistep Methods for the Numerical Solution of Ordinary Differential Equations Made Self-Starting," *Mathematics Research Center Report No.* 495.

Scraton, R. E. (1964), "The Numerical Solution of Second Order Differential Equations Not Containing the First Derivative Explicitly," *Comp. J.*, **6**, pp. 368–370.

Scraton, R. E. (1964B), "Estimation of the Truncation Error in Runge-Kutta and Allied Processes," *Comp. J.*, **7**, pp. 246–248.

Scraton, E. E. (1965), "The Solution of Linear Differential Equations in Chebyshev Series," *Comp. J.*, **8**, pp. 57–61.

Shampine, L. F. and Watts, H. A. (1969), "Block Implicit One-Step Methods," *Math. Comp.*, **23**, No. 108, pp. 731–740.

Shanks, E. B. (1966), "Solutions of Differential Equations by Evaluations of Functions,", *Math. Comp.*, **20**, No. 93, pp. 21–38.

Shaw, B. (1967), "Modified Multistep Methods Based on Nonpolynomial Interpolants," *JACM*, **14**, pp. 143–154.

Shaw, B. (1967B), "Some Multistep Formulas for Special High Order Ordinary Differential Equations," *Num. Math.*, **9**, pp. 367–378.

Silverberg, M. (1968), "A New Method of Solving State Variable Equations Permitting Large Step Sizes," *Proc. IEEE*, **56**, No. 8, pp. 1343–1352.

Sloat, H. and Bickart, T. A. (1970), "An Implicit Formula for the Integration of Stiff Network Equations," *Proc. of the Third Hawaii Interational Conference on Systems Science.*

Spijker, M. (1966), "Convergence and Stability of Step by Step Methods for the Numerical Solution of Initial-Value Problems," *Num. Math.*, **8**, pp. 161–177.

Squier, D. P. (1969), "One-Step Methods for Ordinary Differential Equations," *Num. Math.*, **13**, pp. 176–179.

Stetter, H. J. (1965), "Stabilizing Predictors for Weakly Stable Correctors," *Math. Comp.*, **19**, pp. 84–89.

Stetter, H. J. (1965B), "A Study of Strong and Weak Stability in Discretization Algorithms," *SINUM*, **2**, pp. 265–280.

Stetter, H. J. (1965C), "Asymptotic Expansions for the Error of Discretization Algorithms for Nonlinear Functional Equations," *Num. Math.*, **7**, pp. 18–31.

Stetter, H. J. (1968), "Improved Absolute Stability of Predictor-Corrector Schemes," *Computing*, **3**, pp. 286–296.

Stineman, R. W. (1965), "Digital Time-Domain Analysis of Systems with Widely Separated Poles," *JACM*, **12**, No. 2, pp. 286–293.

Stoer, J. (1961), "Uberzwei Algorithmen zur Interpolation mit Rationalen Funktionen," *Num. Math.*, **3**, pp. 285–304.

Störmer, C. (1907), "Sur les trajectoires des corpuscules électrisés" *Arch. Sci. Phys. Nat., Genève*, **24**, pp. 5–18, 113–158, 221–247.

Störmer, C. (1921), "Méthodes d'intégration numérique des équations différentielles ordinaires," *C. R. Congr. Intern. Math., Strasbourg*, pp. 243–257.

Tewarson, R. P. (1969), "Projection Methods for Solving Sparse Linear Equations," *Comp. J.*, **12**, No. 1, pp. 77–80.

Tewarson, R. P. (1967), "Solution of a System of Simultaneous Linear Equations with a Sparse Coefficient Matrix by Elimination Methods," *BIT*, **7**, pp. 226–239.

Tinney, W. F. and Walker, J. W. (1967), "Direct Solutions of Sparse Network Equations by Optimally Ordered Triangular Factorization," *Proc. IEEE*, **55**, No. 11, pp. 1801–1809.

Todd, J. (1962), ed. *A Survey of Numerical Analysis.* McGraw-Hill, New York.

Treanor, C. E. (1966), "A Method for the Numerical Integration of Coupled First Order Differential Equations with Greatly Different Time Constants," *Math. Comp.*, **20**, No. 93, pp. 39–45.

Tyson, T. J. (1964), "An Implicit Integration Method for Chemical Kinetics," *TRW Report* No. 9840-6002-RU000, Redondo Beach, Calif.

Van Wyk, R. (1970), "Variable Mesh Multistep Methods for Ordinary Differential Equations," *Jour. Comp. Physics*, **5** pp. 244–264.

Verner, J. H. (1969), "The Order of Some Implicit Runge-Kutta Methods," *Num. Math.*, **13**, pp. 14–23.

Watt, J. M. (1967), "The Asymptotic Discretization Error of a Class of Methods for Solving Ordinary Differential Equations," Proc. Cambridge Phil. Society, pp. 441–472.

Whitney, D. E. (1966), "Propagated Error Bounds for Numerical Solution of Transient Response," Proc. *IEEE*, **54**, No. 8, pp. 1084–1085.

Whitney, D. E., (1969), "More about Similarities Between Runge-Kutta and Matrix Exponential Methods for Evaluating Transient Response," Proc. *IEEE*, **57**, No. 11, pp. 2053–2054.

Widlund, O. (1967), "A Note on Unconditionally Stable Linear Multistep Methods," *BIT*, **7**, pp. 65–70.

Willoughby, R. A. (1969), ed. "Proceedings of the Symposium on Sparse Matrixes and their Applications," *Report RA* 1 (#11707), IBM Watson Research Center, Yorktown Heights, N. Y.

Zajac, E. E. (1964), "Note on Overly-Stable Difference Approximation," *Jour. Math. and Phys.*, **18**, No. 1, pp. 51–54.

Zurmuhl, R. (1948), "Runge-Kutta-Verfahren zur numerischen Integration von Differentialgleichungen *n*-ter Ordnung," *ZAMM*, **28**, pp. 173–182.

INDEX